Graph Embedding for Pattern Analysis

Yun Fu • Yunqian Ma
Editors

Graph Embedding for Pattern Analysis

Springer

Editors
Yun Fu
Department of ECE
College of Engineering
Northeastern University
Boston, Massachusetts, USA

Yunqian Ma
Honeywell Aerospace
Honeywell International Inc.
Golden Valley, USA

ISBN 978-1-4614-4456-5 ISBN 978-1-4614-4457-2 (eBook)
DOI 10.1007/978-1-4614-4457-2
Springer New York Heidelberg Dordrecht London

Library of Congress Control Number: 2012951289

Printed on acid-free paper

Springer is part of Springer Science+Business Media (www.springer.com)

Preface

Graph embedding is a computational methodology aiming at representing data as a graph, along with the attributes attached to its nodes and edges. Derived from topological graph theory and algebra, the subject of graph embedding has become important and prominent worldwide with a wide spectrum of applications in pattern analysis, representation, visualization, and classification.

This book is composed of coherent chapters contributed by experts and researchers from both academia and industry in the fields of machine learning, applied statistics, artificial intelligence, computer vision, and pattern classification. Fundamental theories are presented in each chapter to help readers quickly gain the knowledge or review the essential topics. Case studies, experiments, and applications are also provided to further inspire the readers for insightful understanding.

This book may be used as an excellent reference book for researchers or major textbook for graduate student courses requiring minimal undergraduate prerequisites at academic institutions. Existing courses related or focused on graph embedding include the CSE 704 manifold and subspace learning of SUNY-Buffalo, EE364a convex optimization of Stanford University, and graph embedding for pattern recognition hosted by the ICPR 2010 conference in Istanbul.

Chapter "Multilevel Analysis of Attributed Graphs for Explicit Graph Embedding in Vector Spaces" provides the introduction of the graph embedding and describes the multilevel analysis of attributed graphs for explicit graph embedding. Chapter "Feature Grouping and Selection Over an Undirected Graph" presents a method for simultaneous feature grouping and sparseness structures over a given undirected graph and provides a convex and non-convex penalty function. Chapter "Median Graph Computation by Means of Graph Embedding into Vector Spaces" introduces the graph embedding from its solution on the graph representation for the computational complexity and in particular presents one graph embedding method: the median graph. Chapter "Patch Alignment for Graph Embedding" describes the patch alignment framework, which unifies the existing manifold learning-based dimension reduction algorithm and provides a general platform for specific algorithm design. Chapter "Feature Subspace Transformations for Enhancing K-Means Clustering" presents a feature subspace transformation method to transform the database and use the k-means clustering method after that. Motivated by the sparse representation for high-dimensional data analysis, chapter "Learning with ℓ^1-Graph for High Dimensional Data Analysis" presents a method to construct a robust $\ell 1$ graph and uses the $\ell 1$ graph for various machine learning tasks. Chapter "Graph-Embedding Discriminant Analysis on Riemannian Manifolds for Visual Recognition" presents a graph embedding method to embed Riemannian manifolds into reproducing kernel Hilbert spaces and employs many kernel-based learning algorithms. After introducing several linearization methods for subspace learning, chapter "A Flexible and Effective Linearization Method for Subspace Learning" presents a flexible manifold embedding method for semi-supervised and unsupervised subspace learning. Among many applications for horizontal anomaly detection, chapter "A Multi-graph Spectral Framework for Mining Multi-source Anomalies" presents a method to detect objects with inconsistent behavior using multiple information sources. Chapter "Graph Embedding for Speaker Recognition" presents the application of graph embedding to the speaker recognition.

We would like to sincerely thank all the contributors of this book for presenting their research in an easily accessible manner and for putting such discussion into a historical context. We would like to thank Brett Kurzman from Springer for support to this book project.

Boston, MA — Yun Fu
Golden Valley, MN — Yunqian Ma

Contents

Multilevel Analysis of Attributed Graphs for Explicit Graph Embedding in Vector Spaces

Muhammad Muzzamil Luqman, Jean-Yves Ramel, and Josep Lladós

1 Introduction

Ability to recognize patterns is among the most crucial capabilities of human beings for their survival, which enables them to employ their sophisticated neural and cognitive systems [1], for processing complex audio, visual, smell, touch, and taste signals. Man is the most complex and the best existing system of pattern recognition. Without any explicit thinking, we continuously compare, classify, and identify huge amount of signal data everyday [2], starting from the time we get up in the morning till the last second we fall asleep. This includes recognizing the face of a friend in a crowd, a spoken word embedded in noise, the proper key to lock the door, smell of coffee, the voice of a favorite singer, the recognition of alphabetic characters, and millions of more tasks that we perform on regular basis.

The scientific domains of artificial intelligence (AI) and pattern recognition (PR) can be seen as the transportation of the human capability of analyzing—to compare, to classify, and to identify—the audio and visual signals to computers, so that computers may assist humans for pattern recognition tasks and to replace humans for some of them. Pattern recognition has emerged as an important research domain and has supported the development of numerous applications in many different areas of activity: robot assisted manufacture, medical diagnostic systems,

M.M. Luqman (✉)
François Rabelais University of Tours, France,
Autonòma University of Barcelona, Spain
e-mail: mluqman@univ-tours.fr; mluqman@cvc.uab.es

J.Y. Ramel
François Rabelais University of Tours, France
e-mail: ramel@univ-tours.fr

J. Lladós
Autonòma University of Barcelona, Spain
e-mail: josep@cvc.uab.es

Y. Fu and Y. Ma (eds.), *Graph Embedding for Pattern Analysis*,
DOI 10.1007/978-1-4614-4457-2_1,

forecast of economic variables, exploration of earth resources, analysis of satellite data, face detection, verification and recognition, object detection and recognition, handwritten digit and character recognition, speech and speaker verification and recognition, information and image retrieval, text detection and categorization, gender classification and prediction [3–5], being some of the important to mention.

The problems of pattern recognition are often very complex and it is nearly impossible to write an explicitly programmed solution for them. For example, it is impossible to write an analytic program for recognizing a face in a photo [6]. The pattern recognition research community has overcome this problem by adapting a learning methodology that is highly inspired by human ability to learn. A learning methodology refers to the approach where instead of precisely defining a set of specifications for solving a problem analytically, the machine is trained on data and it learns the concept of a class by discriminating between groups of similar objects. Based on the inferred rules and learning performed during training, the machine is able to make predictions about new and unseen data. More formally, the machine acquires generalization power through learning [7].

A pattern recognition task for computers can be looked upon as consisting of two main steps: (1) the representation of the signal data by a data structure and (2) the computation of desired operation (pattern recognition). Each of the two important subdomains of pattern recognition—namely the structural pattern recognition and the statistical pattern recognition—has its strength only in one of the two aforementioned steps, respectively. The structural pattern recognition offers the most powerful relational data structure of graph. For the last three decades, graphs have been used for pattern recognition and image analysis, for extracting and representing complex relations in underlying data. However, there is still a lack of efficient computational tools and learning models that can process this data structure. On the other side, the statistical pattern recognition offers highly efficient computational models of machine learning, classification, and clustering, by employing the well-established theory of statistics. But these computational models can work only on simple numeric vectors and cannot process complex high-dimensional relational data structures.

Over decades of parallel research in structural and statistical subdomains of pattern recognition powerful representations and efficient computational models (respectively) have been built. But little progress has been made towards the long desired objective of pattern recognition research community, of joining the advantages of the structural and statistical pattern recognition approaches for building more powerful and efficient algorithms. Recently, an important step forward has been achieved by the emerging field of graph embedding in pattern recognition. Graph embedding joins the advantages of structural and statistical pattern recognition approaches by acting as a bridge between powerful structural pattern recognition representations and efficient computational models of statistical pattern recognition. Graph embedding achieves this by embedding the graph-based structural representations into numeric vector spaces, thus enabling them to employ the computational efficient models of statistical pattern recognition.

In this chapter, first we will briefly discuss the merits and limitations of structural and statistical subdomains of pattern recognition in Sect. 2. Then we will present an overview along with the literature review on the emerging field of graph embedding in pattern recognition in Sect. 3. This will be followed by a detailed discussion on the multilevel analysis of attributed graphs for explicit graph embedding in vector spaces in Sect. 4.

2 Structural and Statistical Pattern Recognition

Structural pattern recognition is characterized by the utilization of symbolic data structures. The widely used symbolic data structures are graphs, strings, and trees. Overtime the use of graph representations has become very popular in structural pattern recognition. This is because of the fact that both strings and trees are special instances of graphs [8], and thus we can safely term graphs to be the representative of symbolic data structures.

Graph-based representations have their application to a wide range of domains, as graphs provide a convenient and powerful representation of relational information. They are able to represent not only the values of both symbolic and numeric properties of an object, but they can also explicitly model the spatial, temporal, and conceptual relations that exist between its parts. Graphs do not suffer from the constraint of fixed dimensionality. For example, the number of nodes and edges in a graph is not limited a priori and depends on the size and the complexity of the actual object to be modeled [9]. The most important advantage that graphs have is that they have foundations in strong mathematical formulation and have a mature theory at their basis. However, along with the various advantages of graphs, they have a serious drawback. Graph-based representations are computational expensive. The much needed operations of exact and inexact graph matching are NP-complete. A second serious drawback of graphs is that they are sensitive to noise. We recommend [8–12] for an in-depth reading on structural pattern recognition.

Statistical pattern recognition is characterized by the utilization of numeric feature vectors. The feature vectors are very basic representations. A very important advantage of these representations is that because of their simple structure, the basic operations that are needed in machine learning can easily be executed on them. This makes a large number of mature algorithms for pattern analysis and classification immediately available to statistical pattern recognition. And as a result of this fact, the statistical pattern recognition offers state-of-the-art computational efficient tools of learning, classification, and clustering. However, feature vector-based representations have associated representational limitations. These limitations arise from their simple structure and the fact that feature vectors have same length and structure regardless of the complexity of object to be modeled [13]. We recommend [1] for a more detailed reading on statistical pattern recognition and classification.

Table 1 Structural vs statistical pattern recognition

	Pattern recognition	
	Structural	Statistical
Data structure	Symbolic data structure	Numeric feature vector
Representational strength	+	
Fixed dimensionality	+	
Sensitive to noise		+
Efficient computational tools		+

Table 1 summarizes the above-discussed properties of structural and statistical pattern recognition methodologies.

3 Graph Embedding

Over decades of research in pattern recognition, the research community has developed a range of expressive and powerful approaches for diverse problem domains. Graph-based structural representations are usually employed for extracting the structure, topology, and geometry, in addition to the statistical details of underlying data. During the next step in the processing chain, generally these representations could not be exploited to their full strength because of limited availability of computational tools for them. On the other hand, the efficient and mature computational models, offered by statistical approaches, work only on vector data and cannot be directly applied to these high-dimensional representations. Recently, this problem has been addressed by the emerging research domain of graph embedding in pattern recognition.

Definition 1 (Graph embedding). Graph embedding is a methodology aimed at representing a whole graph, along with the attributes attached to its nodes and edges, as a point in a suitable vector space.

Graph embedding is a natural outcome of parallel advancements in structural and statistical pattern recognition. It offers a straightforward solution, by employing the representational power of symbolic data structures and the computational superiority of feature vectors [11]. It acts as a bridge between structural and statistical approaches [14, 15] and allows a pattern recognition method to benefit from computational efficiency of state-of-the-art statistical models and tools [16] along with the convenience and representational power of classical symbolic representations. This permits the last three decades of research on graph-based structural representations in various domains [10], to benefit from the state-of-the-art machine learning models and tools. Graph embedding has its application to the whole variety of domains that are entertained by pattern recognition and where the use of a

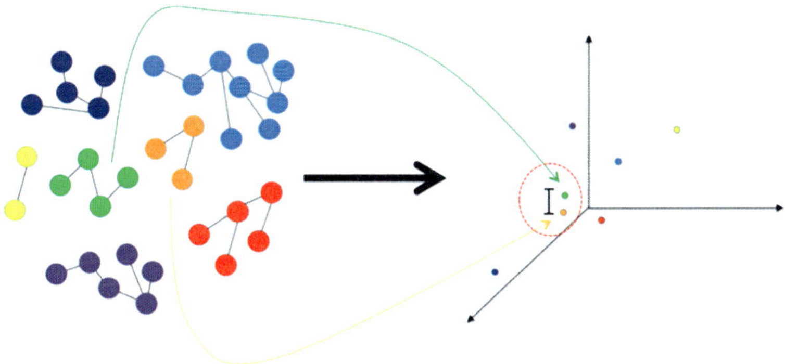

Fig. 1 An illustration of the mapping of a graph into a point in a vector space

relational data structure is mandatory for performing high-level tasks. Apart from reusing the computational efficient methods for vector spaces, another important motivation behind graph embedding methods is to solve the computationally hard problems geometrically [17]. The pattern recognition research community acknowledges the emerging importance of graph embedding methods [18] for realizing the classical idea of using the structural and statistical methods together. We refer the interested reader to [19] for further reading on the applications of graph embedding.

3.1 Implicit and Explicit Graph Embedding

The graph embedding methods are formally categorized as implicit graph embedding or explicit graph embedding. The implicit graph embedding methods are based on graph kernels. A graph kernel is a function that can be thought of as a dot product in some implicitly existing vector spaces. Instead of mapping graphs from graph space to vector space and then computing their dot product, the value of the kernel function is evaluated in graph space. Such an embedding satisfies the main mathematical properties of dot product. Since it does not explicitly map a graph to a point in vector space, a strict limitation of implicit graph embedding is that it does not permit all the operations that could be defined on vector spaces. We refer the interested reader to [18, 20, 21] for further reading on graph kernels and implicit graph embedding. On the other hand, the more useful, explicit graph embedding methods explicitly embed an input graph into a feature vector and thus enable the use of all the methodologies and techniques devised for vector spaces. Figure 1 pictorially illustrates the mapping of a graph to a point in a suitable vector space.

Definition 2 (Attributed graph (AG)). Let A_{V} and A_{E} denote the domains of possible values for attributed vertices and edges, respectively. These domains are assumed to include a special value that represents a null value of a vertex or an edge. An attributed graph AG over $(A_{\mathrm{V}}, A_{\mathrm{E}})$ is defined to be a four-tuple:

$$\mathrm{AG} = \left(V, E, \mu^{\mathrm{V}}, \mu^{\mathrm{E}}\right)$$

where

V is a set of vertices.
$E \subseteq V \times V$ is a set of edges.
$\mu^{\mathrm{V}} : V \longrightarrow A_{\mathrm{V}}^{k}$ is function assigning k attributes to vertices.
$\mu^{\mathrm{E}} : E \longrightarrow A_{\mathrm{E}}^{l}$ is a function assigning l attributes to edges.

Definition 3 (Explicit graph embedding). Explicit graph embedding maps a graph to a point in suitable vector space. It encodes the graphs by equal size vectors and produces one vector per graph. Mathematically, for a given graph $\mathrm{AG} = (V, E, \mu^{\mathrm{V}}, \mu^{\mathrm{E}})$, explicit graph embedding is a function ϕ, which maps graph AG from graph space G to a point $(f_1, f_2, \ldots, f_n)$ in n-dimensional vector space $\mathbb{R}^n$. It is given as

$$\begin{aligned} \phi : G &\longrightarrow \mathbb{R}^n \\ \mathrm{AG} &\longmapsto \phi(\mathrm{AG}) = (f_1, f_2, \ldots, f_n) \end{aligned}$$

The vectors obtained by an explicit graph embedding method can also be employed in a standard dot product for defining an implicit graph embedding function between two graphs [8]. An interesting property of explicit graph embedding is that the graphs are embedded in pattern spaces in a manner that similar structures come close to each other and different structures go far away, i.e., an implicit clustering is achieved [22]. Another important property of explicit graph embedding is that the graphs of different size and order need to be embedded into a fixed size feature vector. This means that for constructing the feature vector, an important step is to mark the important details that are available in all the graphs and are applicable to a broad range of graph types. We refer the interested reader to [18] for further reading on explicit graph embedding.

3.2 A Quick Literature Review on Explicit Graph Embedding

Recent research surveys on graph embedding are presented by [8, 18, 19]. In the literature the problem of explicit graph embedding has been approached by three important families of algorithms.

- Graph probing based methods.
- Spectral based explicit graph embedding.
- Dissimilarity based explicit graph embedding.

3.2.1 Graph Probing Based Methods

The first family of graph embedding methods is based on the frequencies of appearance of specific knowledge-dependent substructures in graph. This embedding is based on feature extraction on structural data. These numeric features capture both topology information (number of nodes, arcs, degree of nodes) and the contents of the graph (histogram of the labels for example). This embeds a graph into a feature vector in Euclidean space.

One of the first methods of graph probing dates back to the 1940s [23]. In this method, each graph is represented by an index called Wiener index. The index is a topological descriptor defined by the sum of all shortest paths in the graph, such that if $G = (G(V), G(E))$ is a graph, then the Wiener index $\mathrm{W}(G)$ of G is

$$\mathrm{W}(G) = \sum\nolimits_{v_i \in G(V)} \sum\nolimits_{v_j \in G(V)} l(v_i, v_j)$$

where $l(v_i, v_j)$ is a function that defines the length of the shortest path between the node and the nodes v_i and v_j in the graph G.

Another approach is introduced in [24]. For undirected and unattributed graphs, the authors calculate the degree of each node and form a sorted histogram of degrees. So they get the vector $\phi(G) \in \mathbb{R}^n$ such that if $G = (V(G), E(G))$ is a graph then $\phi(G) = (n_0, n_1, n_2, \ldots)$ where $n_i = \{v \subset V | d_G(v) = i\}$. The features of the resulting vector are the number of nodes of degree 1, the number of nodes of degree 2, and so on.

An extension of this technique is proposed in [25]. The authors also suggest a representation based on local descriptors of nodes and generalize the method for all types of graphs. This representation is based on the degree of vertices in the case of undirected and unlabeled graphs. In the case of directed graphs, the representation uses in-degree and out-degree of nodes. This representation can be adapted to take into consideration labeled graphs, taking into account, in addition to the degree of the nodes, the labels of the connected edges. On the other hand, the authors present an interesting relationship between the graph edit distance and the distance between the embedding of two graphs. They have shown that the distance between the embedding of two graphs g_1 and g_2, given by $d_{\mathrm{GEM}}(g_1, g_2)$, is less than 4 times of the graph edit distance between them, given by $d_{\mathrm{GED}}(g_1, g_2)$. Mathematically this is given as $d_{\mathrm{GEM}}(g_1, g_2) \leq 4 \times d_{\mathrm{GED}}(g_1, g_2)$.

Recently, Gibert et al. in [26] have proposed a graph embedding method by counting the frequency of appearance of specific set of representatives of labels of nodes and their corresponding edges. In [27] the authors propose an improvement of their graph embedding technique by dimensionality reduction of the obtained feature vector.

The algorithms presented in the cited works provide an embedding of graph into feature vector, in linear time complexity. Their simplicity of implementation is an important advantage of this family of methods. However, the features they use are very localized to nodes and arcs. The graph embedding contains little information on the topology, which can have a negative impact on the classification results.

A subfamily of this category of graph embedding methods is based on the frequencies of appearance of specific knowledge-dependent substructures in graph. These works are mostly proposed for chemical compounds and molecular structures. Graph representations of molecules are assigned feature vectors whose components are the frequencies of appearance of specific knowledge-dependent substructures in graphs [28, 29, for example].

Recently, Sidere et al. in [30] have proposed a graph embedding method. The method is based on the extraction of substructures of 2 nodes, 3 nodes , 4 nodes, and so on, from a graph. The feature vector representation is then obtained by counting the frequencies of these substructures in the graph. The method proposed by Sidere is rich in topological information and is very interesting for pattern recognition.

The methods based on the frequencies of appearance of specific knowledge-dependent substructures in graph are based on finding subgraphs in graph and are capable of exploiting domain knowledge. However, they have a drawback that finding substructures in graphs is computationally challenging.

3.2.2 Spectral Based Explicit Graph Embedding

The second family of graph embedding algorithms is spectral based embedding. Spectral based embedding is a very prominent class of graph embedding methods and is proposed by lots of works in literature. In order to embed graphs into feature vectors, this family of methods extracts features from graphs by eigen decomposition of adjacency and Laplacian matrices and then apply a dimensionality reduction technique on the eigen features. Many works for graph embedding exploit the spectral theory of graphs [31] and are interested in the properties of the spectrum of a graph and the characterization of the topology graph using eigenvalues and eigenvectors of the adjacency matrix or Laplacian matrix.

In [32], graph embedding using the spectral approach is proposed. The authors use the leading eigenvectors of the adjacency matrix to define the eigenspaces of adjacency matrices. Spectral properties are then calculated for each eigenmode. In [33], the authors also use the adjacency matrices of graphs as a support and then compute their eigenvectors and thus obtain modes to define the vector space (perimeter, volume, Cheeger number, etc.). Construction of the vector is completed by applying the dimensionality reduction using principal component analysis (PCA), independent component analysis, or multidimensional scaling.

Graph embedding has also been approached by using Laplacian matrix. For example, in [34], a spectral approach is proposed using the Laplace–Beltrami operator to embed the graphs in a Riemannian manifold or a metric where one can define the length of a path (called a geodesic) between two points of the manifold. This length of the path is then used to calculate similarity between graphs. In [35], the embedding of the nodes of a graph is performed on an eigenspace defined by the first Eigenvectors. In [22], the authors propose to use a spectral decomposition of Laplacian matrix and construct the symmetric polynomials. The coefficients of these polynomials are then used for graph embedding.

As for graph matching techniques, the spectral theory of graphs shows interesting properties that can be used for graph embedding. Its main benefit is the linear complexity associated with operations on matrices. The spectral family of graph embedding methods provides solid theoretical insight into the meaning and significance of extracted features. However, these approaches have some limitations. Spectral methods are sensitive to noise, removing a node, for example, changes the matrices (adjacency or Laplacian) and these errors affect the embedding functions. In addition, their use is restricted to unlabeled graphs making it difficult to use these approaches in most applications of pattern recognition.

3.2.3 Dissimilarity Based Explicit Graph Embedding

Finally, the third family of graph embedding algorithms is based on dissimilarity of a graph from a set of prototypes. The dissimilarity based graph embedding can handle arbitrary graphs. The dissimilarity based graph embedding methods usually use graph edit distance and exploit domain knowledge.

Definition 4 (Graph edit distance (GED)). Let $g_1 = (V(g_1), E(g_1))$ and $g_2 = (V(g_2), E(g_2))$ be two graphs. The graph edit distance between g_1 and g_2 is defined as

$$d(g_1, g_2) = \min_{(e_1,\ldots,e_k) \in \gamma(g_1,g_2)} \sum_{i=1}^{k} c(e_i)$$

where

$\gamma(g_1, g_2)$ denotes the set of edit paths transforming g_1 into g_2.
c denotes the cost function measuring the strength $c(e_i)$ of edit operation e_i.
The edit operations include, for example, addition and deletion of node/edge.

But since graph edit distance is computationally expensive, the dissimilarity based graph embedding methods may become computationally challenging. Unlike the aforementioned approaches of graph embedding, the dissimilarity based graph embedding techniques do not focus on the extraction of a vector from the graph. Rather, the idea is to construct a vector by comparing a graph with a selection of graphs called prototypes.

In [36], the authors present dissimilarity based representation as an alternative to the feature vectors. Starting from the postulate that the dissimilarity is used to separate a class of objects from another, the authors offer to characterize objects not by absolute attributes but as a vector of dissimilarity from objects of other classes. An object is defined by its embedding in a dissimilarity space.

Bunke et al. [9, 37–39] propose to adapt this dissimilarity space for the construction of a graph embedding function. The main idea of this work is to construct a vector of graph edit distances from the graph to be embedded and a set of k prototypes selected in the graph database. The embedding of the graph is thus a vector of k distances. Formally, let $\Gamma = g_1, \ldots, g_n$ be a set of graphs and $p =$

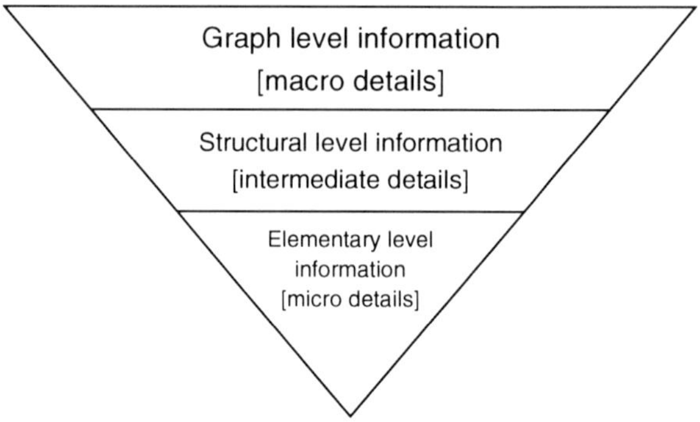

Fig. 2 Multi-facet view of discriminatory information in graph

$p_1, \ldots, p_k \subset \Gamma$ be a subset of selected prototypes from Γ. The graph embedding is defined as the function $\Phi : \Gamma \longmapsto (\mathbb{R})^k$, such that $\Phi(g) = [d(g, p_1), \ldots, d(g, p_k)]$ where $d(g, p_i)$ is the graph edit distance between graph g and the ith prototype graph in p. In [8,40], the authors propose an improvement of the graph embedding method by using feature selection method.

This type of projection is very interesting. However, the limitation of setting the edit distance is found in this method. In addition, the choice of prototype graphs is also a significant parameter as it determines the size of the vector and its capacity to effectively represent the graph in the vector space. Also, it remains highly dependent on the application and its learning set. In [7, 9], the authors have given some indications on the choice of the prototype graphs.

4 Multilevel Analysis of Attributed Graphs for Explicit Graph Embedding

The vector representation of graphs by an explicit graph embedding algorithm seems to be the track to meet the technical obstacles for pattern recognition. This is because of the fact that explicit graph embedding can effectively combine the power and flexibility of graph representation, with the diversity, learning abilities, and computational efficiency of statistical tools. However, no efficient and simple method to adapt to the data is yet available. Most of the existing works on graph embedding can handle only the graphs that are comprised of edges with a single attribute and vertices with either no or only symbolic attributes. These methods are only useful for specific application domains for which they are designed. In this section we will outline our method of explicit graph embedding for attributed graphs with many symbolic as well as numeric attributes on both nodes and edges [41–43]. The method is named fuzzy multilevel graph embedding and is abbreviated as FMGE. It preserves multi-facet information from global, topological, and local point of views and is based on multilevel analysis of a graph for embedding it into a feature vector. FMGE employs fuzzy overlapping trapezoidal intervals for

Graph order	Graph size	Embedding of node degree	Embedding(s) of subgraph(s) homogenity	Embedding(s) of node attribute(s)	Embedding(s) of edge attribute(s)

Fig. 3 Feature vector of fuzzy-interval based approach for explicit graph embedding

minimizing the information loss while mapping from continuous graph space to discrete feature vector space. FMGE has build-in unsupervised learning abilities and thus is inexpensively deployable to a wide range of application domains.

4.1 Multilevel Information in Graph

FMGE performs multilevel analysis of graph to extract discriminatory information of three different levels from a graph. These include the graph level information, structural level information, and the elementary level information. The three levels of information represent three different views of graph for extracting global details, details on topology of graph, and details on elementary building units of graph.

FMGE encodes the numeric part of each of the different levels of the multilevel information by fuzzy histograms and symbolic part by crisp histograms. Once the histograms are constructed, FMGE employs the histogram representation of the multilevel information of a graph for embedding it into a numeric feature vector. The feature vector of FMGE is named fuzzy structural multilevel feature vector and abbreviated as FSMFV. Figure 3 present details on the features in FSMFV. It contains features extracted from three levels of information in graph (viz. graph level information, structural level information, elementary level information). The features for graph level information represent a coarse view of graph and give a general information about the graph. These features include the graph order and graph size. The features for structural level information represent a deeper view of graph and are extracted from the node degrees and subgraph homogeneity in graph. The third level of information is extracted by penetrating into further depth and more granular view of graph and employing details of the elementary building blocks of graph. These features represent the information extracted from the node and edge attributes.

4.1.1 Graph Level Information

The graph level information in FSMFV is embedded by two numeric features, encoding the order and the size of graph.

Graph order: A graph vertex is an abstract representation of the primitive components of underlying content. The order of a graph provides very important discriminatory topological information on the graph.

Histogram of node degrees h^d	Histograms of node attributes resemblance h_d^{nr} and $h_1^{nr}, h_2^{nr}, \ldots, h_k^{nr}$	Histograms of edge attributes resemblance $h_1^{er}, h_2^{er}, \ldots, h_l^{er}$

Fig. 4 Embedding of structural level information

Graph size: An edge is an abstract representation of the relationship between the primitive components of underlying content. Graph size also provides important discriminatory information on the topological details of graph. It enables to differentiate between two equal ordered graphs and define similarity between two equal sized graphs.

4.1.2 Structural Level Information

The embedding of structural level information is a novelty and the most critical part of FMGE. Very few existing works on graph embedding clearly use this information. We use node degree's information and subgraph homogeneity measure embedded by histograms of node attribute's resemblance and edge attribute's resemblance, for embedding structural level information. Figure 4 outlines this part of FSMFV.

Node degrees: The degrees of nodes represent the distribution of edges in graph and provide complementary discriminatory information on the structure and topology of graph. It permits to discriminate between densely connected graphs and sparsely connected graphs. Node degree's information is encoded by a histogram of s_i fuzzy intervals. The fuzzy intervals are learned during a prior learning phase, which employs degrees of all the nodes of all graphs in dataset. Node degree's features (h^d in Fig. 4), for an attributed graph, embed the histogram of its nodes for the s_i fuzzy intervals. In Fig. 4, the histogram h^d is a fuzzy histogram as node degrees is a numeric information.

For directed graphs this feature is represented by two sub-features of in-degree and out-degree, i.e., a fuzzy histogram for encoding the distribution of in-degree and another fuzzy histogram for encoding the distribution of out-degree of nodes.

Node attribute's resemblance for edges: The resemblance between two primitive components that have a relationship between them, in a graph, is a supplementary information available in the graph. The node attribute's resemblance for an edge encodes structural information for the respective node-couple. To compute resemblance information for an edge, the node degrees of its two nodes and the list of node attributes as given by μ^V are employed for extracting additional information. This additional information is represented as new edge attributes and is processed like other edge attributes.

Given an edge between two nodes, say $node_1$ and $node_2$ in a graph, the resemblance between a numeric node attribute a is computed by (1) and the resemblance between a symbolic node attribute b is computed by (2).

For each numeric node attribute, the resemblance attribute nr is represented by s_{nr} features in FSMFV. This resemblance information is encoded by a fuzzy histogram of s_{nr} fuzzy intervals. The fuzzy intervals are learned during a prior learning phase, which employs resemblance attribute nr of all the edges of all graphs in dataset:

$$\text{resemblance}(a_1, a_2) = \frac{\min(|a_1|, |a_2|)}{\max(|a_1|, |a_2|)} \tag{1}$$

where

$a \in \{\text{node degree}, \mu^{\mathrm{V}}\}$
a_1 is the value of the attribute a for $node_1$
a_2 is the value of the same attribute a for $node_2$

$$\text{resemblance}(b_1, b_2) = \begin{vmatrix} 1 & b_1 = b_2 \\ 0 & \text{otherwise} \end{vmatrix} \tag{2}$$

where

$b \in \mu^{\mathrm{V}}$.
b_1 is the value of the attribute b for $node_1$.
b_2 is the value of the same attribute b for $node_2$.

For each symbolic node attribute, we represent the resemblance attribute nr by exactly two possible numeric features. The resemblance for symbolic attributes can either be 0 or 1 (2). The cardinalities of the two resemblance values in input graph are encoded by a crisp histogram which is used as features in FSMFV.

In Fig. 4 the histogram $h_{\mathrm{d}}^{\mathrm{nr}}$ represents the features for encoding resemblance attribute for node degrees. Whereas, the histograms $h_1^{\mathrm{nr}}, h_2^{\mathrm{nr}}, \ldots, h_k^{\mathrm{nr}}$ represent the features for encoding resemblance attributes for k node attributes μ^{V}. The histogram $h_{\mathrm{d}}^{\mathrm{nr}}$ is a fuzzy histogram since node degree is a numeric information. Each of the histograms $h_1^{\mathrm{nr}}, h_2^{\mathrm{nr}}, \ldots, h_k^{\mathrm{nr}}$ is a crisp histogram if its corresponding attribute is symbolic and is a fuzzy histogram if the attribute that it is encoding is a numeric attribute.

Edge attribute's resemblance for nodes: The resemblance among the relationships associated to a primitive component is a supplementary information available in the graph. The edge attribute's resemblance for a node encodes the structural information for the respective edges of node and brings more topological information to FSMFV. To compute resemblance information for a node, each attribute of its edges (as given by μ^{E}) is employed for extracting additional information. This additional information is represented as new node attributes and is processed like other node attributes.

Given a node, say *node* in a graph, the resemblance for the edges connected to *node* is computed as the mean of the resemblances between all the pair of edges connected to *node*. For a pair of edges, say $edge_1$ and $edge_2$ connected to *node*, the resemblance for a numeric attribute c is computed by (3) and the resemblance between a symbolic edge attribute d is computed by (4):

$$\text{resemblance}(c_1, c_2) = \frac{\min(|c_1|, |c_2|)}{\max(|c_1|, |c_2|)} \tag{3}$$

where

$c \in \mu^{\mathrm{E}}$.
c_1 is the value of the attribute c for $edge_1$.
c_2 is the value of the same attribute c for $edge_2$.

$$\text{resemblance}(d_1, d_2) = \begin{vmatrix} 1 & d_1 = d_2 \\ 0 \text{ otherwise} \end{vmatrix} \tag{4}$$

where

$d \in \mu^{\mathrm{E}}$.
d_1 is the value of the attribute d for $edge_1$.
d_2 is the value of the same attribute d for $edge_2$.

For each symbolic edge attribute, the resulting resemblance attribute er is treated as a numeric attribute and is embedded by a fuzzy histogram. The resemblance for symbolic edge attributes is computed as mean of the resemblances between all the pair of edges connected to a node. Although the resemblance for a pair of edges is always either 0 or 1 but mean resemblance for all the pair of edges of a node can be any numeric value (this is different from the symbolic node attribute's resemblance which is always either 0 or 1). Therefore, for each symbolic and numeric edge attribute, the resemblance attribute er is represented by s_{er} features in FSMFV. This resemblance information is encoded by a fuzzy histogram of s_{er} fuzzy intervals. The fuzzy intervals are learned during a prior learning phase, which employs resemblance attribute er of all the nodes of all graphs in dataset.

In Fig. 4 the histograms $h_1^{\mathrm{er}}, h_2^{\mathrm{er}}, \ldots, h_l^{\mathrm{er}}$ represent the features for encoding resemblance attributes for l edge attributes μ^{E}. Each of the histograms $h_1^{\mathrm{er}}, h_2^{\mathrm{er}}, \ldots, h_l^{\mathrm{er}}$ is a fuzzy histogram.

4.1.3 Elementary Level Information

Embedding of elementary level information allows FMGE to extract discriminatory information from individual nodes and edges of the graph. The symbolic attributes are encoded by crisp histograms and the numeric attributes by fuzzy histograms. FMGE can embed attributed graphs with many symbolic and numeric attributes on both nodes and edges.

For every symbolic node attribute, each modality that can be taken by this attribute is represented by exactly one numeric feature in FSMFV. This feature encodes the cardinality of this modality in an input graph.

Each numeric node attribute is encoded by a fuzzy histogram of its s_i fuzzy intervals. The fuzzy intervals are learned for each of the numeric node attributes

Fig. 5 Embedding of elementary level information

Histograms of node attributes $h_1^n, h_2^n, \ldots, h_k^n$	Histograms of edge attributes $h_1^e, h_2^e, \ldots, h_l^e$

in the input graph dataset during a prior learning phase. The features for a numeric attribute of an input graph embed the histogram for these s_i fuzzy intervals. Figure 5 outlines this part of FSMFV.

Node attributes: The node attributes provide additional details on primitive components of underlying content and aid FSMFV to discriminate between two similar structured graphs. In Fig. 5 the histograms $h_1^n, h_2^n, \ldots, h_k^n$ represent the features for encoding k node attributes μ^{V}. Each of the histograms $h_1^n, h_2^n, \ldots, h_k^n$ is a crisp histogram if its corresponding attribute is symbolic and is a fuzzy histogram if the attribute that it is encoding is a numeric attribute.

The node attributes are very important information for discriminating between two equal ordered and equal sized graphs, which are representing quite similar structure as well (for example the graph of a small square can be differentiated from that of a big square by using a length attribute on the nodes).

Edge attributes: The edge attributes provide supplementary details on the relationships between the primitive components of the underlying content and aid FSMFV to discriminate between two similar structured graphs. In Fig. 5 the histograms $h_1^e, h_2^e, \ldots, h_l^e$ represent the features for encoding l edge attributes μ^{E}. Each of the histograms $h_1^e, h_2^e, \ldots, h_l^e$ is a crisp histogram if its corresponding attribute is symbolic and is a fuzzy histogram if the attribute that it is encoding is a numeric attribute.

The edge attributes are very important information for discriminating between two equal ordered and equal sized graphs, which are representing quite similar structure as well (for example the graph of a square can be differentiated from that of a rhombus by using angle attribute on the edges).

4.2 *Embedding Multilevel Information of Graphs into Vector Spaces*

In FMGE framework, the mapping of input collection of graphs to appropriate points in a suitable vector space $\mathbb{R}^n$ is achieved in two phases, i.e., the off-line unsupervised learning phase and the online graph embedding phase. We would like to again highlight the important fact that in FMGE the structural level information is represented by new resemblance attributes on the nodes and edges of the graphs. This section will detail the procedure of encoding the attribute information in graphs (including original attributes and the new resemblance attributes) for embedding them into feature vector spaces.

Fig. 6 Learning fuzzy intervals for a numeric attribute i

4.2.1 Unsupervised Learning

The unsupervised learning phase learns a set of fuzzy intervals for features linked to distribution analysis of the input graphs, i.e., features for node degree, numeric node, and edge attributes and the corresponding resemblance attributes. We refer each of them as an attribute i. For symbolic node and edge attributes and the corresponding resemblance attributes, the unsupervised learning phase employs the modalities taken by this attribute and treats them as crisp intervals. The graph embedding phase then employs these intervals for computing different bins of the respective fuzzy or crisp histograms.

Input and output: The input to unsupervised learning phase of FMGE is the collection of m attributed graphs, given by

$$\{\mathrm{AG}_1, \mathrm{AG}_2, \ldots, \mathrm{AG}_e, \ldots, \mathrm{AG}_m\}$$

where the eth graph is denoted by

$$\mathrm{AG}_e = \left(V_e, E_e, \mu^{\mathrm{V}_e}, \mu^{\mathrm{E}_e}\right)$$

A second input to this phase could be if necessary for each feature the desired number of fuzzy intervals. This is referred by s_i for an attribute i. Some methods of discretization are able to find the "optimal" number of intervals by themselves (for example, equal frequency bins).

As output the unsupervised learning phase of FMGE produces s_i fuzzy overlapping trapezoidal intervals for an attribute i in input graphs (example in Fig. 7).

Description: The main steps for learning of fuzzy intervals for an attribute i are outlined in Fig. 6 and are explained in subsequent paragraphs.

The first step is the computation of crisp intervals from the list of values of attribute i for all the graphs in input collection of graphs. This is straightforward and is achieved by any standard data discretization technique. A survey of popular discretization techniques is presented by Liu et al. [44]. We propose to use equally spaced bins for obtaining an initial set of crisp intervals, as they avoid over-fitting and offers FMGE a better generalization capability to unseen graphs during graph embedding phase. Algorithm 4.1 outlines the pseudo-code for computing an initial set of crisp intervals for an attribute i. It uses a pseudo-call "*GetEqualSpacBin*" for getting an initial set of equally spaced bins. This pseudo-call refers to an appropriate data discretization function (available in underlying implementation

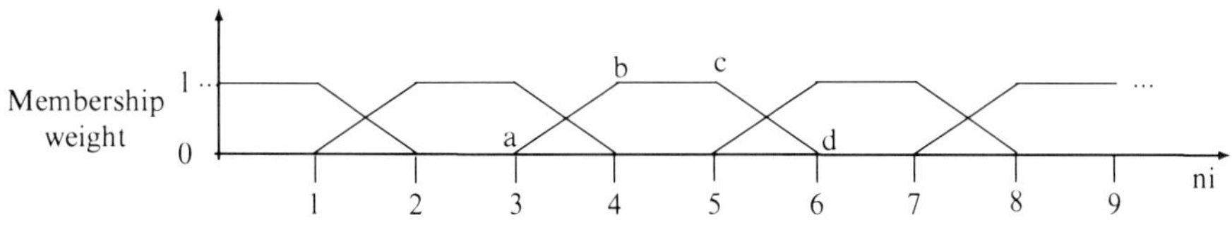

Fig. 7 Five fuzzy overlapping trapezoidal intervals (s_i) defined over nine equally spaced crisp intervals (n_i)

platform). Another possibility is to use the equal frequency bins. An example of this type of discretization is the technique proposed by [45] for discretization of continuous data. It is based on use of Akaike Information Criterion (AIC). It starts with an initial histogram of data and finds optimal number of bins for underlying data. The adjacent bins are iteratively merged using an AIC-based cost function until the difference between AIC-beforemerge and AIC-aftermerge becomes negative. Thus, we get an optimal set of crisp intervals for the underlying data. This set of intervals is representative of the distribution of underlying data.

The initial set of equally spaced bins (or equal frequency bins) is used to construct a data structure, which stores the crisp intervals. This data structure is employed for computing s_i fuzzy intervals for attribute i.

After computing the initial set of crisp intervals, in the next step, these crisp intervals are arranged in an overlapping fashion to get fuzzy overlapping intervals. Normally trapezoidal, triangular, and Gaussian arrangements are popular choices for fuzzy membership functions [46]. We propose to use the trapezoidal membership function, which is the generally used fuzzy membership function. It allows a range of instances to lie under full membership and assigns partial membership to the boundary instances. Figure 7 outlines a trapezoidal interval defined over crisp intervals. A trapezoidal interval is defined by four points, as is given by points a,b,c,d in Fig. 7.

It is very important to highlight here that the first fuzzy overlapping trapezoidal interval covers all values till $-\infty$ and the last fuzzy overlapping trapezoidal interval is limited by ∞. This makes sure that during the graph embedding phase every attribute instance falls under the range of fuzzy overlapping trapezoidal intervals and furthermore it strengthens the generalization abilities of the method to unseen graphs.

A fuzzy interval defined in trapezoidal fashion assigns a membership weight of 1 (full membership) between points b and c. The membership weight gradually approaches 0 as we move from b to a and from c to d. This trapezoidal behavior allows to assign full membership, partial membership, and no membership to attribute instances. This is important to highlight here that the total membership assigned to an instance is always exactly equal to 1, i.e., either one full membership or two partial memberships are assigned to each attribute instance.

Algorithm 4.1: GETINITCRISPINTERVAL($listvaluesAttribute_i, n_i$)

comment: Computes equally spaced crisp intervals.

comment: Requires: List of values of an attribute i

comment: Requires: Number of crisp intervals for attribute i ($n_i \geq 2$)

comment: Returns: 'n_i' crisp intervals for attribute i

$equallySpacedBins \leftarrow$ GETEQUALSPACBIN($listvaluesAttribute_i, n_i$)

$crispIntervals \leftarrow empty$
$st \leftarrow -\infty$
$en \leftarrow equallySpacedBins[1]$

$j \leftarrow 1$
repeat

$$\begin{cases} crispIntervals[j].start \leftarrow st \\ crispIntervals[j].end \leftarrow en \\ st \leftarrow en \\ en \leftarrow equallySpacedBins[j+1] \\ j \leftarrow j+1 \end{cases}$$

until $j > n_i$

return ($crispIntervals$)

Algorithm 4.2 outlines the pseudo-code for computing fuzzy overlapping trapezoidal intervals from an initial set of crisp intervals for an attribute i in input collection of graphs. It first computes the initial set of crisp intervals using Algorithm 4.1 and then arranges them in an overlapping trapezoidal fashion for obtaining a set of fuzzy overlapping trapezoidal intervals for attribute i. The number of fuzzy intervals for attribute i depends upon the number of features desired for attribute i in FSMFV and is controlled by parameter s_i which can either be manually specified, automatically learned by using an equal frequency based discretization or empirically learned and optimized on validation set.

The number of crisp intervals n_i for desired number of fuzzy intervals s_i is computed by (5):

$$n_i = 2 \times s_i - 1 \quad (5)$$

For the sake of continuity and readability, we have used the terms fuzzy intervals, fuzzy overlapping intervals, and fuzzy overlapping trapezoidal intervals interchangeably.

Algorithm 4.2: GETFUZZYOVLAPTRAPZINTERVAL($listvaluesAttribute_i, s_i$)

comment: Computes fuzzy intervals for an attribute i.

comment: Requires: List of values of an attribute i

comment: Requires: Number of fuzzy intervals for attribute i ($s_i \geq 2$)

comment: Returns: 's_i' fuzzy intervals for attribute i

```
n_i ← 2 * s_i − 1
crispIntervals ← GETINITCRISPINTERVAL(listvaluesAttribute_i, n_i)

fuzzyIntervals ← empty

fuzzyIntervals[1].a ← −∞
fuzzyIntervals[1].b ← −∞
fuzzyIntervals[1].c ← crispIntervals[1].end
fuzzyIntervals[1].d ← crispIntervals[2].end

j ← 1
jcrisp ← 0
repeat
    j ← j + 1
    jcrisp ← jcrisp + 2
    fuzzyIntervals[j].a ← fuzzyIntervals[j − 1].c
    fuzzyIntervals[j].b ← fuzzyIntervals[j − 1].d

    if (jcrisp + 1 ≥ n_i)
      then  fuzzyIntervals[j].c ← ∞
            fuzzyIntervals[j].d ← ∞
            break

    fuzzyIntervals[j].c ← crispIntervals[jcrisp + 1].end
    fuzzyIntervals[j].d ← crispIntervals[jcrisp + 2].end
until j ≥ s_i

return (fuzzyIntervals)
```

4.2.2 Graph Embedding

The graph embedding phase of FMGE employs the learned crisp intervals and fuzzy intervals to compute respective histograms for embedding an input attributed graph into a feature vector. This achieves the mapping of the input graphs to appropriate points in a suitable vector space $\mathbb{R}^n$. The graph embedding phase of FMGE produces a feature vector FSMFV_e for input graph AG_e. The length of the feature vector is strictly dependent on the size of histograms used for encoding the three levels of information. The length of this feature vector is uniform for all graphs in an input collection and is given by (6):

$$\text{Length of FSMFV} = 2 + \sum s_i + \sum c_i \tag{6}$$

where

- 2 refers to the features for graph order and graph size.
- $\sum s_i$ refers to the sum of number of bins in fuzzy interval encoded. histograms for numeric information in graph (i.e., attribute i).
- $\sum c_i$ refers to the sum of number of bins in crisp interval encoded. histograms for symbolic information in graph.

Numeric attributes: The values of each numeric attribute i in input graph AG_e are fuzzified by employing its s_i fuzzy intervals and trapezoidal membership function. Mathematically, the membership function α defined over a trapezoidal interval x is given by (7):

$$\alpha(x) = \begin{vmatrix} (x-a)/(b-a) & a \le x < b \\ 1 & b \le x \le c \\ (x-d)/(c-d) & c < x \le d \\ 0 & x < a \\ 0 & x > d \end{vmatrix} \tag{7}$$

In (7), x refer to an instance of attribute i to be fuzzified and a,b,c,d refers to the limits of a trapezoidal fuzzy interval for attribute i. Function $\alpha(x)$ computes the degree of membership of an instance x with the trapezoidal interval defined by a,b,c,d. The possible memberships can be *a full membership* if x is between b and c, *a partial membership* if x is between a and b or is between c and d, or it can be *a no membership* if x is outside the interval a,b,c,d.

We represent the fuzzy histogram of attribute i as h_i^{num}, which actually refers to the fuzzy histogram for node degrees (h^d in Fig. 4), the fuzzy histograms for numeric node attribute's resemblance (h_{d}^{nr} and $h_1^{\text{nr}}, h_2^{\text{nr}}, \ldots, h_k^{\text{nr}}$ in Fig. 4), the fuzzy histograms for numeric and symbolic edge attribute's resemblance ($h_1^{\text{er}}, h_2^{\text{er}}, \ldots, h_l^{\text{er}}$ in Fig. 4), the fuzzy histograms for numeric node attributes ($h_1^n, h_2^n, \ldots, h_k^n$ in Fig. 5), and the fuzzy histograms for numeric edge attributes ($h_1^e, h_2^e, \ldots, h_l^e$ in Fig. 5). The fuzzy histogram of an attribute i represents the embedding of attribute i for input graph AG_e. For an input graph AG_e, the fuzzy histogram h_i^{num} for attribute i is constructed by first computing the degree-of-memberships of all instances of attribute i in AG_e for the s_i fuzzy intervals, and then summing the memberships for each of the s_i fuzzy intervals.

Symbolic attributes: Each symbolic attribute j in input graph AG_k is encoded by a histogram of all its possible modalities (or labels). This histogram encodes the number of instances for each possible label of the symbolic attribute. We call this histogram as a crisp histogram (in contrary to fuzzy histogram for numeric attributes). We call a crisp histogram for a symbolic attribute j as h_j^{sym}, which actually refers to the crisp histograms for symbolic node attribute's resemblance ($h_1^{\mathrm{nr}}, h_2^{\mathrm{nr}}, \ldots, h_k^{\mathrm{nr}}$ in Fig. 4), the crisp histograms for symbolic node attributes ($h_1^n, h_2^n, \ldots, h_k^n$ in Fig. 5), and the crisp histograms for symbolic edge attributes ($h_1^e, h_2^e, \ldots, h_l^e$ in Fig. 5).

After constructing the fuzzy interval encoded histograms for each numeric attribute i and crisp histogram for each symbolic attribute j, the FSMFV_e for input graph AG_e is constructed from the value of graph order, the value of graph size and fuzzy interval encoded histograms h_i^{num} of its node degree, numeric node and edge attributes appended by the crisp histograms h_j^{sym} for symbolic node and edge attributes. This gives an embedding of the input graph AG_e into a feature vector FSMFV_e.

4.3 Application to Graph Classification

As an application of FMGE, we report some experimental results on three datasets from "IAM Graph Database Repository for Graph Based Pattern Recognition and Machine Learning." The IAM graph database repository is publicly available from the web site of IAPR technical committee on graph-based representations (TC-15)[1] and contains graph datasets from the field of document image analysis and graphics recognition, describing both synthetic and real data [39].

Datasets: The summary of the letter, GREC and fingerprint datasets, together with some characteristic properties, is given in Table 2. The letter graph dataset is comprised of graphs extracted from drawings of 15 capital letters of Roman alphabet that consists of straight lines only. The prototype drawing of letters are converted into prototype graphs by representing lines by undirected edges and ending points of lines by nodes. Each node is labeled with a two-dimensional attribute giving its position relative to a reference coordinate system. The GREC graph dataset is comprised of graphs representing 22 symbols from architectural and electronic drawings. Graphs are extracted from the denoised images by representing ending points, corners, intersections, and circles by nodes and labeled with a two-dimensional attribute giving their position. The nodes are connected by undirected edges which are labeled as line or arc and have the angle with respect to the horizontal direction as attribute. Fingerprint images are converted into graphs by representing the ending points and bifurcation points of the skeletonized regions as nodes. Each node is labeled with a two-dimensional attribute giving its position. The edges are attributed with an angle denoting the orientation of the edge with respect to the horizontal direction.

Experimental setup and results: The datasets consist of training, validation, and test sets. The experiments are performed by tuning the parameters on the validation

[1] http://www.greyc.ensicaen.fr/iapr-tc15/index.php.

Table 2 IAM graph database

		Letter LOW	GREC	Fingerprint
Size	Train	750	836	500
	Valid	750	836	300
	Test	750	1,628	2,000
Classes		15	22	4
Average	$\|V\|$	4.7	11.5	5.4
	$\|E\|$	3.1	12.2	4.4
Maximum	$\|V\|$	8	25	26
	$\|E\|$	6	30	25
Numeric attribute	$\|V\|$	2	2	2
	$\|E\|$	0	1	1
Symbolic attribute	$\|V\|$	0	1	0
	$\|E\|$	0	1	0

set and then testing with the best found configuration of the parameters on the test set. The considered parameters are the number of fuzzy intervals for embedding numeric information in graph, i.e., the node degree, numeric node attributes, and numeric edge attributes. Starting from 2 intervals, the number of fuzzy intervals for embedding the numeric information is increased until 25 (in steps of 1). Out of these different configurations, we selected the one which gave highest recognition rate on validation set. The selected parameter configuration is employed for embedding the graphs in test set.

After embedding the training and test sets into feature vectors, we performed PCA in vector space, for reducing the dimensionality of feature vectors. PCA requires to adopt a criterion for selecting the number of dimensions to keep. We have used eigenvalue-based intrinsic dimensionality estimation on training set for finding the number of interesting dimensions to keep for our experiments. The eigenvalue-based intrinsic dimensionality estimation performs PCA with two dimensions and evaluates the eigenvalues corresponding to the principal components (of the high-dimensional feature vectors in training set). The eigenvalues provide insight into the amount of variance that is described by their corresponding eigenvectors. The estimation of the intrinsic dimensionality is performed by counting the number of normalized (and ordered) eigenvalues that is higher than a very small threshold value. In our experimentation we kept all the eigenvalues greater than 0.025. The number of dimensions obtained from intrinsic dimensionality reduction is used for reducing the dimensionality of the feature vectors in training and test sets. The dimensions of original and PCA reduced feature vectors are given in Table 3. The reference system [39] employs a graph edit distance based nearest-neighbor classifier, because of the fact that there is a lack of general classification algorithms that can be applied to graphs. One of the few classifiers directly applicable to arbitrary graphs is the k-nearest-neighbor classifier (k-NN). Given a labeled set of training graphs, an unknown graph is assigned to the class that occurs most frequently among the k nearest graphs (in terms of edit distance) from the training

Table 3 Results on test sets for original vectors and PCA reduced vectors. Recognition rate (%) obtained by *1-NN* classifier and DIM is the dimensionality of feature vector

	Letter LOW		GREC		Fingerprint	
	Recognition rate	DIM	Recognition rate	DIM	Recognition rate	DIM
Classification in graph space	99.6		95.5		76.6	
Original FSMFV	96.5	58	97.2	79	76.6	127
PCA reduced FSMFV	96.3	10	96.8	14	77.2	16

set. The decision boundary of this classifier is a piecewise linear function which makes it very flexible.

Table 3 presents the recognition rates on test set, before and after the dimensionality reduction. The application of PCA successfully reduces the dimensionality of the feature vector without degrading the performance of the original graph embedding technique. The new low-dimensional feature vectors are very interesting for being efficiently processed by complex machine learning algorithms (sophisticated classification and clustering tools). In case of letter and GREC datasets the recognition rate on PCA reduced vectors is a little less than that of original vectors but the drastic decrease in dimensionality is very interesting achievement. For fingerprint dataset, as it was desired, the PCA reduced vector has very low dimensionality and it manages to achieve a little more recognition rate than original vectors.

The results clearly demonstrate that graph embedding followed by classification in vector space permits to employ the best of structural and statistical pattern recognition. The low dimensionality of FSMFV feature space gives a big gain in computational efficiency as compared to original graph space. The recognition rates are not true representative of FMGE's capabilities, since the datasets do not have many node and edge attributes. We have reported these results only to show the applicability of the explicit graph embedding approaches to the problem of graph classification. By embedding the graphs into feature vector spaces, the graph embedding methods open new horizons of computational tools (including learning, classification and clustering), for various application domains where the use of graph-based relational data structures is mandatory for performing high-level semantic operations.

5 Summary

In this chapter we first presented the new emerging research domain of graph embedding along with recent developments in explicit graph embedding methods. The discussion on the multilevel analysis of attributed graphs for explicit graph embedding in vector spaces outlined our explicit graph embedding method FMGE (the fuzzy multilevel graph embedding).

FMGE is a straightforward, simple, and computational efficient solution for facilitating the use of graph-based powerful representations together with learning and computational strengths of state-of-the-art machine learning, classification, and clustering. The method exploits multilevel analysis of graphs for extracting three levels of information from graphs (i.e. graph level, structural level, and elementary level) and embedding them into numeric feature vectors. It represents the graph topology information by new node and edge attributes and uses an unsupervised learning based algorithm to embed it into feature vector spaces. The method offers the embedding of attributed graphs with numeric as well as symbolic attributes on both nodes and edges. It has built-in learning abilities for adapting its parameters to underlying graph repositories. Computational time complexity of FMGE is linear to size of graphs and the number of attributes in graphs. However, the method is strongly dependent on the attributes of the nodes and edges in the graph. Apart from the elementary level details, the structural level details which are extracted by the homogeneity of subgraphs in the graph are also defined in terms of the node and edge attributes. The proposed feature vector lacks in information on the topology of the graph. This makes this method very useful for the application domains which extract attribute rich graphs from data, i.e., there are lots of meaningful attributes. The method is less useful for application domains where graphs have less number of attributes and topological details are required to be extracted (without using attributes) for graph embedding.

We conclude this chapter with a remark that graph embedding is an interesting approximate solution for addressing the problem of in-exact graph matching which belongs to the class of NP-complete problems. By mapping a high-dimensional graph into a point in suitable vector space, graph embedding permits to perform the basic mathematical computations which are required by various statistical pattern recognition techniques and offers interesting solutions to the problems of graph clustering and classification. However, in our opinion because of the strict limitation of the resulting feature vector not being capable of preserving the matching between nodes of graphs, graph embedding always lacks the capabilities to address the problem of graph isomorphism (i.e., exact graph matching).

References

1. Duda R, Hart P, Stork D (2000) Pattern classification, vol 2. Wiley Interscience, New York
2. Kuncheva L (2004) Combining pattern classifiers: Methods and algorithms. Wiley, New York
3. Byun H (2003) A survey on pattern recognition applications of support vector machines. Int J Pattern Recogn Artif Intell 17(3):459–486
4. De Sa J (2001) Pattern recognition: Concepts, methods, and applications. Springer, Berlin
5. Friedman M, Kandel A (1999) Introduction to pattern recognition. World Scientific, Singapore
6. Shawe-Taylor J, Cristianini N (2004) Kernel methods for pattern analysis. Cambridge University Press, Cambridge

7. Riesen K (2010) Classification and clustering of vector space embedded graphs. PhD thesis, University of Bern, Switzerland
8. Bunke H, Riesen K (2011) Recent advances in graph-based pattern recognition with applications in document analysis. Pattern Recogn 44:1057–1067
9. Riesen K, Bunke H (2009) Graph classification based on vector space embedding. Int J Pattern Recogn Artif Intell 23(6):1053–1081
10. Conte D, Foggia P, Sansone C, Vento M (2004) Thirty years of graph matching in pattern recognition. Int J Pattern Recogn Artif Intell 18(3):265–298
11. Bunke H, Irniger C, Neuhaus M (2005) Graph matching – challenges and potential solutions. In: International conference on image analysis and processing, Springer-Verlag Berlin, Heidelberg, pp 1–10
12. Shokoufandeh A, Macrini D, Dickinson S, Siddiqi K, Zucker S (2005) Indexing hierarchical structures using graph spectra. IEEE Trans Pattern Anal Mach Intell 27(7):1125–1140
13. Ferrer M, Valveny E, Serratosa F, Riesen K, Bunke H (2010) Generalized median graph computation by means of graph embedding in vector spaces. Pattern Recogn 43:1642–1655
14. Bunke H, Gunter S, Jiang X (2001) Towards bridging the gap between statistical and structural pattern recognition: Two new concepts in graph matching. In: International conference on advances in pattern recognition. Springer, Berlin, pp 1–11
15. Roth V, Laub J, Kawanabe M, Buhmann J (2003) Optimal cluster preserving embedding of nonmetric proximity data. IEEE Trans Pattern Anal Mach Intell 25(12):1540–1551
16. Chen T, Yang Q, Tang X (2007) Directed graph embedding. In: International joint conference on artificial intelligence, Morgan Kaufmann Publishers Inc. San Francisco, CA, USA, pp 2707–2712
17. Shaw B, Jebara T (2009) Structure preserving embedding. In: International conference on machine learning, ACM New York, NY, USA, pp 1–8
18. Foggia P, Vento M (2010) Graph Embedding for Pattern Recognition. In Ünay D, Çataltepe Z, Aksoy S (eds) Recognizing patterns in signals, speech, images and videos. Lecture notes in computer science, vol 6388. Springer, Berlin, pp 75–82
19. Lee G, Madabhushi A (2010) Semi-supervised graph embedding scheme with active learning (SSGEAL): Classifying high dimensional biomedical data. In: Pattern recognition in bioinformatics. Lecture notes in computer science, vol 6282. Springer, Berlin, pp 207–218
20. Riesen K, Bunke H (2010) Graph classification and clustering based on vector space embedding. World Scientic, Singapore
21. Riesen K, Bunke H (2010) Graph classification and clustering based on vector space embedding. World Scientific, Singapore
22. Wilson RC, Hancock ER, Luo B (2005) Pattern vectors from algebraic graph theory. IEEE Trans Pattern Anal Mach Intell 27:1112–1124
23. Wiener H (1947) Structural determination of paraffin boiling points. J Am Chem Soc 69(17)
24. Papadopoulos Y, et Manolopoulos AN (1999) Structure-based similarity search with graph histograms. In: International workshop on database and expert systems applications. IEEE Computer Society Press
25. Lopresti D, Wilfong G (2003) A fast technique for comparing graph representations with applications to performance evaluation, International Journal on Document Analysis and Recognition, 6: 219–229,
26. Gibert J, Valveny E, Bunke H (2011) Vocabulary selection for graph of words embedding. In: 5th Iberian conference on pattern recognition and image analysis. LNCS, 6669 ed. Springer, Berlin, pp 216–223
27. Gibert J, Valveny E, Bunke H (2011) Dimensionality reduction for graph of words embedding. In: LNCS 6658, Springer, pp 22–31
28. Kramer S, Raedt L (2001) Feature construction with version spaces for biochemical application. In: 18th international conference on machine learning, Morgan Kaufmann Publishers Inc. pp 258–265
29. Inokuchi A, Washio T, Motoda H (2000) An apriori-based algorithm for mining frequent substructures from graph data. Lect Notes Comput Sci 1910:13–23

30. Sidère N, Héroux P, Ramel J-Y (2009) Vector representation of graphs: application to the classification of symbols and letters. In: International conference on document analysis and recognition, IEEE Computer Society Press pp 681–685
31. Chung FRK (1997) Spectral graph theory. American Mathematical Society, Providence
32. Harchaoui Z (2007) Image classification with segmentation graph kernels. In: IEEE conference on computer vision and pattern recognition. IEEE Computer Society Press
33. Luo B, Wilson R, Hancock E (2003) Spectral embedding of graphs. Pattern Recogn 36:2213–2230
34. Robles-Kelly A, Hancock E (2007) A Riemannian approach to graph embedding. Pattern Recogn 40:1042–1056
35. Kosinov S, Caelli T (2002) Inexact multisubgraph matching using graph eigenspace and clustering models. In: SSPR/SPR, Springer, pp 133–142
36. Pekalska E, Duin RPW (2005) The dissimilarity representation for pattern recognition: Foundations and applications. World Scientific Publishing, Singapore
37. Riesen K, Neuhaus M, Bunke H (2007) Graph embedding in vector spaces by means of prototype selection. In: International conference on graph-based representations in pattern recognition. Springer, Berlin, pp 383–393
38. Ferrer M, Valveny E, Serratosa F, Riesen K, Bunke H (2008) An approximate algorithm for median graph computation using graph embedding. In: International conference on pattern recognition. IEEE, New York, pp 1–4
39. Riesen K, Bunke H (2010) IAM graph database repository for graph based pattern recognition and machine learning. In: Structural, syntactic, and statistical pattern recognition. Springer, Berlin, pp 287–297
40. Bunke H, Riesen K (2011) Improving vector space embedding of graphs through feature selection algorithms. Pattern Recogn 44:1928–1940
41. Luqman MM, Lladós J, Ramel J-Y, Brouard T (2010) A fuzzy-interval based approach for explicit graph embedding. In: Recognizing patterns in signals, speech, images and videos, vol 6388, Springer, pp 93–98
42. Luqman MM, Lladós J, Ramel J-Y, Brouard T (2011) Dimensionality reduction for fuzzy-interval based explicit graph embedding. In: GREC, pp 117–120
43. Luqman MM, Ramel J-Y, Lladós J, Brouard T (2013) Fuzzy multilevel graph embedding. Pattern recogn. **46**(2)551–565. ISSN 0031-3203, 10.1016/j.patcog.2012.07.029
44. Liu H, Hussain F, Tan C, Dash M (2002) "Discretization: An enabling technique. Data mining and knowledge, Springer, pp 393–423
45. Colot O, Courtellemont P, El-Matouat A (1994) Information criteria and abrupt changes in probability laws. In: Signal processing VII: theories and applications, pp 1855–1858
46. Ishibuchi H, Yamamoto T (2003) Deriving fuzzy discretization from interval discretization. In: International conference on fuzzy systems. IEEE, New York, pp 749–754

Feature Grouping and Selection Over an Undirected Graph

Sen Yang, Lei Yuan, Ying-Cheng Lai, Xiaotong Shen, Peter Wonka, and Jieping Ye

1 Introduction

High-dimensional regression/classification is challenging due to the *curse of dimensionality*. Lasso [18] and its various extensions [10], which can simultaneously perform feature selection and regression/classification, have received increasing attention in this situation. However, in the presence of highly correlated features lasso tends to only select one of those features resulting in suboptimal performance [25]. Several methods have been proposed to address this issue in the literature. Shen and Ye [15] introduce an adaptive model selection procedure that corrects the estimation bias through a data-driven penalty based on generalized degrees of freedom. The Elastic Net [25] uses an additional l_2 regularizer to encourage highly correlated features to stay together. However, these methods do not incorporate prior knowledge into the regression/classification process, which is critical in many applications. As an example, many biological studies have suggested that genes tend to work in groups according to their biological functions, and there are some regulatory relationships between genes [9]. This biological knowledge can be represented as a graph, where the nodes represent the genes, and the edges imply the regulatory relationships between genes. Therefore, we want to study how estimation accuracy can be improved using dependency information encoded as a graph.

Given feature grouping information, the group lasso [1, 6, 11, 21] yields a solution with grouped sparsity using l_1/l_2 penalty. The original group lasso does

S. Yang • L. Yuan • Y.-C. Lai • P. Wonka • J. Ye (✉)
Arizona State University, Tempe, AZ 85287, USA
e-mail: senyang@asu.edu; lei.yuan@asu.edu; ying-cheng.lai@asu.edu; peter.wonka@asu.edu
jieping.ye@asu.edu

X. Shen
University of Minnesota, Minneapolis, MN 55455, USA
e-mail: xshen@stat.umn.edu

Y. Fu and Y. Ma (eds.), *Graph Embedding for Pattern Analysis*,
DOI 10.1007/978-1-4614-4457-2_2,

not consider the overlaps between groups. Zhao et al. [22] extend the group lasso to the case of overlapping groups. Jacob et al. [6] introduce a new penalty function leading to a grouped sparse solution with overlapping groups. Yuan et al. [20] propose an efficient method to solve the overlapping group lasso. Other extensions of group lasso with tree structured regularization include [7, 11]. Prior works have demonstrated the benefit of using feature grouping information for high-dimensional regression/classification. However, these methods need the feature groups to be pre-specified. In other words, they only utilize the grouping information to obtain solutions with grouped sparsity, but lack the capability of identifying groups.

There are also a number of existing methods for feature grouping. Fused lasso [19] introduces an l_1 regularization method for estimating subgroups in a certain serial order, but pre-ordering features is required before using fused lasso. A study about parameter estimation of the fused lasso can be found in [12]; Shen et al. [13] propose a non-convex method to select all possible homogenous subgroups, but it fails to obtain sparse solutions. OSCAR [2] employs an l_1 regularizer and a pairwise l_∞ regularizer to perform feature selection and automatic feature grouping. Li and Li [9] suggest a grouping penalty using a Laplacian matrix to force the coefficients to be similar, which can be considered as a graph version of the Elastic Net. When the Laplacian matrix is an identity matrix, Laplacian lasso [5,9] is identical to the Elastic Net. GFlasso employs an l_1 regularization over a graph, which penalizes the difference $|\beta_i - \text{sign}(r_{ij})\beta_j|$, to encourage the coefficients β_i, β_j for features i, j connected by an edge in the graph to be similar when $r_{ij} > 0$, but dissimilar when $r_{ij} < 0$, where r_{ij} is the sample correlation between two features [8]. Although these grouping penalties can improve the performance, they would introduce additional estimation bias due to strict convexity of the penalties or due to possible graph misspecification. For example, additional bias may occur when the signs of coefficients for two features connected by an edge in the graph are different in Laplacian lasso [5,9], or when the sign of r_{ij} is inaccurate in GFlasso [8].

In this chapter, we focus on simultaneous estimation of grouping and sparseness structures over a given undirected graph. Features tend to be grouped when they are connected by an edge in a graph. When features are connected by an edge in a graph, the absolute values of the model coefficients for these two features should be similar or identical. We propose a convex and non-convex penalty to encourage both sparsity and equality of absolute values of coefficients for connected features. The convex penalty includes a pairwise l_∞ regularizer over a graph. The non-convex penalty improves the convex penalty by penalizing the difference of absolute values of coefficients for connected features. These penalties are designed to resolve the aforementioned issues of Laplacian lasso and GFlasso. Several recent works analyze their theoretical properties [14, 24]. Through ADMM and DC programming, we develop computational methods to solve the proposed formulations. The proposed methods can combine the benefit of feature selection and that of feature grouping to improve regression/classification performance. Due to the equality of absolute values of coefficients, the model complexity of the learned model can be reduced. We have performed experiments on synthetic data and a real dataset. The results demonstrate the effectiveness of the proposed methods.

The rest of the chapter is organized as follows. We introduce the proposed convex method in Sect. 2 and the proposed non-convex method in Sect. 3. Experimental results are given in Sect. 4. We conclude the chapter in Sect. 5.

2 A Convex Formulation

Consider a linear model in which response y_i depends on a vector of p features:

$$\mathbf{y} = \mathbf{X}\beta + \varepsilon, \tag{1}$$

where $\beta \in \mathbb{R}^p$ is a vector of coefficients, $\mathbf{X} \in \mathbb{R}^{n\times p}$ is the data matrix, and ε is random noise. Given an undirected graph, we try to build a prediction model (regression or classification) incorporating the graph structure information to estimate the nonzero coefficients of β and to identify the feature groups when the number of features p is larger than the sample size n. Let (N,E) be the given undirected graph, where $N = \{1,2,\ldots,p\}$ is a set of nodes, and E is the set of edges. Node i corresponds to feature $\mathbf{x}_i$. If nodes i and j are connected by an edge in E, then features $\mathbf{x}_i$ and $\mathbf{x}_j$ tend to be grouped. The formulation of graph OSCAR (GOSCAR) is given by

$$\min_{\beta} \frac{1}{2}\|\mathbf{y} - \mathbf{X}\beta\|^2 + \lambda_1\|\beta\|_1 + \lambda_2 \sum_{(i,j)\in E} \max\{|\beta_i|,|\beta_j|\}, \tag{2}$$

where λ_1, λ_2 are regularization parameters. We use a pairwise l_∞ regularizer to encourage the coefficients to be equal [2], but we only put grouping constraints over the nodes connected over the given graph. The l_1 regularizer encourages sparseness. The pairwise l_∞ regularizer puts more penalty on the larger coefficients. Note that $\max\{|\beta_i|,|\beta_j|\}$ can be decomposed as

$$\max\{|\beta_i|,|\beta_j|\} = \frac{1}{2}(|\beta_i + \beta_j| + |\beta_i - \beta_j|).$$

$\frac{1}{2}(|\beta_i + \beta_j| + |\beta_i - \beta_j|)$ can be represented by

$$|\mathbf{u}^{\mathrm{T}}\beta| + |\mathbf{v}^{\mathrm{T}}\beta|,$$

where $\mathbf{u},\mathbf{v}$ are sparse vectors, each with only two nonzero entries $u_i = u_j = \frac{1}{2}$, $v_i = -v_j = \frac{1}{2}$. Thus (2) can be rewritten in a matrix form as

$$\min_{\beta} \frac{1}{2}\|\mathbf{y} - \mathbf{X}\beta\|^2 + \lambda_1\|\beta\|_1 + \lambda_2\|\mathbf{T}\beta\|_1, \tag{3}$$

where $\mathbf{T}$ is a sparse matrix constructed from the edge set E.

The proposed formulation is closely related to OSCAR [2]. The penalty of OSCAR is $\lambda_1\|\beta\|_1 + \lambda_2\sum_{i<j}\max\{|\beta_i|,|\beta_j|\}$. The l_1 regularizer leads to a sparse solution, and the l_∞ regularizer encourages the coefficients to be equal. OSCAR can be efficiently solved by accelerated gradient methods, whose key projection can be solved by a simple iterative group merging algorithm [23]. However, OSCAR assumes each node is connected to all the other nodes, which is not sufficient for many applications. Note that OSCAR is a special case of GOSCAR when the graph is complete. GOSCAR, incorporating an arbitrary undirected graph, is much more challenging to solve.

2.1 Algorithm

We propose to solve GOSCAR using the alternating direction method of multipliers (ADMM) [3]. ADMM decomposes a large global problem into a series of smaller local subproblems and coordinates the local solutions to identify the globally optimal solution. ADMM attempts to combine the benefits of dual decomposition and augmented Lagrangian methods for constrained optimization [3]. The problem solved by ADMM takes the form of

$$\begin{aligned}&\min_{\mathbf{x},\mathbf{z}} f(\mathbf{x}) + g(\mathbf{z})\\ &s.t.\ \mathbf{A}\mathbf{x} + \mathbf{B}\mathbf{z} = \mathbf{c}.\end{aligned}$$

ADMM uses a variant of the augmented Lagrangian method and reformulates the problem as follows:

$$L_\rho(\mathbf{x},\mathbf{z},\mu) = f(\mathbf{x}) + g(\mathbf{z}) + \mu^{\mathrm{T}}(\mathbf{A}\mathbf{x} + \mathbf{B}\mathbf{z} - \mathbf{c}) + \frac{\rho}{2}\|\mathbf{A}\mathbf{x} + \mathbf{B}\mathbf{z} - \mathbf{c}\|^2,$$

with μ being the augmented Lagrangian multiplier and ρ being the nonnegative dual update step length. ADMM solves this problem by iteratively minimizing $L_\rho(\mathbf{x},\mathbf{z},\mu)$ over $\mathbf{x}$, $\mathbf{z}$, and μ. The update rule for ADMM is given by

$$\begin{aligned}\mathbf{x}^{k+1} &:= \arg\min_{\mathbf{x}} L_\rho(\mathbf{x},\mathbf{z}^k,\mu^k),\\ \mathbf{z}^{k+1} &:= \arg\min_{\mathbf{z}} L_\rho(\mathbf{x}^{k+1},\mathbf{z},\mu^k),\\ \mu^{k+1} &:= \mu^k + \rho(\mathbf{A}\mathbf{x}^{k+1} + \mathbf{B}\mathbf{z}^{k+1} - \mathbf{c}).\end{aligned}$$

Consider the unconstrained optimization problem in (3), which is equivalent to the following constrained optimization problem:

$$\min_{\beta,\mathbf{q},\mathbf{p}} \frac{1}{2}\|\mathbf{y}-\mathbf{X}\beta\|^2+\lambda_1\|\mathbf{q}\|_1+\lambda_2\|\mathbf{p}\|_1$$
$$s.t.\ \beta-\mathbf{q}=\mathbf{0},$$
$$\mathbf{T}\beta-\mathbf{p}=\mathbf{0}, \tag{4}$$

where $\mathbf{q},\mathbf{p}$ are slack variables. Equation (4) can then be solved by ADMM. The augmented Lagrangian is

$$L_\rho(\beta,\mathbf{q},\mathbf{p},\mu,\upsilon)=\frac{1}{2}\|\mathbf{y}-\mathbf{X}\beta\|^2+\lambda_1\|\mathbf{q}\|_1+\lambda_2\|\mathbf{p}\|_1+\mu^{\mathrm{T}}(\beta-\mathbf{q})$$
$$+\upsilon^{\mathrm{T}}(\mathbf{T}\beta-\mathbf{p})+\frac{\rho}{2}\|\beta-\mathbf{q}\|^2+\frac{\rho}{2}\|\mathbf{T}\beta-\mathbf{p}\|^2,$$

where μ,υ are augmented Lagrangian multipliers.

Update β: In the $(k+1)$-th iteration, β^{k+1} can be updated by minimizing L_ρ with $\mathbf{q},\mathbf{p},\mu,\upsilon$ fixed:

$$\beta^{k+1}=\arg\min_{\beta}\frac{1}{2}\|\mathbf{y}-\mathbf{X}\beta\|^2+(\mu^k+\mathbf{T}^{\mathrm{T}}\upsilon^k)^{\mathrm{T}}\beta+\frac{\rho}{2}\|\beta-\mathbf{q}^k\|^2+\frac{\rho}{2}\|\mathbf{T}\beta-\mathbf{p}^k\|^2. \tag{5}$$

The above optimization problem is quadratic. The optimal solution is given by $\beta^{k+1}=\mathbf{F}^{-1}\mathbf{b}^k$, where

$$\mathbf{F}=\mathbf{X}^{\mathrm{T}}\mathbf{X}+\rho(\mathbf{I}+\mathbf{T}^{\mathrm{T}}\mathbf{T}),$$
$$\mathbf{b}^k=\mathbf{X}^{\mathrm{T}}\mathbf{y}-\mu^k-\mathbf{T}^{\mathrm{T}}\upsilon^k+\rho\mathbf{T}^{\mathrm{T}}\mathbf{p}^k+\rho\mathbf{q}^k.$$

The computation of β^{k+1} involves solving a linear system, which is the most time-consuming part in the whole algorithm. To compute β^{k+1} efficiently, we compute the Cholesky factorization of $\mathbf{F}$ at the beginning of the algorithm:

$$\mathbf{F}=\mathbf{R}^{\mathrm{T}}\mathbf{R}.$$

Note that $\mathbf{F}$ is a constant and positive definite matrix. Using the Cholesky factorization we only need to solve the following two linear systems at each iteration:

$$\mathbf{R}^{\mathrm{T}}\hat{\beta}=\mathbf{b}^k,\ \mathbf{R}\beta=\hat{\beta}. \tag{6}$$

Since $\mathbf{R}$ is an upper triangular matrix, solving these two linear systems is very efficient.

Update $\mathbf{q}$: $\mathbf{q}^{k+1}$ can be obtained by solving

$$\mathbf{q}^{k+1}=\arg\min_{\mathbf{q}}\frac{\rho}{2}\|\mathbf{q}-\beta^{k+1}\|^2+\lambda_1\|\mathbf{q}\|_1-(\mu^k)^{\mathrm{T}}\mathbf{q},$$

which is equivalent to the following problem:

$$\mathbf{q}^{k+1} = \arg\min_{\mathbf{q}} \frac{1}{2}\|\mathbf{q} - \beta^{k+1} - \frac{1}{\rho}\mu^k\|^2 + \frac{\lambda_1}{\rho}\|\mathbf{q}\|_1. \tag{7}$$

Equation (7) has a closed-form solution, known as *soft-thresholding*

$$\mathbf{q}^{k+1} = S_{\lambda_1/\rho}\left(\beta^{k+1} + \frac{1}{\rho}\mu^k\right), \tag{8}$$

where the *soft-thresholding operator* is defined as:

$$S_\lambda(x) = \operatorname{sign}(x)\max(|x| - \lambda, 0).$$

Update p: Similar to updating $\mathbf{q}$, $\mathbf{p}^{k+1}$ can also be obtained by soft-thresholding:

$$\mathbf{p}_i^{k+1} = S_{\lambda_2/\rho}\left(\mathbf{T}\beta^{k+1} + \frac{1}{\rho}\upsilon^k\right). \tag{9}$$

Update μ, υ:

$$\begin{aligned} \mu^{k+1} &= \mu^k + \rho(\beta^{k+1} - \mathbf{q}^{k+1}), \\ \upsilon^{k+1} &= \upsilon^k + \rho(\mathbf{T}\beta^{k+1} - \mathbf{p}^{k+1}). \end{aligned} \tag{10}$$

A summary of GOSCAR is shown in Algorithm 1.

Algorithm 1: The GOSCAR algorithm

Input: $\mathbf{X}, \mathbf{y}, E, \lambda_1, \lambda_2, \rho$
Output: β
Initialization: $\mathbf{p}^0 \leftarrow \mathbf{0}, \mathbf{q}^0 \leftarrow \mathbf{0}, \mu^0 \leftarrow \mathbf{0}, \upsilon^0 \leftarrow \mathbf{0}$;
Compute the Cholesky factorization of $\mathbf{F}$;
do
  Compute β^{k+1} according to Eq. (6).
  Compute $\mathbf{q}^{k+1}$ according to Eq. (8).
  Compute $\mathbf{p}^{k+1}$ according to Eq. (9).
  Compute $\mu^{k+1}, \upsilon^{k+1}$ according to Eq. (10).
Until *Convergence*;
return β;

In Algorithm 1, the Cholesky factorization only needs to be computed once, and each iteration involves solving one linear system and two soft-thresholding operations. The time complexity of the soft-thresholding operation in (8) is $O(p)$. The other one in (9) involves a matrix–vector multiplication. Due to the sparsity of $\mathbf{T}$, its time complexity is $O(n_e)$, where n_e is the number of edges. Solving the linear system involves computing $\mathbf{b}^k$ and solving (6), whose total time complexity is $O(p(p+n)+n_e)$. Thus the time complexity of each iteration is $O(p(p+n)+n_e)$.

3 A Non-convex Formulation

The grouping penalty of GOSCAR overcomes the limitation of Laplacian lasso that the different signs of coefficients can introduce additional penalty. However, under the l_∞ regularizer, even if $|\beta_i|$ and $|\beta_j|$ are close to each other, the penalty on this pair may still be large due to the property of the max operator, resulting in the coefficient β_i or β_j being over penalized. The additional penalty would result in biased estimation, especially for large coefficient, as in the Lasso case [18]. Another related grouping penalty is GFlasso, $|\beta_i - \text{sign}(r_{ij})\beta_j|$, where r_{ij} is the pairwise sample correlation. GFlasso relies on the pairwise sample correlation to decide whether β_i and β_j are enforced to be close or not. When the pairwise sample correlation wrongly estimates the sign between β_i and β_j, an additional penalty on β_i and β_j would occur, introducing estimation bias. This motivates our non-convex grouping penalty, $||\beta_i| - |\beta_j||$, which shrinks only small differences in absolutes values. As a result, estimation bias is reduced as compared to these convex grouping penalties. The proposed non-convex method performs well even when the graph is wrongly specified, unlike GFlasso. Note that the proposed non-convex grouping penalty does not assume that the sign of an edge is given; it only relies on the graph structure.

The proposed non-convex formulation (ncFGS) solves the following optimization problem:

$$\min_{\beta} f(\beta) := \frac{1}{2}\|\mathbf{y} - \mathbf{X}\beta\|^2 + \lambda_1\|\beta\|_1 + \lambda_2 \sum_{(i,j)\in E} ||\beta_i| - |\beta_j||, \tag{11}$$

where the grouping penalty $\sum_{(i,j)\in E} ||\beta_i| - |\beta_j||$ controls only magnitudes of differences of coefficients ignoring their signs over the graph. Through the l_1 regularizer and grouping penalty, simultaneous feature grouping and selection are performed, where only large coefficients as well as pairwise differences are shrunk.

A computational method for the non-convex optimization in (11) is through DC programming. We will first give a brief review of DC programming.

A particular DC program on $\mathbb{R}^p$ takes the form of

$$f(\beta) = f_1(\beta) - f_2(\beta)$$

with $f_1(\beta)$ and $f_2(\beta)$ being convex on $\mathbb{R}^p$. Algorithms to solve DC programming based on the duality and local optimality conditions have been introduced in [17]. Due to their local characteristic and the non-convexity of DC programming, these algorithms cannot guarantee the computed solution to be globally optimal. In general, these DC algorithms converge to a local solution, but some researchers observed that they converge quite often to a global one [16].

To apply DC programming to our problem we need to decompose the objective function into the difference of two convex functions. We propose to use:

$$f_1(\beta) = \frac{1}{2}\|\mathbf{y} - \mathbf{X}\beta\|^2 + \lambda_1\|\beta\|_1 + \lambda_2 \sum_{(i,j)\in E} (|\beta_i + \beta_j| + |\beta_i - \beta_j|),$$

$$f_2(\beta) = \lambda_2 \sum_{(i,j)\in E} (|\beta_i| + |\beta_j|).$$

The above DC decomposition is based on the following identity: $||\beta_i| - |\beta_j|| = |\beta_i + \beta_j| + |\beta_i - \beta_j| - (|\beta_i| + |\beta_j|)$. Note that both $f_1(\beta)$ and $f_2(\beta)$ are convex functions.

Denote $f_2^k(\beta) = f_2(\beta^k) + \langle \beta - \beta^k, \partial f_2(\beta^k) \rangle$ as the affine minorization of $f_2(\beta)$, where $\langle \cdot, \cdot \rangle$ is the inner product. Then DC programming solves (11) by iteratively solving a subproblem as follows:

$$\min_{\beta} f_1(\beta) - f_2^k(\beta). \tag{12}$$

Since $\langle \beta^k, \partial f_2(\beta^k) \rangle$ is constant, (12) can be rewritten as

$$\min_{\beta} f_1(\beta) - \langle \beta, \partial f_2(\beta^k) \rangle. \tag{13}$$

Let $\mathbf{c}^k = \partial f_2(\beta^k)$. Note that

$$c_i^k = \lambda_2 d_i \mathrm{sign}(\beta_i^k) \mathbb{I}(\beta_i^k \neq 0), \tag{14}$$

where d_i is the degree of node i and $\mathbb{I}(\cdot)$ is the indicator function. Hence, the formulation in (13) is

$$\min_{\beta} \frac{1}{2}\|\mathbf{y} - \mathbf{X}\beta\|^2 + \lambda_1\|\beta\|_1 - (\mathbf{c}^k)^{\mathrm{T}}\beta + \lambda_2 \sum_{(i,j)\in E} (|\beta_i + \beta_j| + |\beta_i - \beta_j|), \tag{15}$$

which is convex. Note that the only differences between the problems in Eq. (2) and Eq. (15) are the linear term $(\mathbf{c}^k)^{\mathrm{T}}\beta$ and the second regularization parameter. Similar to GOSCAR, we can solve (15) using ADMM, which is equivalent to the following optimization problem:

$$\begin{aligned} &\min_{\beta,\mathbf{q},\mathbf{p}} \frac{1}{2}\|\mathbf{y} - \mathbf{X}\beta\|^2 - (\mathbf{c}^k)^{\mathrm{T}}\beta + \lambda_1\|\mathbf{q}\|_1 + 2\lambda_2\|\mathbf{p}\|_1 \\ &s.t\ \beta - \mathbf{q} = \mathbf{0}, \\ &\mathbf{T}\beta - \mathbf{p} = \mathbf{0}. \end{aligned} \tag{16}$$

There is an additional linear term $(\mathbf{c}^k)^{\mathrm{T}}\beta$ in updating β compared to Algorithm 1. Hence, we can use Algorithm 1 to solve Eq. (15) with a small change in updating β:

$$\mathbf{F}\beta - \mathbf{b}^s - \mathbf{c}^k = 0,$$

where s represents the iteration number in Algorithm 1.

Table 1 The algorithms and their associated penalties

Algorithm	Penalty
Lasso	$\lambda_1 \|\|\beta\|\|_1$
OSCAR	$\lambda_1 \|\|\beta\|\|_1 + \lambda_2 \sum_{i<j} \max\{\|\beta_i\|, \|\beta_j\|\}$
GFlasso	$\lambda_1 \|\|\beta\|\|_1 + \lambda_2 \sum_{(i,j)\in E} \|\beta_i - \text{sign}(r_{ij})\beta_j\|$
GOSCAR	$\lambda_1 \|\|\beta\|\|_1 + \lambda_2 \sum_{(i,j)\in E} \max\{\|\beta_i\|, \|\beta_j\|\}$
ncFGS	$\lambda_1 \|\|\beta\|\|_1 + \lambda_2 \sum_{(i,j)\in E} \|\|\beta_i\| - \|\beta_j\|\|$

The key steps of ncFGS are shown in Algorithm 2.

Algorithm 2: The ncFGS algorithm

Input: $\mathbf{X}, \mathbf{y}, E, \lambda_1, \lambda_2, \varepsilon$
Output: β
Initialization: $\beta^0 \leftarrow \mathbf{0}$;
while $f(\beta^k) - f(\beta^{k+1}) > \varepsilon$ **do**
  Compute $\mathbf{c}^k$ according to Eq. (14).
  Compute β^{k+1} using Algorithm 1 with $\mathbf{c}^k$ and $\lambda_1, 2\lambda_2$ as regularization parameters.
end
return β;

4 Numerical Results

We compare GOSCAR and ncFGS against lasso, GFlasso, and OSCAR on synthetic datasets and a real dataset: Breast Cancer.[1] The experiments are performed on a PC with dual-core Intel 3.0GHz CPU and 4GB memory. The code is written in MATLAB. The algorithms and their associated penalties are summarized in Table 1:

4.1 Efficiency

To evaluate the efficiency of the proposed methods, we conduct experiments on a synthetic dataset with a sample size of 100 and dimensions varying from 100 to 3,000. The regression model is $\mathbf{y} = \mathbf{X}\beta + \varepsilon$, where $\mathbf{X} \sim \mathcal{N}(0, \mathbf{I}_{p\times p})$, $\beta_i \sim \mathcal{N}(0,1)$, and $\varepsilon_i \sim \mathcal{N}(0, 0.01^2)$. The graph is randomly generated. The number of edges n_e varies from 100 to 3,000. The regularization parameters are set as $\lambda_1 = \lambda_2 = 0.8\max\{|\beta_i|\}$ with n_e fixed. Since the graph size affects the penalty, λ_1 and λ_2 are scaled by $\frac{1}{n_e}$ to avoid trivial solutions with dimension p fixed. The average computational time based on 30 repetitions is reported in Fig. 1. As can be seen in Fig. 1, GOSCAR can achieve 1e-4 precision in less than 10 s when the dimension and the number of edges are 1,000. The computational time of ncFGS

[1] http://cbio.ensmp.fr/~jvert/publi/

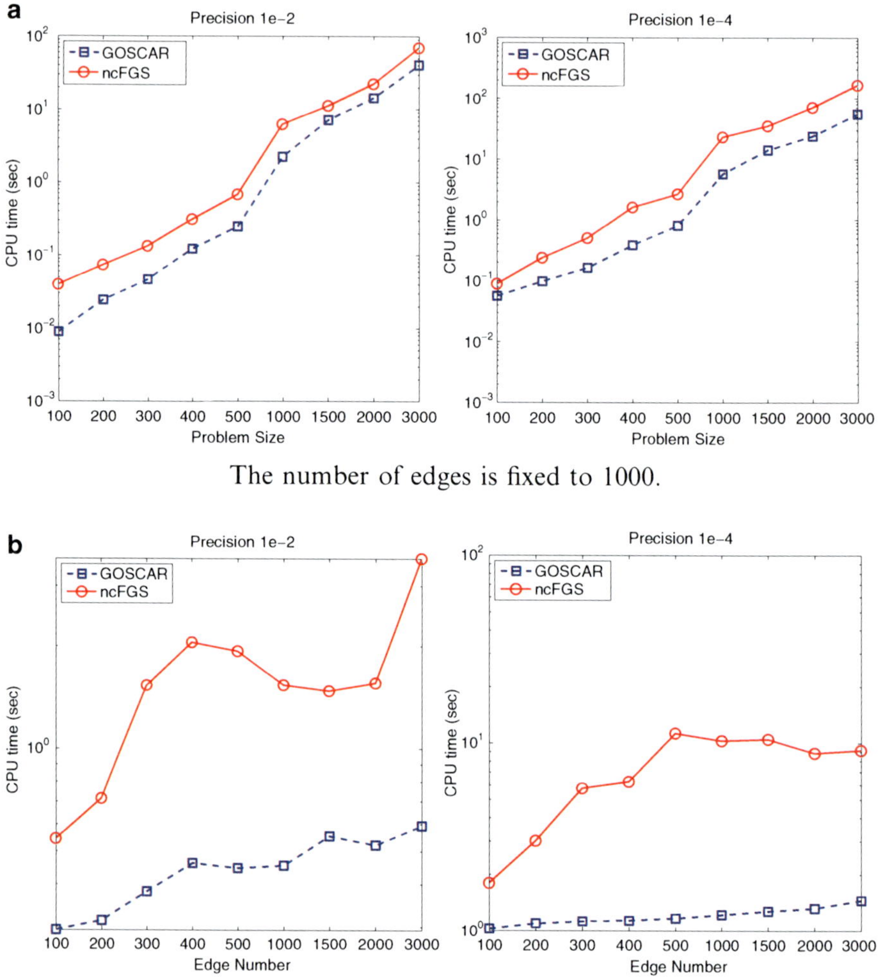

Fig. 1 Comparison of GOSCAR and ncFGS in terms of computation time with different dimensions, precisions, and the numbers of edges (in seconds and in logarithmic scale)

is about seven times higher than that of GOSCAR in this experiment. We can also observe that the proposed methods scale very well to the number of edges. The computational time of the proposed method increases less than 4 times when the number of edges increases from 100 to 3,000. It is not surprising because the complexity of each iteration in Algorithm 1 is linear with respect to n_e, and the sparsity of $\mathbf{T}$ makes the algorithm much more efficient. The increase of dimension is more costly than that of the number of edges, as the complexity is quadratic with respect to p.

4.2 Simulations

We use five synthetic problems that have been commonly used in the sparse learning literature [2, 9] to compare the performance of different methods. The data is generated from the regression model $\mathbf{y} = \mathbf{X}\beta + \varepsilon,\ \varepsilon_i \sim \mathcal{N}(0, \sigma^2)$. The five problems are given by:

1. $n = 100,\ p = 40$, and $\sigma = 2, 5, 10$. The true parameter is given by

$$\beta = (\underbrace{0,\ldots,0}_{10}, \underbrace{2,\ldots,2}_{10}, \underbrace{0,\ldots,0}_{10}, \underbrace{2,\ldots,2}_{10})^{\mathrm{T}}.$$

   $\mathbf{X} \sim \mathcal{N}(\mathbf{0}, \mathbf{S}_{p\times p})$ with $s_{ii} = 1, \forall i$ and $s_{ij} = 0.5$ for $i \neq j$.
2. $n = 50,\ p = 40,\ \beta = (\underbrace{3,\ldots,3}_{15}, \underbrace{0,\ldots,0}_{25})^{\mathrm{T}}$, and $\sigma = 2, 5, 10$. The features are generated as

$$\begin{aligned} \mathbf{x}_i &= Z_1 + \varepsilon_i^x,\ Z_1 \sim \mathcal{N}(0,1),\ \ i = 1,\ldots,5 \\ \mathbf{x}_i &= Z_2 + \varepsilon_i^x,\ Z_2 \sim \mathcal{N}(0,1),\ \ i = 6,\ldots,10 \\ \mathbf{x}_i &= Z_3 + \varepsilon_i^x,\ Z_3 \sim \mathcal{N}(0,1),\ \ i = 11,\ldots,15 \\ \mathbf{x}_i &\sim \mathcal{N}(0,1)\ \ i = 16,\ldots,40 \end{aligned}$$

   with $\varepsilon_i^x \sim \mathcal{N}(0, 0.16)$, and $\mathbf{X} = [\mathbf{x}_1, \ldots, \mathbf{x}_{40}]$.
3. Consider a regulatory gene network [9], where an entire network consists of n_{TF} subnetworks, each with one transcription factor (TF) and its 10 regulatory target genes. The data for each subnetwork can be generated as $\mathbf{X}_i^{\mathrm{TF}} \sim \mathcal{N}(0, \mathbf{S}_{11\times 11})$ with $s_{ii} = 1, s_{1i} = s_{i1} = 0.7, \forall i, i \neq 1$ and $s_{ij} = 0$ for $i \neq j, j \neq 1, i \neq 1$. Then $\mathbf{X} = [\mathbf{X}_1^{\mathrm{TF}}, \ldots, \mathbf{X}_{n_{\mathrm{TF}}}^{\mathrm{TF}}]$, $n = 100,\ p = 110$, and $\sigma = 5$. The true parameters are

$$\beta = \left(\underbrace{\frac{5}{\sqrt{11}},\ldots,\frac{5}{\sqrt{11}}}_{11}, \underbrace{\frac{-3}{\sqrt{11}},\ldots,\frac{-3}{\sqrt{11}}}_{11}, \underbrace{0,\ldots,0}_{p-22}\right)^{\mathrm{T}}.$$

4. Same as Problem 3 except that

$$\beta = \left(5, \underbrace{\frac{5}{\sqrt{10}},\ldots,\frac{5}{\sqrt{10}}}_{10}, -3, \underbrace{\frac{-3}{\sqrt{10}},\ldots,\frac{-3}{\sqrt{10}}}_{10}, \underbrace{0,\ldots,0}_{p-22}\right)^{\mathrm{T}}$$

5. Same as Problem 3 except that

$$\beta = \Bigg(\underbrace{5, \frac{5}{\sqrt{10}}, \dots, \frac{5}{\sqrt{10}}}_{10}, \underbrace{-5, \frac{-5}{\sqrt{10}}, \dots, \frac{-5}{\sqrt{10}}}_{10},$$

$$\underbrace{3, \frac{3}{\sqrt{10}}, \dots, \frac{3}{\sqrt{10}}}_{10}, \underbrace{-3, \frac{-3}{\sqrt{10}}, \dots, \frac{-3}{\sqrt{10}}}_{10} \underbrace{0, \dots, 0}_{p-44}\Bigg)^{\mathrm{T}}$$

We assume that the features in the same group are connected in a graph, and those in different groups are not connected. We use MSE to measure the performance of estimation of β, which is defined as

$$\mathrm{MSE}(\beta) = (\beta - \beta^*)^{\mathrm{T}}\mathbf{X}^{\mathrm{T}}\mathbf{X}(\beta - \beta^*).$$

For feature grouping and selection, we introduce two separate metrics to measure the accuracy of feature grouping and selection. Denote $I_i, i = 0, 1, 2, \dots, K$ as the index of different groups, where I_0 is the index of zero coefficients. Then the metric for feature selection is defined as

$$s_0 = \frac{\sum_{i \in I_0} \mathbb{I}(\beta_i = 0) + \sum_{i \notin I_0} \mathbb{I}(\beta_i \neq 0)}{p},$$

and the metric for feature grouping is defined as

$$s = \frac{\sum_{i=1}^{K} s_i + s_0}{K+1},$$

where

$$s_i = \frac{\sum_{i \neq j, i, j \in I_i} \mathbb{I}(|\beta_i| = |\beta_j|) + \sum_{i \neq j, i \in I_i, j \notin I_i} \mathbb{I}(|\beta_i| \neq |\beta_j|)}{|I_i|(p-1)}.$$

s_i measures the grouping accuracy of group i under the assumption that the absolute values of entries in the same group should be the same, but different from those in different groups. s_0 measures the accuracy of feature selection. It is clear that $0 \leq s_0, s_i, s \leq 1$.

For each dataset, we generate n samples for training, as well as n samples for testing. To make the synthetic datasets more challenging, we first randomly select $\lfloor n/2 \rfloor$ coefficients, and change their signs, as well as those of the corresponding features. Denote $\tilde{\beta}$ and $\tilde{\mathbf{X}}$ as the coefficients and features after changing signs. Then $\tilde{\beta}_i = -\beta_i$, $\tilde{\mathbf{x}}_i = -\mathbf{x}_i$, if the i-th coefficient is selected; otherwise, $\tilde{\beta}_i = \beta_i$, $\tilde{\mathbf{x}}_i = \mathbf{x}_i$, so that $\tilde{\mathbf{X}}\tilde{\beta} = \mathbf{X}\beta$. We apply different approaches on $\tilde{\mathbf{X}}$. The covariance matrix of $\mathbf{X}$ is used in GFlasso to simulate the graph misspecification. The results of β converted from $\tilde{\beta}$ are reported.

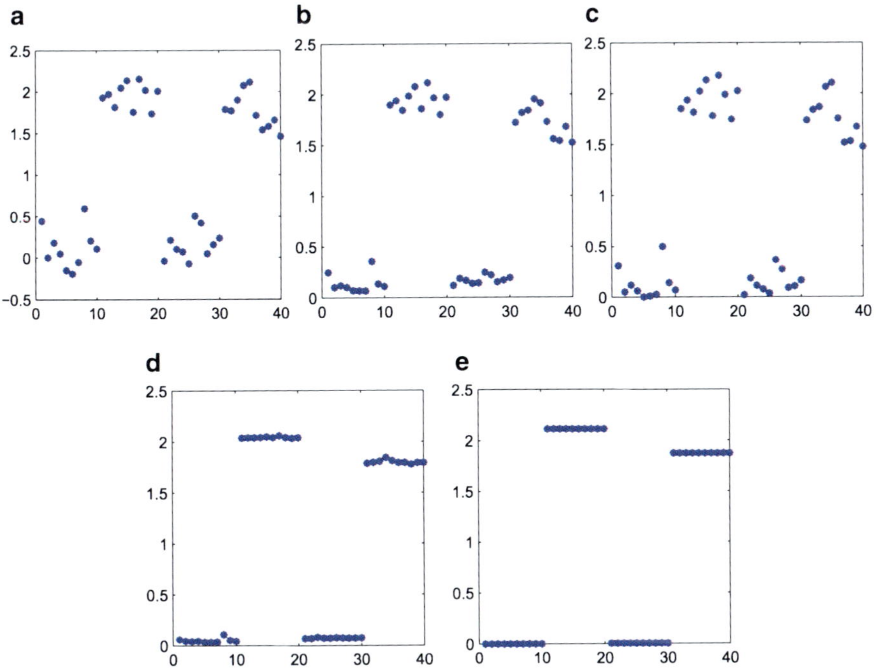

Fig. 2 The average nonzero coefficients obtained on dataset 1 with $\sigma = 2$: (**a**) Lasso; (**b**) GFlasso; (**c**) OSCAR; (**d**) GOSCAR; (**e**) ncFGS

Table 2 Comparison of performance in terms of MSEs and estimated standard deviations (in parentheses) for different methods based on 30 simulations on different synthetic datasets

Datasets	Lasso	OSCAR	GFlasso	GOSCAR	ncFGS
Data1 ($\sigma = 2$)	1.807 (0.331)	1.441 (0.318)	1.080 (0.276)	0.315 (0.157)	**0.123 (0.075)**
Data1 ($\sigma = 5$)	5.294 (0.983)	5.328 (1.080)	3.480 (1.072)	1.262 (0.764)	**0.356 (0.395)**
Data1 ($\sigma = 10$)	12.628 (3.931)	13.880 (4.031)	13.411 (4.540)	6.061 (4.022)	**1.963 (1.600)**
Data2 ($\sigma = 2$)	1.308 (0.435)	1.084 (0.439)	0.623 (0.250)	0.291 (0.208)	**0.226 (0.175)**
Data2 ($\sigma = 5$)	4.907 (1.496)	4.868 (1.625)	2.538 (0.656)	0.781 (0.598)	**0.721 (0.532)**
Data2 ($\sigma = 10$)	18.175 (6.611)	18.353 (6.611)	6.930 (2.858)	4.601 (2.623)	**4.232 (2.561)**
Data3 ($\sigma = 5$)	5.163 (1.708)	4.503 (1.677)	4.236 (1.476)	3.336 (1.725)	**0.349 (0.282)**
Data4 ($\sigma = 5$)	7.664 (2.502)	7.167 (2.492)	7.516 (2.441)	7.527 (2.434)	**5.097 (0.780)**
Data5 ($\sigma = 5$)	9.893 (1.965)	7.907 (2.194)	9.622 (2.025)	9.810 (2.068)	**7.684 (1.1191)**

Figure 2 shows the average nonzero coefficients obtained on dataset 1 with $\sigma = 2$. As can be seen in Fig. 2, GOSCAR and ncFGS are able to utilize the graph information and achieve good parameter estimation. Although GFlasso can use the graph information, it performs worse than GOSCAR and ncFGS due to the graph misspecification.

The performance in terms of MSEs averaged over 30 simulations is shown in Table 2. As indicated in Table 2, among existing methods (Lasso, GFlasso,

Table 3 Accuracy of feature grouping and selection based on 30 simulations for five feature grouping methods: the first row for each dataset corresponds to the accuracy of feature selection; the second row corresponds to the accuracy of feature grouping. The numbers in parentheses are the standard deviations

Datasets	OSCAR	GFlasso	GOSCAR	ncFGS
Data1 ($\sigma = 2$)	0.675 (0.098)	0.553 (0.064)	0.513 (0.036)	**0.983 (0.063)**
	0.708 (0.021)	0.709 (0.017)	0.702 (0.009)	**0.994 (0.022)**
Data1 ($\sigma = 5$)	0.565 (0.084)	0.502 (0.009)	0.585 (0.085)	**1.000 (0.000)**
	0.691 (0.011)	0.709 (0.016)	0.708 (0.017)	**1.000 (0.000)**
Data1 ($\sigma = 10$)	0.532 (0.069)	0.568 (0.088)	0.577 (0.061)	**0.983 (0.063)**
	0.675 (0.031)	0.725 (0.022)	0.708 (0.020)	**0.994 (0.022)**
Data2 ($\sigma = 2$)	0.739 (0.108)	0.544 (0.272)	**1.000 (0.000)**	0.958 (0.159)
	0.625 (0.052)	0.823 (0.029)	**0.837 (0.014)**	0.831 (0.052)
Data2 ($\sigma = 5$)	0.763 (0.114)	0.717 (0.275)	**0.999 (0.005)**	0.979 (0.114)
	0.650 (0.066)	0.741 (0.062)	0.833 (0.011)	**0.845 (0.030)**
Data2 ($\sigma = 10$)	0.726 (0.101)	0.818 (0.149)	0.993 (0.024)	**1.000 (0.000)**
	0.597 (0.058)	0.680 (0.049)	0.829 (0.025)	**0.851 (0.015)**
Data3 ($\sigma = 5$)	0.886 (0.135)	0.736 (0.103)	0.382 (0.084)	**0.992 (0.026)**
	0.841 (0.056)	0.739 (0.041)	0.689 (0.013)	**0.995 (0.017)**
Data4 ($\sigma = 5$)	0.875 (0.033)	0.881 (0.026)	**0.882 (0.037)**	0.796 (0.245)
	0.834 (0.030)	0.805 (0.035)	0.805 (0.036)	**0.895 (0.114)**
Data5 ($\sigma = 5$)	0.760 (0.203)	0.802 (0.153)	0.861 (0.051)	**0.881 (0.174)**
	0.858 (0.031)	0.821 (0.037)	0.805 (0.037)	**0.920 (0.056)**

OSCAR), GFlasso is the best, except in the two cases where OSCAR is better. GOSCAR is better than the best existing method in all cases except for two, and ncFGS outperforms all the other methods.

Table 3 shows the results in terms of accuracy of feature grouping and selection. Since Lasso does not perform feature grouping, we only report the results of the other four methods: OSCAR, GFlasso, GOSCAR, and ncFGS. Table 3 shows that ncFGS achieves higher accuracy than other methods in most cases.

4.3 Real Data

We conduct experiments on the breast cancer dataset. The metrics to measure the performance of different algorithms include accuracy (acc.), sensitivity (sen.), specificity (spe.), degrees of freedom (dof.), and the number of nonzero coefficients (nonzero coeff.). The dof. of lasso is the number of nonzero coefficients [18]. For the algorithms capable of feature grouping, we use the same definition of dof. in [2], which is the number of estimated groups.

The breast cancer dataset consists of gene expression data for 8,141 genes in 295 breast cancer tumors (78 metastatic and 217 non-metastatic). The network described in [4] is used as the input graph in this experiment. Figure 3 shows a subgraph

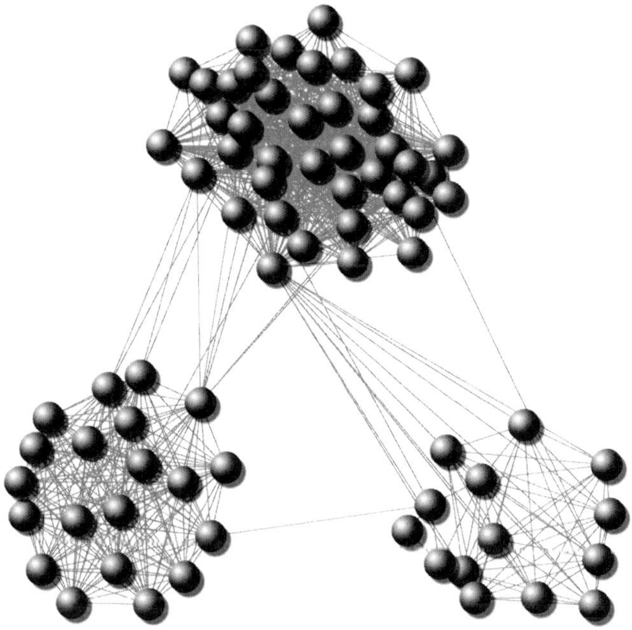

Fig. 3 A subgraph of the network in breast cancer dataset [4]. The subgraph consists of 80 nodes

Table 4 Comparison of classification accuracy, sensitivity, specificity, degrees of freedom, and the number of nonzero coefficients averaged over 30 replications for various methods on breast cancer dataset

Metrics	Lasso	OSCAR	GFlasso	GOSCAR	ncFGS
acc.	0.739 (0.054)	0.755 (0.055)	0.771 (0.050)	**0.783 (0.042)**	0.779 (0.041)
sen.	0.707 (0.056)	0.720 (0.060)	0.749 (0.060)	**0.755 (0.050)**	**0.755 (0.055)**
pec.	0.794 (0.071)	0.810 (0.068)	0.805 (0.056)	**0.827 (0.061)**	0.819 (0.058)
dof.	239.267	165.633	108.633	70.267	**57.233**
nonzero coeff.	239.267	243.867	144.867	140.667	**79.833**

The numbers in parentheses are the standard deviations

consisting of 80 nodes of the used graph. We restrict our analysis to the 566 genes most correlated to the output, but also connected in the graph. About 2/3 data is randomly chosen as training data, and the remaining 1/3 data is used as testing data. The tuning parameter is estimated by fivefold cross validation. Table 4 shows the results averaged over 30 replications. As indicated in Table 4, GOSCAR and ncFGS outperform the other three methods.

5 Summary

In this chapter, we consider simultaneous feature grouping and selection over a given undirected graph. We propose a convex and non-convex penalty to encourage both sparsity and equality of absolute values of coefficients for features connected in the graph. We employ ADMM and DC programming to solve the proposed formulations. Numerical experiments on synthetic and real data demonstrate the effectiveness of the proposed methods. Our results also demonstrate the benefit of simultaneous feature grouping and feature selection through the proposed non-convex method. In this chapter, we focus on undirected graphs. A possible future direction is to extend the formulations to directed graphs.

Acknowledgements This work was supported in part by NSF (IIS-0953662, MCB-1026710, CCF-1025177, DMS-0906616) and NIH (R01LM010730, 2R01GM081535-01, R01HL105397).

References

1. Bach F, Lanckriet G, Jordan M (2004) Multiple kernel learning, conic duality, and the SMO algorithm. In: ICML ACM New York, NY, USA. DOI 10.1145/1015330.1015424
2. Bondell H, Reich B (2008) Simultaneous regression shrinkage, variable selection, and supervised clustering of predictors with oscar. Biometrics 64(1):115–123
3. Boyd S, Parikh N, Chu E, Peleato B, Eckstein J (2011) Distributed optimization and statistical learning via the alternating direction method of multipliers Foundations and Trends ® in Machine Learning, Now Publishers Inc 3(1):1–122
4. Chuang H, Lee E, Liu Y, Lee D, Ideker T (2007) Network-based classification of breast cancer metastasis. Mol Syst Biol 3(1)
5. Fei H, Quanz B, Huan J (2010) Regularization and feature selection for networked features. In: CIKM, ACM New York, NY, USA pp 1893–1896. DOI 10.1145/1871437.1871756
6. Jacob L, Obozinski G, Vert J (2009) Group lasso with overlap and graph lasso. In: ICML, ACM New York, NY, USA pp 433–440. DOI 10.1145/1553374.1553431
7. Jenatton R, Mairal J, Obozinski G, Bach F (2010) Proximal methods for sparse hierarchical dictionary learning. In: ICML ACM New York, NY, USA
8. Kim S, Xing E (2009) Statistical estimation of correlated genome associations to a quantitative trait network. PLoS Genet 5(8):e1000587
9. Li C, Li H (2008) Network-constrained regularization and variable selection for analysis of genomic data. Bioinformatics 24(9):1175–1182
10. Liu J, Ji S, Ye J (2009) SLEP: Sparse learning with efficient projections. Arizona State University, http://www.public.asu.edu/~jye02/Software/SLEP/
11. Liu J, Ye J (2010) Moreau-Yosida regularization for grouped tree structure learning. In: NIPS
12. Rinaldo A (2009) Properties and refinements of the fused lasso. Ann Stat 37(5B):2922–2952
13. Shen X, Huang H (2009) Grouping pursuit through a regularization solution surface. J Am Stat Assoc 105(490):727–739
14. Shen X, Huang H, Pan W (2012) Simultaneous supervised clustering and feature selection over a graph. Biometrika, to appear
15. Shen X, Ye J (2002) Adaptive model selection. J Am Stat Assoc 97(457):210–221
16. Tao P, An L (1997) Convex analysis approach to DC programming: Theory, algorithms and applications. Acta Math Vietnam 22(1):289–355

17. Tao P, El Bernoussi S (1988) Duality in DC (difference of convex functions) optimization. Subgradient methods. Trends Math Optimiz 84:277–293
18. Tibshirani R (1996) Regression shrinkage and selection via the lasso. J Roy Stat Soc Ser B, 58(1): 267–288
19. Tibshirani R, Saunders M, Rosset S, Zhu J, Knight K (2005) Sparsity and smoothness via the fused lasso. J Roy Stat Soc Ser B 67(1):91–108
20. Yuan L, Liu J, Ye J (2011) Efficient methods for overlapping group lasso. In: NIPS
21. Yuan M, Lin Y (2006) Model selection and estimation in regression with grouped variables. J Roy Stat Soc Ser B 68(1):49–67
22. Zhao P, Rocha G, Yu B (2009) The composite absolute penalties family for grouped and hierarchical variable selection. Ann Stat 37(6A):3468–3497
23. Zhong L, Kwok J (2011) Efficient sparse modeling with automatic feature grouping. In: ICML
24. Zhu Y, Shen X, Pan W (2012) Simultaneous grouping pursuit and feature selection in regression over an undirected graph. Manuscript
25. Zou H, Hastie T (2005) Regularization and variable selection via the elastic net. J Roy Stat Soc Ser B 67(2):301–320

Median Graph Computation by Means of Graph Embedding into Vector Spaces

Miquel Ferrer, Itziar Bardají, Ernest Valveny, Dimosthenis Karatzas, and Horst Bunke

1 Introduction

In pattern recognition [8, 14], a key issue to be addressed when designing a system is how to represent input patterns. Feature vectors is a common option. That is, a set of numerical features describing relevant properties of the pattern are computed and arranged in a vector form. The main advantages of this kind of representation are computational simplicity and a well sound mathematical foundation. Thus, a large number of operations are available to work with vectors and a large repository of algorithms for pattern analysis and classification exist. However, the simple structure of feature vectors might not be the best option for complex patterns where nonnumerical features or relations between different parts of the pattern become relevant.

In this context, graphs comprise an attractive alternative to represent complex and structured objects. One of the main advantages of graphs over feature vectors is that they can explicitly model the relations between the different parts of the

M. Ferrer (✉)
DAMA-UPC, Universitat Politècnica de Catalunya - BarcelonaTech, Barcelona, Spain
e-mail: mferrer@ac.upc.edu

I. Bardají
Institut de Robòtica i Informàtica Industrial, CSIC-UPC, Universitat Politècnica de Catalunya - BarcelonaTech, Barcelona, Spain
e-mail: ibardaji@iri.upc.edu

E. Valveny • D. Karatzas
Centre de Visió per Computador, Universitat Autònoma de Barcelona, Bellaterra, Spain
e-mail: ernest@cvc.uab.cat; dimos@cvc.uab.cat

H. Bunke
NICTA, Vistoria Research Laboratory, The University of Melbourne, Parkville, Victoria, Australia
e-mail: bunke@iam.unibe.ch

Y. Fu and Y. Ma (eds.), *Graph Embedding for Pattern Analysis*,
DOI 10.1007/978-1-4614-4457-2_3,

object, whereas feature vectors are only able to describe an object as an aggregation of numerical properties. In addition, graphs permit to associate any kind of label (not only numbers) to both edges and nodes. Furthermore, the dimensionality of graphs, i.e., the number of nodes and edges, can be different for every object. Thus, the more complex an object is, the larger the number of nodes and edges used to represent it can be. However, the main drawback of using graphs arises from the difficulty and complexity of computational manipulation. Even the simple task of comparing two graphs, which is commonly referred to as graph matching, turns out to be very complex under some conditions [5]. In addition, most mathematical operations required in many learning and classification algorithms are not possible in the graph domain.

In order to overcome these limitations graph embedding techniques have gained popularity recently. They can combine the strengths of both domains, that is the high representational power of graphs together with all the mathematical foundation and algorithms available for the feature vectors. Graph embedding [20] aims to convert graphs into real vectors and then operate in the associated vector space. Thus, it emerges as a powerful way to provide graph-based representations with access to the rich repository of algorithmic tools available in statistical pattern analysis [7, 16]. To this end, different graph embedding procedures have been proposed in the literature so far. Some of them [4, 25, 33, 39, 42, 46] are based on the spectral graph theory. Graphs are converted into a vector representation using some spectral features extracted from the adjacency or the Laplacian matrix of a graph. Another family of graph embedding approaches is based on the similarity between graphs. For instance, in [22] graph features are extracted out of the dissimilarity matrix among a set of graphs. Alternatively, another embedding inspired in the work proposed in [31] is presented in [38]. Given a set of some a priori selected graph prototypes each point is embedded into a vector space by taking the distance (in this case the graph edit distance is used) to all these prototypes. The basic intuition of this work is that the description of the regularities in observations of classes and objects is the basis to perform pattern classification. Thus, assuming that the prototypes have been chosen appropriately, each class will form a compact zone in the vector space. Finally, there is a family of embedding approaches based on computing frequencies of certain substructures of the graphs [15, 24].

Graph embedding permits to go from the graph domain to the vector domain. The reverse problem, going from the vector space back to the graph space, is not so easy since it implies recovering the structural information that is usually lost with the embedding. The ability to translate a vectorial result calculated in the embedding space back to a graph is a condition sine qua non for linking statistical and structural pattern recognition through graph embedding.

Formally, the median graph [21] is defined as the graph that has the minimum sum of distances (SOD) to all graphs in a given set. Thus, it can be taken as the representative of the set and, therefore, it has a large number of potential applications including classical algorithms for learning, clustering, and classification that are normally used in the vector domain. As a matter of fact, it can be potentially applied to any graph-based algorithm where a representative of a set of graphs is needed.

However, the computation of the median graph is exponential both in the number of input graphs and their size [21]. The only exact algorithm proposed up to now [27] is based on an A^* algorithm using a data structure called multimatch. As the computational cost of this algorithm is very high, a set of approximate algorithms have also been presented in the past based on different approaches such as genetic search [21,27], greedy algorithms [19], and spectral graph theory [11,45]. However, all these algorithms can only be applied to restricted sets of graphs, regarding either the type or the size of the graphs.

Graph embedding can help in finding more efficient methods to compute the median graph and has already been explored in the past [12, 13]. The hypothesis underlying these works is that embedding the graphs into vectors and computing the median in the vector space can lead to a corresponding graph in the graph space is a good approximation of the median graph. Therefore, this procedure relies on some method to reconstruct the graph corresponding to the median vector. In these previous works [12, 13] an approximate reconstruction is obtained using some heuristics that permit to recover a graph based on the concept of the weighted mean of a pair of graphs using the median vector and the graphs corresponding to the 2 or 3 closest points to it.

In this chapter we present a generic procedure to convert a point in a vector space into a graph, given that we have a set of n graphs that have been previously embedded in the vector space (with n being the dimension of the vector space) and that the point we want to convert lies inside the convex hull of such embedded vectors, which is for example the case with the median and barycenter of the set of points. The basic idea of this approach is as follows. Given n graphs mapped to their corresponding points in the n-dimensional real space and a point inside the convex hull of such set a of points, we iteratively project the point into subspaces of lower dimensionality until a projected point is obtained lying on a line that connects the maps of two graphs of the given set. The graph corresponding to this point can be approximately reconstructed by means of the weighted mean. Next, we recursively consider all other projected points obtained before in higher dimensional spaces and apply the same reconstruction principle until the graph corresponding to the desired point is obtained. Ideally, this procedure would permit to recover the graph that corresponds exactly to the input point. However, due to the complexity of graph matching problems, we are forced to use approximate algorithms at different steps and therefore we will only be able to obtain partial approximations of the input point.

We show the application of this procedure to the computation of the median graph. As in the previous works, the median graph is obtained after embedding a set of graphs into a vector space, computing the median of such a set in the vector domain and then recovering the graph corresponding to median vector. For this last step, the proposed generic procedure is used. The application of this procedure raises some considerations about the order in which graphs have to be taken along the procedure. In this sense, we analyze four additional variations of the method which take into account different sorting schemes of the original set of graphs. These variations can help to understand the influence of the approximations assumed across the procedure. It is also important to remark that this generic framework could be potentially used in conjunction with any embedding technique and with any method to compute the representative of the set in the vector space.

In order to test the feasibility of applying this procedure to the computation of the median graph, we evaluate the final SOD of the median graph obtained to all the graphs of the set as a measure of the quality of median graph. We have also made clustering experiments on three different graph databases, one semi-artificial and two containing real-world data. In these clustering experiments, the proposed algorithms for the median graph computation are used to obtain the centers of the clusters. The underlying graphs have no constraints regarding the number of nodes and edges. The results are evaluated, according to four different clustering measures, namely, the Rand index, the Dunn index, the bipartite index, and the mutual information index. We will show that the clusters obtained using the proposed method are better than those using the set median graph or previous approaches also based on graph embedding.

The rest of this chapter is organized as follows. In the next section we define the basic concepts and we introduce the notation we will use later in the chapter. Then, in Sect. 3 the proposed generic method for recovering a graph from a point in the vector space is described. After that, Sect. 4 presents the practical implementation of the proposed generic framework for the computation of the median graph. Section 5 reports a number of experiments and presents the results achieved with our method. Also a comparison with several reference systems is provided. Finally, in Sect. 6 we draw some conclusions and we point out to possible future work.

2 Basic Concepts

This section introduces the basic terminology and notation we will use throughout the chapter.

2.1 Graph

Given L, a finite alphabet of labels for nodes and edges, a graph g is defined by the four-tuple $g = (V, E, \mu, \nu)$ where V is a finite set of nodes, $E \subseteq V \times V$ is the set of edges, $\mu : V \longrightarrow L$ is the node labeling function, and $\nu : V \times V \longrightarrow L$ is the edge labeling function. The alphabet of labels is not constrained in any way. For example, L can be defined as a vector space (i.e., $L = \mathbb{R}^n$) or simply as a set of discrete labels (i.e., $L = \{\Delta, \Sigma, \Psi, \dots\}$). Edges are defined as ordered pairs of nodes, i.e.,, an edge is defined by (u, v) where $u, v \in V$. The edges are directed in the sense that if the edge is defined as (u, v) then $u \in V$ is the source node and $v \in V$ is the target node.

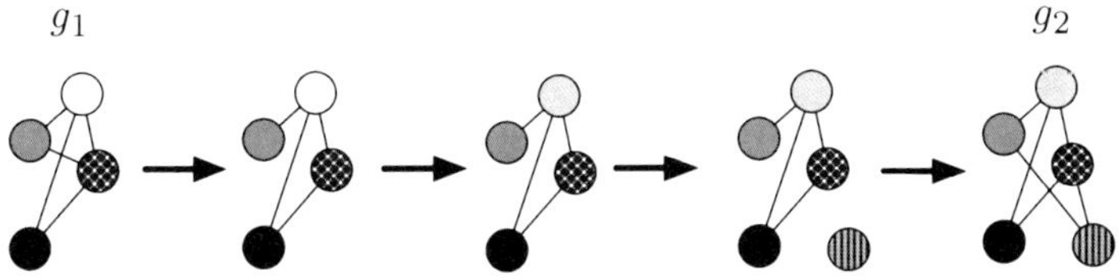

Fig. 1 A possible edit path between two graphs g_1 and g_2. Note that node labels are indicated by different colors

2.2 *Graph Edit Distance*

The process of evaluating the structural similarity of two graphs is commonly referred to as graph matching. This issue has been addressed by a large number of works. For an extensive review of different graph matching methods and applications, the reader is referred to [5]. In this work, we will use the graph edit distance [2, 40], one of the most widely used methods to compute the dissimilarity between two graphs.

The basic idea behind the graph edit distance [2, 40] is to define the dissimilarity of two graphs as the minimum amount of change required to transform one graph into the other. To this end, a number of edit operations e, consisting of the insertion, deletion, and substitution of both nodes and edges, are defined. Given these edit operations, for every pair of graphs, g_1 and g_2, there exists a sequence of edit operations, or edit path $p(g_1, g_2) = (e_1, \ldots, e_k)$ (where each e_i denotes an edit operation) that transforms g_1 into g_2 (see Fig. 1 for example). In general, several edit paths may exist between two given graphs. This set of edit paths is denoted by $\wp(g_1, g_2)$. To evaluate which edit path is the best one, edit costs are introduced through a cost function. The basic idea is to assign a cost $c(e)$ to each edit operation according to the amount of distortion it introduces in the transformation. Then, the edit distance between two graphs g_1 and g_2, denoted by $d(g_1, g_2)$, is the minimum cost edit path over all edit paths that transform g_1 into g_2:

$$d(g_1, g_2) = \min_{(e_1, \ldots, e_k) \in \wp(g_1, g_2)} \sum_{i=1}^{k} c(e_i) \qquad (1)$$

Different optimal and approximate algorithms for the computation of the graph edit distance have been proposed so far. Optimal algorithms are usually based on combinatorial search procedures that explore all the possible mappings of nodes and edges of one graph to the nodes and edges of the second graph [40]. The major drawback of such an approach is its computational complexity, which is exponential in the number of nodes of the involved graphs. As a result, the application of these methods is restricted to graphs of rather small size in practice. As an alternative, a number of suboptimal methods have been proposed to make the graph edit distance less computationally demanding and therefore usable in real applications. Some of these methods are based on local optimization [28]. A linear programming method to compute the graph edit distance with unlabeled edges is presented in [23]. Such a method can be used to obtain lower and upper edit distance bounds in polynomial time. In [29] simple variants of the standard method are proposed to derive two fast suboptimal algorithms for graph edit distance, which make the computation substantially faster. Finally, a new efficient algorithm is presented based on a fast suboptimal bipartite optimization procedure [36]. We will use these two last approximate methods for the graph-edit distance computation.

In [2] it was shown that $d(g_1, g_2)$ is a metric if the underlying cost function is a metric. Under the approximation algorithms of [29, 36] used in this work, however, the metric property is no longer guaranteed. But this does not have any negative impact on the approach proposed in this work because, firstly, the embedding procedure that maps each graph onto an n-dimensional vector can be applied regardless if the underlying distance function is a metric or not [30] and, secondly, after embedding all points, which represent the graph, are located in a Euclidean (and in particular a metric) space.

2.3 Weighted Mean of a Pair of Graphs

For the purpose of median graph computation, the weighted mean of a pair of graphs [3] is a crucial tool. For this reason we include its definition in the following.

Let g and g' be two graphs. The *weighted mean* of g and g' is a graph g'' such that

$$\begin{aligned} d(g, g'') &= a \\ d(g, g') &= a + d(g'', g') \end{aligned}$$

That is, the graph g'' is a graph in between the graphs g and g' along the edit path between them. Furthermore, if the distance between g and g'' is a and the distance between g'' and g' is b, then the distance between g and g' is $a+b$. Figure 2 illustrates this idea.

Observe that g'' is not necessarily unique. Consider, for example, a graph g consisting of only a single node with label A and a graph g' consisting of three isolated nodes labeled with A, B, and C, respectively. Assume that the insertion and deletion of a node has a cost equal to 1, regardless of the label of the affected node. Then we have $d(g, g') = 2$. Obviously, for $a = 1$ there exist two 1-mean graphs: g_1, which consists of two isolated nodes, one with label A and the other with label B, and g_2, which also consists of two isolated nodes, one with label A and the other with label C. In this work, we will assume that two graphs that are at the same distance from the original graphs are equivalent.

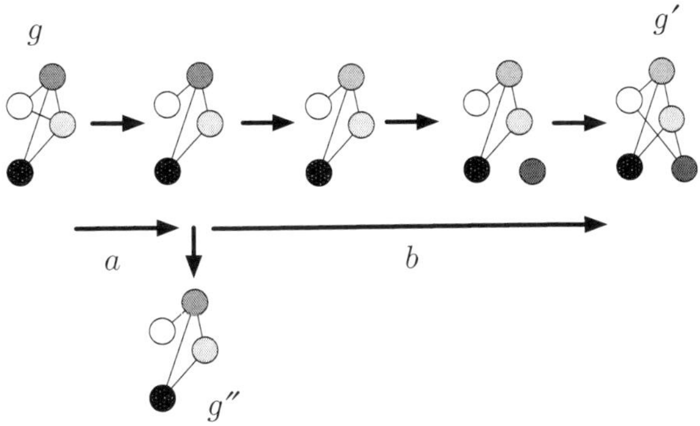

Fig. 2 Weighted mean of a pair of graphs

2.4 Median Graph

Given L, a finite alphabet of labels for nodes and edges, let U be the set of all graphs that can be constructed using labels from L. Given $S = \{g_1, g_2, \ldots, g_n\} \subseteq U$, the generalized median graph $\bar{g}$ of S is defined as

$$\bar{g} = \arg\min_{g \in U} \sum_{g_i \in S} d(g, g_i) \tag{2}$$

That is, the generalized median graph $\bar{g}$ of S is a graph $g \in U$ that minimizes the SOD (SOD) to all the graphs in S. Notice that $\bar{g}$ is usually not a member of S, and in general more than one generalized median graph may exist for a given set S. It can be seen as the representative of the set. Consequently, it can be potentially used by any graph-based algorithm where a representative of a set of graphs is needed.

Despite its simple mathematical definition (2), the computation of the median graph is extremely complex. As shown in (2) some distance measure $d(g, g_i)$ between the candidate median g and every graph $g_i \in S$ must be computed. However, since the computation of the graph edit distance is a well-known NP-complete problem, the computation of the generalized median graph can only be done in exponential time, both in the number of graphs in S and their size (even in the special case of strings, the time required is exponential in the number of input strings [18]). As a consequence, in real applications we are forced to use suboptimal methods in order to obtain approximate solutions for the generalized median graph in reasonable time. Such approximate methods [11, 19, 21, 27, 45] apply some heuristics in order to reduce the complexity of the graph edit distance computation and the size of the search space. Two different approaches that use graph embedding have already been explored in the past [12, 13]. The basic idea underlying these works is that embedding the graphs into vectors and computing the median in the vector space can lead to a median vector whose corresponding graph in the graph space can be a good approximation of the median graph. In these previous works [12, 13] an approximate reconstruction is obtained using, respectively, the 2 or 3 closest points to the median vector. We will use these methods (referred as E2P and E3P) as reference embedding methods for comparison later in the experiments.

Another alternative to reduce the computation time is to use the set median graph instead of the generalized median graph. The difference between the two concepts is only the search space where the median is looked for. As it is shown in (2), the search space for the generalized median graph is U, i.e., the whole universe of graphs. In contrast, the search space for the set median graph is simply S, i.e., the set of graphs in the given set. It makes the computation of set median graph exponential in the size of the graphs, due to the complexity of graph edit distance, but polynomial with respect to the number of graphs in S, since it is only necessary to compute pairwise distances between the graphs in the set. The set median graph is usually not the best representative of a set of graphs, but it is often a good starting point towards the search of the generalized median graph. As a matter of fact, we will use the set median graph as a baseline for the experiments presented later.

3 From Vectors to Graphs

In this section we will propose a generic procedure to compute the graph corresponding to a point in the vector space associated to a particular graph embedding. For this procedure to be applied we require to know the points corresponding to the embedding of a set $S = \{g_1, g_2, \ldots, g_n\}$ of n graphs, where n is the dimension of the vector space. Theoretically, any embedding could be used, as long as the distance relationships in the graph space are maintained in the vector space. In reality, this depends on the embedding method used, and it is only approximately true.

Once we have the set of n points corresponding to the embedding of S, the procedure can be applied to recover any point M lying inside the convex hull defined by these n points. It is based on recursively projecting the point M into hyperplanes of decreasing dimensionality and recovering the graph corresponding to the projected points by means of the weighted mean of a pair of graphs. It is important to note that all geometric operations needed in the reconstruction are carried out in the n-dimensional real space using the Euclidean distance. Hence, all the operations take place in a metric space. Thus, if we were able to compute the exact edit distance and the optimal edit path between two graphs, we would be able to obtain the graph that corresponds to the original point M. However, we are forced to use several approximations in practice. As a result we will only be able to obtain approximations of the corresponding graph.

Let us introduce some important aspects before explaining the method.

1. Given a set of n linearly independent points in $\mathbb{R}^n$ we can define a hyperplane H_{n-1} of dimensionality n-1 (e.g. in the case of n=2, two points define a unique 1D line, in the case of n=3, three points define a unique 2D plane, etc.). See Fig. 3 for example.

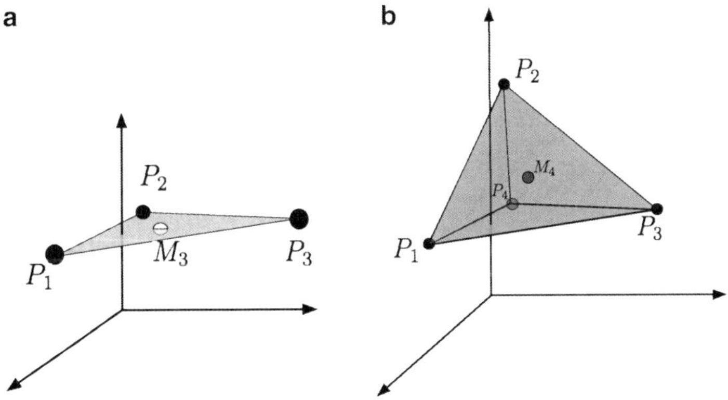

Fig. 3 (**a**) The 2D-hyperplane H_2 is defined by the 3 points $P_i = \{P_1, P_2, P_3\}$. The Euclidean median M_3 falls in the 2D space defined by the 3 points and specifically within the triangle (2D simplex) with vertices P_i $(i = 1, \ldots, 3)$. (**b**) The 3D-hyperplane H_3 is defined by the 4 points $P_i = \{P_1, P_2, P_3, P_4\}$. The Euclidean median M_4 falls in the 3D space defined by the 4 points and specifically within the pyramid (3D simplex) with vertices P_i $(i = 1, \ldots, 4)$

2. Assume that we can define a line segment in the vector space that connects two points P_1 and P_2 corresponding to known graphs g_1 and g_2, such that the point M in a $2D$ space M_2 lies on this line segment. We can then calculate the graph g_{M_2} corresponding to the point M_2 as the weighted mean of g_1 and g_2.[1]

From the second point we can observe that, given n embedded points $\{P_1, P_2, \ldots, P_n\}$ and the original point M_n, in order to obtain the graph corresponding to M_n, the problem is to find two points in the vector space, whose corresponding graphs are known, such that the median M_n lies on the line defined by these two points. In this way, we can then apply the weighted mean of these two points in order to find the graph corresponding to M_n. In the following we will describe how we can obtain such two points and, thus, such a graph. We will illustrate this procedure with the example shown in Fig. 4 with four points. Figure 4a shows the four points $\{P_1, P_2, P_3, P_4\}$ and the original point M_4.

Given $P_1, P_2, \ldots, P_n$, we can choose without loss of generality, any one of them, say P_n, and create the vector $(\mathbf{P_n} - \mathbf{M_n})$ (vector $(\mathbf{P_4} - \mathbf{M_4})$ in Fig. 4b). This vector will lie fully on the hyperplane H_{n-1} defined by these n points. Then, if we call H_{n-2} the hyperplane of dimensionality $n-2$ defined by the set of the remaining $n-1$ points $\{P_1, P_2, \ldots, P_{n-1}\}$, i.e., all the original points except P_n, then the intersection of the line defined by the vector $(\mathbf{P_n} - \mathbf{M_n})$ and the new hyperplane H_{n-2} will be a single point. We will call this new point M_{n-1} (M_3 in Fig. 4b which lies on the hyperplane H_2 (plane) defined by P_1, P_2, and P_3).

[1] For clarity, in the remainder, we will refer to the projection of M into n-dimensional space as M_n.

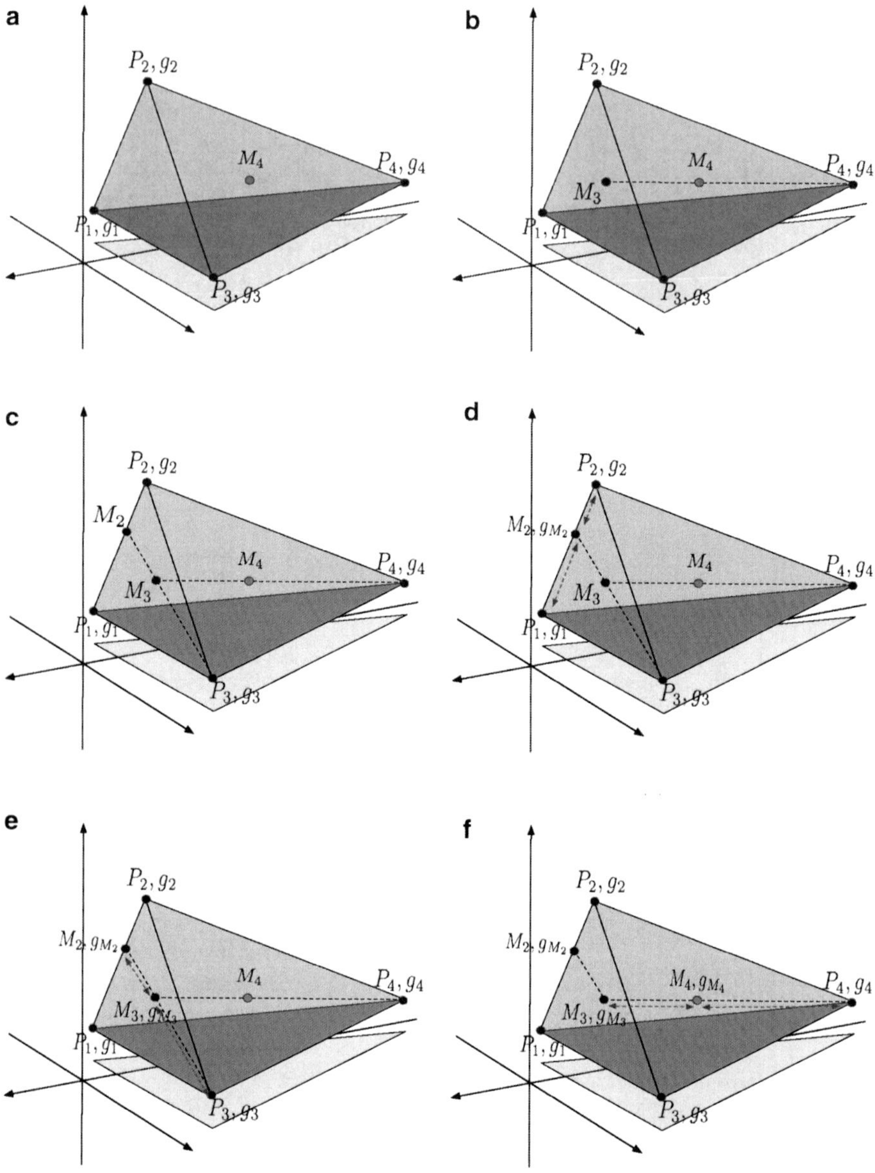

Fig. 4 Complete example of the median recovering with four points $\{P_1, P_2, P_3, P_4\}$

As mentioned before, in order to use the weighted mean of a pair of graphs to calculate the graph corresponding to M_n, we need to first find a point (whose corresponding graph is known) that lies on the line defined by the vector $(\mathbf{P_n} - \mathbf{M_n})$ and specifically on the ray extending M_n (so that M_n lies between P_n and the new point). Now we have two points (P_n and M_{n-1}) and the original point M_n

falling on the line defined by them. However, although we already know the graph corresponding to the point P_n (P_n comes from the graph g_n), we do not know yet the graph corresponding to the point M_{n-1}. Therefore, we cannot apply the weighted mean to find the graph corresponding to M_n. However, we can follow exactly the same procedure as before and consider a new line defined by the vector $(\mathbf{P_{n-1}} - \mathbf{M_{n-1}})$ ($(\mathbf{P_3} - \mathbf{M_3})$ in Fig. 4c). Again, as we did for M_{n-1}, we can define the point of intersection of the above line with the $n-3$ dimensional hyperplane H_{n-3} which is defined by the $n-2$ remaining points $\{P_1, P_2, \ldots, P_{n-2}\}$. Then, we will get a new point M_{n-2} (M_2 in Fig. 4c which lies on the line defined by points P_1 and P_2).

This process is recursively repeated until M_2 is obtained. The case of M_2 is solvable using the weighted mean of a pair of graphs, as M_2 will lie on the line segment defined by P_1 and P_2 which correspond to the known graphs g_1 and g_2 (we obtain g_{M_2} corresponding to M_2 in Fig. 4d).

Having calculated the graph g_{M_2} corresponding to the point M_2, the inverse process can be followed all the way up to M_n. Once g_{M_2} is found, in the next step, the graph g_{M_3} corresponding to M_3 can be calculated as the weighted mean of the graphs corresponding to M_2 and P_3 (Fig. 4e). Generally the graph g_{M_k} corresponding to the point M_k will be given as the weighted mean of the graphs corresponding to M_{k-1} and P_k. The weighted mean algorithm can be applied repeatedly until the graph g_{M_n} corresponding to M_n is obtained (g_{M_4} in Fig. 4f).

In this procedure we claim that the method to recover the graph from a vector should permit to obtain the exact graph in case that:

- The embedding preserves the distance structure.
- We were able to perform exact computations of the graph edit distance.

In general, these two conditions are not easy to satisfy. Concerning the first condition, the procedure simply requires that the edit path between two graphs follows a path along the straight line joining the two corresponding vectors in the vector space. Although there are some cases where using the selected embedding procedure this can be shown to be true, in general, it is not always satisfied. Regarding the second condition, the exact computation of the edit distance is a well-known NP-problem. So, we are forced to use some approximation. For these reasons, we are only able to get approximations of the graph corresponding to the point.

4 Median Graph Computation

In this section we will make use of the procedure to recover a graph from an embedded vector to propose a generic procedure to compute the median graph via embedding. In our case, the embedding of graphs into points in a suitable vector space will permit us to carry the median computation in the vector domain instead

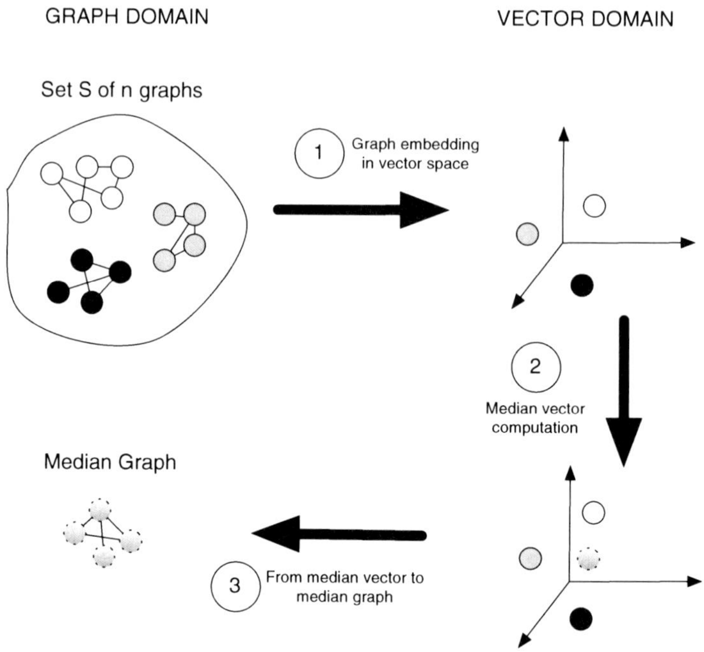

Fig. 5 Overview of the approximate procedure for median graph computation

of performing this operation in the graph domain, which is considerably more complex. This generic procedure we present is composed by three main steps (see Fig. 5).

- Given a set $S = \{g_1, g_2, \ldots, g_n\}$ of n graphs, the first step is to embed every graph in S into the real n-dimensional space. That is, each graph in S becomes a point in $\mathbb{R}^n$. Theoretically, any embedding which fulfils this condition, i.e., each graph becomes an n-dimensional point, could be used in this step. However, it is expected to obtain better results if the distance relationships resemble as much as possible both in the original graph space and the vector space.
- The second step consists of computing a representative of the set in the vector space. As in the case of the first step, several solutions could be applied here. However, the median vector **M** arises as a natural solution [13], since it is the vectorial counterpart of the median graph but in the vector domain: given a set $\wp = \{P_1, P_2, \ldots, P_m\}$ of m points with $P_i \in \mathbb{R}^n$ for $i = 1, \ldots, m$, the median vector is a point $M_n \in \mathbb{R}^n$ that minimizes the sum of the distances to all the points in $\wp$. Thus, if the embedding preserves the distance structure of the graph domain, the median vector should be a good representation of the median graph in the vector space.
- Finally, the resulting median vector has to be mapped back to a corresponding graph. For this, we will use the procedure described in the previous section.

It is important to notice that the three above mentioned steps are generic and independent of each other. That means that different approaches and solutions can be used in each of them and combined. In the following section we will propose a particular implementation of each step in order to compute the median graph. We explain the choices we have made for the first two steps, which are basically the same as in [13], and we discuss some practical considerations about the application of the recursive procedure to recovering the median graph from the median vector in the last step.

4.1 Step I: Graph Embedding

In our proposal, we will use a class of graph embedding procedures based on the selection of some prototypes and graph edit distance computation. This approach was first presented in [37], and it is based on the work proposed in [31]. The basic intuition of this work is that the description of the regularities in observations of classes and objects is the basis to perform pattern classification. Thus, based on the selection of concrete prototypes, each point is embedded into a vector space by taking its distance to all these prototypes. Assuming these prototypes have been chosen appropriately, each class will form a compact zone in the vector space. For the sake of completeness, we briefly describe this approach in the following.

Assume we have a set of training graphs $T = \{g_1, g_2, \ldots, g_n\}$ and a graph dissimilarity measure $d(g_i, g_j)$ $(i, j = 1, \ldots, n; g_i, g_j \in T)$. Then, a set $P = \{p_1, \ldots, p_m\} \subseteq T$ of m prototypes is selected from T (with $m \leq n$). After that, the dissimilarity between a given graph $g \in T$ and every prototype $p \in P$ is computed. This leads to m dissimilarity values, $d_1, \ldots, d_m$ where $d_k = d(g, p_k)$. These dissimilarities can be arranged in a vector $(d_1, \ldots, d_m)$. In this way, we can transform any graph of the training set T into an m-dimensional vector using the prototype set P.

For our purposes, given a set of graphs $S = \{g_1, g_2, \ldots, g_n\}$, we use the graph embedding method described above to obtain the corresponding n-dimensional points $\{P_1, P_2, \ldots, P_n\}$ in $\mathbb{R}^n$. However, in our case, we set $P = S$, i.e., we avoid the problem of selecting a proper subset $P \subseteq S$ of prototypes and use the whole set of graphs.

It is important to mention that, as long as there are no identical graphs in the set S, the vectors $\mathbf{v_i} = (\mathbf{P_i} - \mathbf{O})$, where O is the origin of the n-dimensional space defined, can be assumed to be linearly independent. This arises from the way the coordinates of the points were defined during graph embedding (Fig. 6).
An important relation that has been shown in [37] is

$$\| \phi(g) - \phi(g') \| \leq \sqrt{n} \cdot d(g, g') \tag{3}$$

where $\phi(g)$ and $\phi(g')$ denote the mappings in the vector space of graphs g and g', respectively, after embedding. That is, the upper bound of the Euclidean distance of

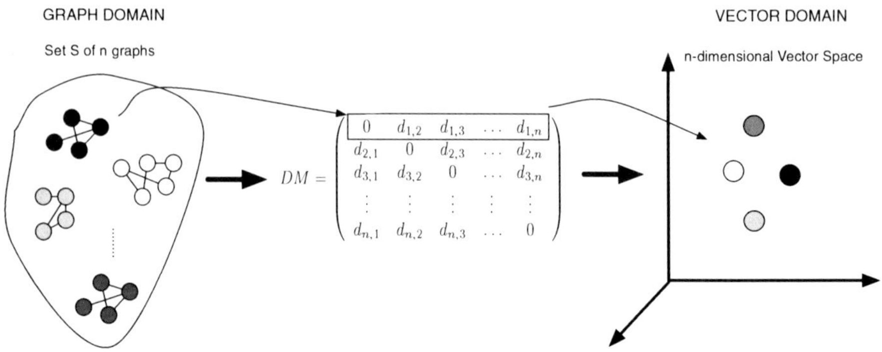

Fig. 6 Step 1. Graph embedding

a pair of graph maps $\phi(g)$ and $\phi(g')$ is given by $\sqrt{n} \cdot d(g,g')$. In other words, if g and g' have a small distance in the graph domain, they will have a small distance after embedding in the Euclidean space as well.

Therefore, at the end of this first step we will have a collection of points in an n-dimensional space, each of them corresponding to one of the original graphs.

4.2 Step II: Median Vector Computation

To obtain the representative of the set in the vector domain, we will use the concept of median vector as we already commented at the beginning of Sect. 3.

The median vector cannot be calculated in a straightforward way. The exact location of the median vector can not be found when the number of elements in $\wp$ is greater than 5 [1]. No algorithm in polynomial time is known, nor has the problem been shown to be NP-hard [17]. In this work we will use the most common approximate algorithm for the computation of the median vector, i.e., Weiszfeld's algorithm [44]. It is a form of iteratively re-weighted least squares that converge to the median vector. To this end, the algorithm first selects an initial estimate solution M'_{n_0} (this initial solution is often chosen randomly). Then, the algorithm defines a set of weights that are inversely proportional to the distances from the current estimate M'_{n_i} to the samples x and creates a new estimate M'_{n_i+1} that is the weighted average of the samples according to these weights.

$$M'_{n_{i+1}} = \frac{\sum_{j=1}^{m} \frac{x_j}{\|x_j - M'_{n_i}\|}}{\sum_{j=1}^{m} \frac{1}{\|x_j - M'_{n_i}\|}} \tag{4}$$

The algorithm may finish when a predefined number of iterations are reached, or under some other criteria, such as that the difference between the current estimate and the previous one is less than a predefined threshold.

Note that the median vector will always fall within the volume of the $n-1$ dimensional simplex whose vertices are the set of points used to compute the median. Thus, it fulfills the required constraint to use the recursive procedure to compute the graph associated to that point. Figure 3 shows an example for $n=4$ and $n=3$.

4.3 Step III: Median Graph Recovering

We propose to obtain the graph corresponding to the median vector by means of the recursive application of the weighted mean of a pair of graphs. As it has already been remarked in the previous section, this recursive procedure relies on a set of approximations concerning the metric space and the computation of the graph edit distance, which will result in the calculation of an approximation of the median graph.

In order to analyze the effect of all these approximations in the final result, we can examine the order in which points P_i in the vector space are considered in the recursive procedure. This is an issue not defined in the original procedure as, if computations were exact, the order would not matter. However, in case of approximate computations, the order can be important for the final solution. For instance, if we start the process of recovering the median graph using the points that are further from the optimal solution to define the connecting line in the vector space, we will probably start introducing some approximation errors in the first steps as the quality of the weighed mean is better the shortest the edit path is. However, in the final steps we will consider the points that are closer to the optimal solution and thus, we will probably balance this effect as we will give more weight to these points in the final solution. If we take the reverse order the expected effect would be the contrary. The final result of these opposite effects is not clear.

Therefore, we have defined different sorting schemes to consider the points in the recursive procedure according to the SOD of every point, calculated either in the graph or the vector domain, to the rest of points. Points with a low SOD will correspond to points close to the optimal solution. Thus, we present four variants of the basic recursive scheme presented in Sect. 3 (BRS in short), which include a preprocessing to sort the graphs. Note that to be consistent with the notation and the explanations performed in Sect. 3, the words *ascending* or *descending* used in the following refer to graphs from g_n to g_1. These sorted schemes will be referred as SRS (sorted recursive schemes).

- *Graph-domain-based Sorted Recursive Scheme in descending order* (GSRSD): In this approach, the graphs are ordered in descending order, taking into account the SOD to the rest of the graphs in S of each of them. Consequently, g_n is the graph with maximum SOD and g_1 is the set median graph. Under this sorting, the graph corresponding to the point M_2 is calculated as the weighted mean of g_1 and g_2, the two graphs with lowest SOD, i.e., the set median (g_1) and the next one in terms of the minimum SOD to S (g_2).

- *Graph-domain-based Sorted Recursive Scheme in ascending order* (GSRSA): This sorting is the inverse to the previous one. The graphs are ordered upwards, based on the SOD. This way, the graph corresponding to M_2 is obtained from the two graphs with maximum SOD, and the graph corresponding to M_n is obtained from the weighted mean between the graphs corresponding to M_{n-1} and g_n (the set median).
- *Vector-domain-based Sorted Recursive Scheme in descending order* (VSRSD): Here the criterion for the ordering is still the SOD, but it is evaluated in the Euclidean space. That is, the SOD of each of the points $\{P_n, \ldots, P_1\}$ to the other points of the set. In this case, g_n is the graph, the corresponding point of which has the maximum SOD,

$$P_{\max} = \arg \max_{P \in \{P_n, \ldots, P_1\}} \sum_{i=1}^{n} \|P_i - P\|.$$

- *Vector-domain-based Sorted Recursive Scheme in ascending order* (VSRSA): As before, in this last sorting, the SOD in the Euclidean space is used to sort the points. The points are ordered upwards with respect to the SOD, such that the first two points used to compute the weighted mean are those with maximum SOD.

In addition note that, given n graphs, in the procedure to recover the median graph we obtain $n - 1$ intermediate graphs (from M_2 to M_n). As we go through the process we get closer to the graph corresponding to the median vector. But, at the same time, at every step we are also introducing more approximation in the final solution. As a result, it could happen that some of the intermediate graphs has an SOD better than the final median graph. Given this situation, we have also analyzed the SOD of these intermediate graphs.

In order to see the differences along these five recursive schemes (BRS and the four variations) we computed several medians using the letter dataset [34]. In this dataset, we consider the 15 capital letters of the Roman alphabet that consist of straight lines only (A, E, F, H, I, K, L, M, N, T, V, W, X, Y, Z). For each class, a prototype line drawing is manually constructed. These prototype drawings are then converted into prototype graphs by representing lines by undirected edges and ending points of lines by nodes. Each node is labeled with a two-dimensional attribute giving its position relative to a reference coordinate system. Edges are unlabeled (see [34] for more characteristics of this dataset). More concretely, we took sets of 50 and 100 elements randomly from the dataset and we computed the median with each of the methods. Figure 7 shows the evolution of the SOD of the intermediate median graphs for each recursive method. The x-axis represents the

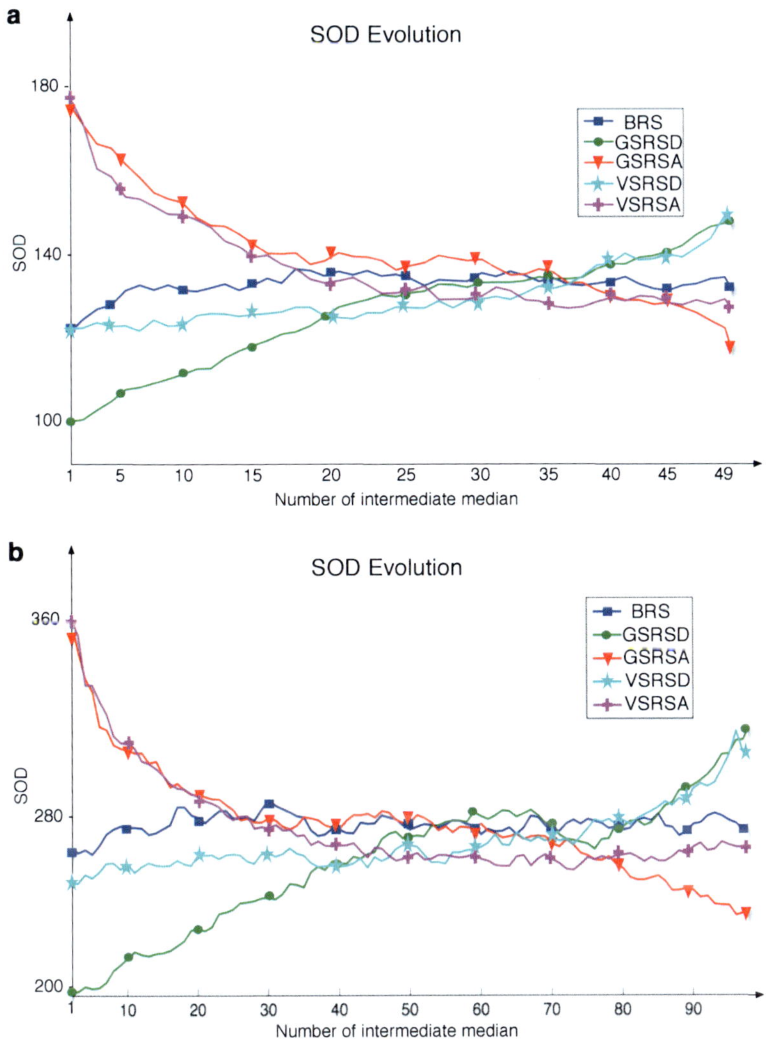

Fig. 7 SOD evolution for the letter dataset using (**a**) sets of 50 elements and (**b**) sets of 100 elements

recursive level, being 1 for the first graph we obtain (i.e., M_2), and the last point representing the last final median (i.e., M_n). The y-axis represents the SOD of each corresponding intermediate graph. Results are the mean over ten repetitions for each size of the set.

First of all, as expected, it is important to note that the results are different for each of the five recursive schemes. As it can be seen in Fig. 7, the evolution of the SOD shows different behavior depending on the initial sorting. However, while the BRS approach shows a random-like behavior (there is no clear tendency in the evolution of the SOD), the sorted schemes show a general tendency in the SOD evolution. Note also that this tendency is independent of the size of the set used to compute the median. One of the most striking facts is that the domain on which the sorting is based is unimportant. That is, in the descending methods (GSRSD and VSRSD), there is a clear tendency in starting with graphs or vectors with lower SODs and terminate with higher SODs. This fact can be explained because in the descending methods, the first intermediate graph (i.e., M_2) is computed using graphs having lower SOD (in the case of GSRSD method, M_2 is computed with the set median and the next graph in terms of the lower SOD). Consequently, M_2 has a low SOD. Then, as we compute more intermediate graphs, they are computed using graphs with higher SODs. This translates into a degradation in terms of the SOD in the intermediate graph. On the contrary, in the ascending schemes (GSRSA and VSRSA) the tendency in the evolution is exactly complementary. Here, we start with graphs having high SODs (and consequently M_2 has a high SOD) and then we use better graphs in terms of their SOD. This translates to a decreasing curve. As a conclusion, we can state that we get better solutions as we consider points that are closer to the optimal solution. However, the behavior of the two sorting schemes is not completely complementary in the sense that the loss in terms of SOD in the descending methods is not the same as the gain obtained in the ascending methods. For this reason, the minimum (or maximum) values of SOD in these evolutions differ. However, the fact that the tendency is kept regardless of the domain of the sorting supports the idea that relative distances are well conserved after mapping graphs into points in the particular embedding considered here.

Another important observation is that if we analyze the SOD of the intermediate graphs we can find intermediate solutions along the recursive path with a lower SOD than the final solution. This fact validates our previous hypothesis that there is a compromise between the amount of approximation and how close we are to the final solution. For this reason, when we compare these methods to other existing approaches for the median graph computation in the next section, we will take into account not only the final solution but also the best solution along the recursive path.

Recursive methods sorted in descending order (specially GSRSD) obtain, in general, the best intermediate graphs. This fact seems to lead to the conclusion that it is better to start the approximation with a graph as closer as possible to the optimal solution. In addition, in these methods, the best median is usually obtained in a very interior call, when few intermediate graphs have been computed. Table 1 shows for each dataset and for each of the five recursive schemes the mean position of the best intermediate median (for 50/100 elements) along all the repetitions. Note that the values obtained by the BRS method are very close to the mid position (i.e., 25 in the case of 50 elements and 50 in the case of 100 elements), while the descending methods have in general lower values than the mid value and the ascending methods have in general higher values than the mid value. This could be used in a future work to improve the method in order to obtain good approximations of the median without need of computing all the intermediate graphs.

Table 1 Average position of the median with minimum SOD

Method	Letter	Molecule	Mutagenicity
BRS	21/50	18/48	23/48
GSRSD	11/18	21/39	15/25
GSRSA	36/56	25/58	31/63
VSRSD	20/50	15/33	14/20
VSRSA	28/48	33/60	33/76

5 Experimental Evaluation

In this section we provide the results of an experimental evaluation of the three proposed methods. To this end, and since the objective of the methods is to find a representative of a set of graphs, we will perform cluster analysis. The median is typically used as a representative of a class, hence it is natural to think that if clustering results are better, then the corresponding theoretical representative is a better prototype of the cluster and the corresponding method to obtain the representatives is better. To this end, different clustering indices will be employed as evaluation measures. The main goal is not only to compare the different methods we propose to compute the representative but also to give an idea of which of them could be a better choice to compute the representative of a given set. Therefore, our reference system in all the experiments will be the set median.

With this objective in mind, we first proceed to provide the basic notions about clustering. In Sect. 5.1 we briefly explain graph clustering with focus on explaining the k-means graph-based clustering method that will be used in the subsequent experiments. In Sect. 5.1.1 we present the four standard clustering quality measures we will use to perform the evaluation of the methods. Finally, we present the results in Sect. 5.2.

5.1 Graph Clustering

Cluster analysis or clustering is the task of assigning a set of objects into groups (called clusters) so that the objects in the same cluster are more similar (in some sense or another) to each other than to those in other clusters. In our case, we use a clustering strategy, the well known k-means algorithm, in which at each iteration of the algorithm a representative for each cluster is needed. For the sake of completeness, the k-means clustering algorithm applied to graphs is presented in Algorithm 3.

The k-means clustering algorithm is one of the most simple and straightforward methods for clustering data [26]. Given k, the desired number of clusters, the k centers of the clusters are randomly initialized picking up k graphs from the original set of n graphs. Then, the remaining graphs are assigned to the cluster corresponding to the closest center. The centers of the clusters are recomputed and the graphs are

Algorithm 3: Graph k-means algorithm

input : A set $S = \{g_1, g_2, \ldots, g_n\}$ of n data items (represented as graphs) and the number of clusters k to create

output: The centers of the clusters and for each data item an integer [1, k] indicating the cluster the item belongs to

begin

1 Assign randomly each graph g_i to a cluster

2 Using this initial assignment, compute the median graph of each cluster.

3 Assign each data item to be in the cluster of its closest center using the graph edit distance.

4 Recompute the centers as in Step 2.

5 Repeat Steps 3 and 4 until the centers do not change.

end

assigned again to the cluster with the closest center until the clusters remain stable or a maximum number of iterations are reached. Note that clustering of data items represented by graphs is then possible by letting the median graph or the barycenter be the center and using graph edit distance whenever a distance is needed.

5.1.1 Clustering Performance Measures

In this work, graph clustering is used as a tool for the evaluation of the different median graph approaches. It is natural to think that if clustering results are better, then the corresponding theoretical representative is a better prototype of the cluster and the corresponding method is better.

Thus let $X = \{g_1, \ldots, g_n\}$ be a set of n graphs belonging to classes $\{C_1, \ldots, C_k\}$, which represents he ground truth, and let $D = \{D_1, \ldots, D_l\}$ be a clustering of X. Let us denote $n_i^j = D_i \cap C_j$ the number of elements of class C_j clustered in D_i.

Before presenting the results obtained for the different datasets, we introduce the clustering performance measures in which we base our evaluation of the clusterings, which have already been used in previous graph-based clustering experiments [10, 35, 41].

We use four standard performance measures. Three of them, the Rand index, the mutual information, and the bipartite index, base their scoring in the comparison between the ground truth and the clustering. In the case of this study, we are aware of the real classification of the data we work with. But it is important to remark that these measures of quality cannot be used when unclassified data are clustered. We compute one more quality measure, which is independent of the ground truth. The Dunn index is based on the assumption that items clustered together must be near each other while being far from items belonging to other clusters. Let us define all of them.

Rand Index: To compute this index, a pairwise comparison between all pairs of items in the dataset is computed. If two elements fall in the same class and belong to the same cluster, it counts as an agreement. Similarly, if the two elements belong to different classes and fall into different clusters it counts as an agreement too. Otherwise, it counts as a disagreement. Let A be the number of agreements and D the number of disagreements. The Rand index [32]

$$R = \frac{A}{A+D} \tag{5}$$

measures the matching of the obtained clusters to the ground truth classes. The Rand index produces measures in the interval [0, 1], with 1 meaning a perfect match between the result of the algorithm and the ground truth.

Mutual Information: The mutual information[6, 43]

$$M = \frac{1}{n}\sum_{j=1}^{l}\sum_{h=1}^{k} n_j^h \log_{lk}\left(\frac{n_j^h n}{\sum_{i=1}^{l} n_i^h \sum_{i=1}^{k} n_j^i}\right) \tag{6}$$

represents the overall degree of agreement between the clustering and the ground truth with a preference for clusters that have high purity. Higher values indicate better performance.

Bipartite Index: Let $\wp_k$ denote the symmetric group, i.e., the set of all the permutations, of the set $1,\ldots,k$. We use permutations to evaluate all the possible assignments of clusters to classes and then compute the bipartite index over the optimal such assignment, as follows [35]:

$$B_I = \max_{\sigma \in \wp} \frac{1}{n} \sum_{i=1}^{k} n_i^{\sigma(i)} \tag{7}$$

This index is also normalized in the $[0,1]$ range, with higher values denoting better performance .

Dunn Index: Let $d_{\min}$ be the minimum distance between any two objects in different clusters and $d_{\max}$ the maximum distance between any two objects in the same cluster. The *Dunn index* [9]

$$D_I = \frac{d_{\min}}{d_{\max}} \tag{8}$$

is a measure of the compactness and separation of the clusters. Higher values of the Dunn index indicate better clustering.

5.2 *Experimental Results*

In this section, the application of the median graph and the barycenter graph for data clustering purposes will be presented.

Table 2 Clustering quality measures for the letter database

	RI	DI	MI	BI
SM	0.805627	0.031315	0.116952	0.222400
E2P	0.885333 •	0.026163	0.197808 •	0.395333 •
E3P	0.895852 •	0.027030	0.214651 •	0.424000 •
BRS	0.854425 •	0.026506 ⋆	0.157418 •	0.288750 •
GSRSD	0.836309 •	0.021861	0.148605 •	0.287500 •
GSRSA	0.901697 •, ∘	0.025869	0.234569 •, ∘	0.454000 •, ⋆
VSRSD	0.840083 •	0.024761	0.148055 •	0.289412 •
VSRSA	0.891671 •, ⋆	0.030308 ∘	0.203391 •, ⋆	0.406889 •
BRS/Best	0.899922 •, ∘	0.028445 ∘	0.292317 •, ∘	0.448000 •, ∘
GSRSD/Best	0.903065 •, ∘	0.030218 ∘	**0.245440** •, ∘	0.465333 •, ∘
GSRSA/Best	**0.905512** •, ∘	0.030634 ∘	0.238033 •, ∘	**0.476833** •, ∘
VSRSD/Best	0.901601 •, ∘	**0.037744** •, ∘	0.236525 •, ∘	0.457810 •, ∘
VSRSA/Best	0.897984 •, ∘	0.031610 •, ∘	0.230356 •, ∘	0.444667 •, ∘

5.2.1 Experimental Setup

The clustering experiments have been applied to the letter, the molecule, and the mutagenicity datasets [34]. Notice that the total number of methods being compared grows up to 11 (set median, E2P, E3P and the five recursive methods), which implies a large number of total instances of the clustering process for each database. For each of these methods, and ten times for each database, 50 elements of each class are randomly picked up and the clustering is carried out. In each instance of the experiments, i.e., for each clustering performed, we compute the value of the four quality measures that are explained in Sect. 5.1.1, and the mean value over the ten repetitions is reported.

5.2.2 Results

The results of these experiments are displayed in three tables, one for each database. Each table contains, for each of the methods, the mean value over the ten repetitions, for each of the four quality indices. Table 2 shows the values of the indices for the experiments made with graphs from the letter database, Table 3 corresponds to Molecules and Table 4 refers to the experiments with mutagen datasets. Results marked with a (•) are those medians performing better as class representatives than the set median. Results marked with a (∘) are the medians performing better than the E2P and E3P methods. Medians marked with the (⋆) are those performing better than E2P or E3P methods. The best median for each index is marked with bold face. Recall that the Rand index (RI), the mutual information (MI), and the bipartite index (BI) are groundtruth-based indices, while the Dunn index (DI) is not. For the sake of simplicity, we will sometimes refer to groundtruth-based indices as GT indices. Recall also that all four indices give higher values to better clusterings.

Table 3 Clustering quality measures for the molecule database

	RI	DI	MI	BI
SM	0.533500	0.367934	0.063922	0.617000
E2P	0.562738 •	0.130584	0.102330 •	0.676923 •
E3P	0.697138 •	0.168914	0.210778 •	0.813846 •
BRS	0.548760 •	0.263230	0.088487 •	0.656000 •
GSRSD	0.534226 •	0.455628 •	0.052444	0.588710
GSRSA	0.676083 •, ⋆	0.144330	0.188662 •, ⋆	0.785833 •, ⋆
VSRSD	0.517486	**0.459920** •	0.034428	0.560952
VSRSA	0.617291 •, ⋆	0.158076	0.147863 •, ⋆	0.737273 •, ⋆
BRS/Best	0.705227 •, ∘	0.126002	0.214477 •, ∘	0.809333 •, ⋆
GSRSD/Best	**0.781244** •, ∘	0.126477	**0.277252** •, ∘	**0.870000** •, ∘
GSRSA/Best	0.651940 •	0.115464	0.176478 •, ∘	0.775000 •
VSRSD/Best	0.701178 •, ∘	0.135930	0.214217 •, ∘	0.813333 •, ⋆
VSRSA/Best	0.770160 •, ∘	0.093471	0.266716 •, ∘	0.864000 •, ∘

Table 4 Clustering quality measures for the mutagenicity database

	RI	DI	MI	BI
SM	0.512620	**0.230830**	0.012476	0.571000
E2P	0.531980 •	0.055990	0.027048 •	0.621000 •
E3P	0.532200 •	0.051731	0.026672 •	0.624000 •
BRS	0.519187 •	0.070817	0.015229 •	0.586875 •
GSRSD	0.500000 •	0.314786	0.022740 •	0.500000
GSRSA	0.527580 •	0.047640	0.022250 •	0.613000 •
VSRSD	0.500000 •	0.314786	0.021190 •	0.500000
VSRSA	0.512053 •	0.051021	0.009267 •	0.573333 •
BRS/Best	0.534200 •, ∘	0.048426	0.027776 •, ∘	0.630000 •
GSRSD/Best	**0.540818** •, ∘	0.050847	**0.034978** •, ∘	**0.642727** •
GSRSA/Best	0.517640 •	0.077794	0.014892 •	0.570000
GSRSD/Best	0.527674 •	0.081238	0.023582 •	0.607895 •
VSRSA/Best	0.527778 •	0.057365	0.023378 •	0.611111 •

Letter: In general, except for the Dunn index, the recursive give better results than the set median graph. Which means that the graph embedding approach turns out to be a useful tool for median graph computation. In addition, the recursive methods give almost always better results than at least one of the non-recursive embedding methods, which means that using the whole set of graphs to obtain the median gives a significant improvement. In all the indices, the best results are given by one of the recursive methods (taking the best intermediate median). In addition to that, regarding the type of ordering used in the recursive methods, the ascending order gives the better results in the general case. However, if we take the best intermediate median along the recursive path, such a difference is less evident and the best results are spread among both ascending and descending schemes.

Molecule: Similar results can be drawn for the molecule dataset. In general, the embedding methods are able to obtain better representatives than the set median. However, only the best intermediate medians give better results than the two of the non-recursive embedding methods, although most of the recursive embedding methods give better results than at least one of the non-recursive embedding methods. Here, the best intermediate medians of the GSRSD methods give the best results in general. In relation to the sorting type, the ascending schemes are those giving the best results in general in the recursive methods when the full recursive path is taken into account. This tendency is not followed when the best intermediate median is taken along the recursive path. In this case, better results are given in three out of the four indices by the GSRSD method.

Mutagenicity: In this case, the differences between all the embedding methods seem to be less than before. Although in general, the embedding methods give better results than the set median, the recursive embedding methods perform only slightly better than the non-recursive embedding methods, although three out of the four best results are given by the SRSGD method using the best intermediate median. Again, the ascending order gives the best results in general when the final graph along the recursive path is taken as the median. But, descending order and especially the GSRSD method give the best results (three out of the four indices) when the best intermediate median is taken as the final median.

These clustering experiments confirm that in general the embedding methods give better results than the set median in terms of their quality as representative points of a class. In addition, the recursive embedding methods give in most cases better results than the non-recursive embedding methods. This may suggest that using the whole set of original graphs to compute the median is important to the final result. It is important to remark that some differences can be seen between the ascending and descending sorting schemes. This suggests that a deep study of the implications of each sorting scheme should be performed in order to try to improve the median computation, for instance stopping the computation before the whole recursive path is computed.

6 Summary

Graph embedding methods have become very popular in the last few years since they permit to use the whole repository of machine learning algorithms with graphs. However an unsolved problem yet is the reverse step, i.e., how to recover the graph that corresponds to a point in the vector space. In this chapter we have described a generic recursive procedure that permits to recover such a graph given that the point lies inside the convex hull of n previously embedded graphs. This procedure is based on recursively projecting the point into hyperplanes of decreasing dimensionality and recovering the graph from these projections using the concept of the weighted mean of a pair of graphs.

One problem where this procedure can be successfully applied is the computation of the median graph. The median graph has been shown to be a good choice to obtain a representative of a set of graphs. However, its computation is extremely complex. As a consequence, in real applications we are forced to use suboptimal methods in order to obtain approximate solutions for the generalized median graph in reasonable time. The procedure proposed in this chapter comprises a new algorithm for the computation of the median graph based on graph embedding. First, the graphs are mapped to points in an n-dimensional vector space using an embedding based on the graph edit distance. Then, the crucial point of obtaining the median of the set is carried out in the vector space, not in the graph domain, which dramatically simplifies this operation. Finally, the new procedure to recover a graph from the vector space permits to obtain the graph corresponding to the median vector. This last step is the main difference with previous methods that also compute the median graph using graph embedding. We analyze four variations of the base algorithm taking into account the order in which the graphs are considered in the recursive path.

In order to evaluate the proposed method (and all its variations), we have made experiments on three different graph databases, one semi-artificial and two containing real-world data. The underlying graphs have no constraints regarding the number of nodes and edges. We compared this approach with state-of-the-art embedding-based methods for median graph computation and also with the set median approach. Results show that with the proposed recursive approach we can obtain, in general, better medians, in terms of their SOD and their clustering performance, than existing embedding-based methods or the set median.

The proposed novel method for median graph computation is approximate in a double sense, namely through the graph embedding and graph recovery step. Nevertheless, as experiments on a number of databases with quite different characteristics have shown, it is able to discover median graphs of better quality than previous approximate methods that use the set median or the closest two or three points as the basis for approximation.

A number of important questions remain open regarding the nature of the proposed procedure. For instance, a deep analysis of the influence of the different approximations in the final graph that is obtained, should be carefully investigated. It would also be interesting to establish some relation between the degree of the exactness of the recovered graph and the characteristics of several embeddings that exist in the literature and, particularly, regarding how well these embeddings preserve the graph distances in the vector space. Finally, applications of this procedure to problems other than the median graph computation should also be investigated.

References

1. Bajaj C (1988) The algebraic degree of geometric optimization problems. Discrete Comput Geom 3(2):177–191
2. Bunke H, Allerman G (1983) Inexact graph matching for structural pattern recognition. Pattern Recogn Lett 1(4):245–253
3. Bunke H, Günter S (2001) Weighted mean of a pair of graphs. Computing 67(3):209–224
4. Caelli T, Kosinov S (2004) Inexact graph matching using eigen-subspace projection clustering. IJPRAI 18(3):329–354. DOI http://dx.doi.org/10.1142/S0218001404003186
5. Conte D, Foggia P, Sansone C, Vento M (2004) Thirty years of graph matching in pattern recognition. Int J Pattern Recogn Artif Intell 18(3):265–298
6. Cover TM, Thomas JA (1991) Elements of information theory. Wiley-Interscience, New York
7. Demirci MF, Shokoufandeh A, Keselman Y, Bretzner L, Dickinson SJ (2006) Object recognition as many-to-many feature matching. Int J Comput Vision 69(2):203–222
8. Duda R, Hart P, Stork D (2000) Pattern classification, 2nd edn. Wiley Interscience, New York
9. Dunn J (1974) Well separated clusters and optimal fuzzy partitions. J Cibernet 4:95–104
10. Ferrer M (2008) Theory and algorithms on the median graph. Application to graph-based classification and clustering. PhD Thesis, Universitat Autònoma de Barcelona
11. Ferrer M, Serratosa F, Sanfeliu A (2005) Synthesis of median spectral graph. 2nd Iberian Conference on Pattern Recognition and Image Analysis. Estoril, Portugal. Lecture Notes in Computer Science, Springer-Verlag, vol 3523, pp 139–146
12. Ferrer M, Valveny E, Serratosa F, Bardají I, Bunke H (2009a) Graph-based -means clustering: A comparison of the set median versus the generalized median graph. In: Jiang X, Petkov N (eds) CAIP. Lecture notes in computer science, vol 5702. Springer, Berlin, pp 342–350
13. Ferrer M, Valveny E, Serratosa F, Riesen K, Bunke H (2010) Generalized median graph computation by means of graph embedding in vector spaces. Pattern Recognition 43(4): pp. 1642–1655.
14. Friedman M, Kandel A (1999) Introduction to pattern recognition. World Scientific, New York
15. Gibert J, Valveny E, Bunke H (2012) Graph embedding in vector spaces by node attribute statistics. Pattern Recogn 45(9):3072–3083. DOI 10.1016/j.patcog.2012.01.009. URL http://dx.doi.org/10.1016/j.patcog.2012.01.009
16. Grauman K, Darrell T (2004) Fast contour matching using approximate earth mover's distance. In: Proceedings of the IEEE conference on Computer Vision and Pattern Recognition Washington, DC, pp 220–227
17. Hakimi SL (2000) Location theory. CRC, West Palm Beach
18. de la Higuera C, Casacuberta F (2000) Topology of strings: Median string is NP-complete. Theor Comput Sci 230(1–2):39–48
19. Hlaoui A, Wang S (2006) Median graph computation for graph clustering. Soft Comput 10(1):47–53
20. Indyk P (2001) Algorithmic applications of low-distortion geometric embeddings. FOCS '01 Proceedings of the 42nd IEEE symposium on Foundations of Computer Science. IEEE Computer Society Washington, DC, USA, pp 10–33
21. Jiang X, Münger A, Bunke H (2001) On median graphs: Properties, algorithms, and applications. IEEE Trans Pattern Anal Mach Intell 23(10):1144–1151
22. Jouili S, Tabbone S (2010) Graph embedding using constant shift embedding. In: Proceedings of the 20th international conference on recognizing patterns in signals, speech, images, and videos, ICPR'10. Springer, Berlin, pp 83–92. URL http://dl.acm.org/citation.cfm?id=1939170.1939183
23. Justice D, Hero AO (2006) A binary linear programming formulation of the graph edit distance. IEEE Trans Pattern Anal Mach Intell 28(8):1200–1214
24. Kramer S, Raedt LD (2001) Feature construction with version spaces for biochemical applications. In: Proceedings of the eighteenth international conference on machine learning, ICML '01. Morgan Kaufmann Publishers Inc., San Francisco, pp 258–265. URL http://dl.acm.org/citation.cfm?id=645530.655667

25. Luo B, Wilson RC, Hancock ER (2003) Spectral embedding of graphs. Pattern Recogn 36(10):2213–2230
26. Mitchell TM (1997) Machine learning. McGraw-Hill, New York
27. Münger A (1998) Synthesis of prototype graphs from sample graphs. Diploma thesis, University of Bern (in German)
28. Neuhaus M, Bunke H (2004) An error-tolerant approximate matching algorithm for attributed planar graphs and its application to fingerprint classification. In: Fred ALN, Caelli T, Duin RPW, Campilho AC, de Ridder D (eds) SSPR/SPR. Lecture notes in computer science, vol 3138. Springer, Berlin, pp 180–189
29. Neuhaus M, Riesen K, Bunke H (2006) Fast suboptimal algorithms for the computation of graph edit distance. In Proc. Joint IAPR Workshops on Structural and Syntactic Pattern Recognition and Statistical Techniques in Pattern Recognition. Lecture Notes on Computer Science, Springer-Verlag, Vol 4109, pp 163–172
30. Pekalska E, Duin R (2005) The dissimilarity representation for pattern recognition – foundations and applications. World Scientific, Singapore
31. Pekalska E, Duin RPW, Paclík P (2006) Prototype selection for dissimilarity-based classifiers. Pattern Recogn 39(2):189–208
32. Rand WM (1971) Objective criteria for the evaluation of clustering methods. J Am Stat Assoc 66:846–850
33. Ren P, Wilson RC, Hancock ER (2011) Graph characterization via ihara coefficients. IEEE Trans Neural Networks 22(2):233–245
34. Riesen K, Bunke H (2008a) IAM graph database repository for graph based pattern recognition and machine learning. In N. da Vitoria Lobo et al., (eds) SSPR&SPR. Lecture Notes in Computer Science, Springer-Verlag, vol 5342, pp 287–297
35. Riesen K, Bunke H (2008b) Kernel k-means clustering applied to vector space embeddings of graphs. In: Prevost L, Marinai S, Schwenker F (eds) ANNPR. Lecture notes in computer science, vol 5064. Springer, Berlin, pp 24–35
36. Riesen K, Bunke H (2009) Approximate graph edit distance computation by means of bipartite graph matching. Image Vision Comput 27(7):950–959
37. Riesen K, Bunke H (2010) Graph classification and clustering based on vector space embedding. World Scientific, Singapore
38. Riesen K, Neuhaus M, Bunke H (2007) Graph embedding in vector spaces by means of prototype selection. In: 6th IAPR-TC-15 international workshop, GbRPR 2007. Lecture notes in computer science, vol 4538. Springer, Berlin, pp 383–393
39. Robles-Kelly A, Hancock ER (2007) A Riemannian approach to graph embedding. Pattern Recogn 40(3):1042–1056
40. Sanfeliu A, Fu K (1983) A distance measure between attributed relational graphs for pattern recognition. IEEE Trans Syst Man Cybernet 13(3):353–362
41. Schenker A, Bunke H, Last M, Kandel A (2005) Graph-theoretic techniques for web content mining. World Scientific, Singapore
42. Shokoufandeh A, Macrini D, Dickinson S, Siddiqi K, Zucker SW (2005) Indexing hierarchical structures using graph spectra. IEEE Trans Pattern Anal Mach Intell 27(7):1125–1140. DOI 10.1109/TPAMI.2005.142. URL http://dx.doi.org/10.1109/TPAMI.2005.142
43. Strehl A, Ghosh J, Mooney R (2000) Impact of similarity measures on web-page clustering. In: Proceedings of the 17th national conference on artificial intelligence: workshop of artificial intelligence for web search, (AAAI 2000), Austin, Texas, 30–31 July 2000, pp 58–64
44. Weiszfeld E (1937) Sur le point pour lequel la somme des distances de n points donnés est minimum. Tohoku Math J (43):355–386
45. White D, Wilson RC (2006) Mixing spectral representations of graphs. In: 18th international conference on pattern recognition (ICPR 2006). IEEE Computer Society, Hong Kong, China, 20–24 August 2006, pp 140–144
46. Wilson RC, Hancock ER, Luo B (2005) Pattern vectors from algebraic graph theory. IEEE Trans Pattern Anal Mach Intell 27(7):1112–1124

Patch Alignment for Graph Embedding

Yong Luo, Dacheng Tao, and Chao Xu

1 Introduction

Dozens of manifold learning-based dimensionality reduction algorithms have been proposed in the literature. The most representative ones are locally linear embedding (LLE) [65], ISOMAP [76], Laplacian eigenmaps (LE) [4], Hessian eigenmaps (HLLE) [20], and local tangent space alignment (LTSA) [102]. LLE uses linear coefficients, which reconstruct a given example by its neighbors, to represent the local geometry, and then seeks a low-dimensional embedding, in which these coefficients are still suitable for reconstruction. ISOMAP preserves global geodesic distances of all the pairs of examples. LE preserves proximity relationships by manipulations on an undirected weighted graph, which indicates neighbor relations of pairwise examples. LTSA exploits the local tangent information as a representation of the local geometry and this local tangent information is then aligned to provide a global coordinate. HLLE obtains the final low-dimensional representations by applying eigenanalysis to a matrix, which is built by estimating the Hessian over neighborhood. All of these algorithms suffer from the out of sample problem [6]. One common response to this problem is to apply a linearization procedure to construct explicit maps over new examples. Examples of this approach include locality preserving projections (LPP) [41], a linearization of LE; neighborhood preserving embedding (NPE) [39], a linearization of LLE; orthogonal neighborhood

Y. Luo • C. Xu
Key Laboratory of Machine Perception, Peking University, No.5 Yihe Road, Haidian District, Beijing 100871, P.R. China
e-mail: yluo180@gmail.com; xuchao@cis.pku.edu.cn

D. Tao (✉)
Centre for Quantum Computation and Intelligent Systems, University of Technology, Sydney, Jones Street, Ultimo, NSW 2007, Sydney, Australia
e-mail: Dacheng.Tao@uts.edu.au

Y. Fu and Y. Ma (eds.), *Graph Embedding for Pattern Analysis*, DOI 10.1007/978-1-4614-4457-2_4,

preserving projections (ONPP) [47], a linearization of LLE with the orthogonal constraint over the projection matrix; and linear local tangent space alignment (LLTSA) [99], a linearization of LTSA.

All the aforementioned algorithms are designed according to specific intuitions, and solutions are given by optimizing intuitive and pragmatic objectives. That is, these algorithms have been developed based on the experience and knowledge of field experts for their own purposes. As a result, common properties and intrinsic differences of these algorithms are not completely clear. This problem is solved by the "patch alignment" framework [98], which can unify different spectral analysis-based dimensionality reduction algorithms. This framework consists of two stages: part optimization and whole alignment. For part optimization, different algorithms have different optimization criteria over patches, each of which is built by one example associated with its related ones. The part optimization procedure can preserve local information of data distribution, i.e., the near neighbors in the original space are still close to each other in the low-dimensional subspace. For whole alignment, all part optimizations are integrated to form the final global coordinate for all independent patches based on the alignment trick which is also used in developing LTSA [102].

Patch alignment framework not only unifies existing manifold learning based dimension reduction algorithms but also provides a general platform for specific algorithm design. In recent years, a series of related frameworks have been developed. Nonnegative patch alignment framework (NPAF) was proposed in [35] by incorporating the nonnegativity constraint in the patch alignment formulation. NPAF unified popular nonnegative matrix factorization (NMF) related dimensionality reduction algorithms and offered a new viewpoint to better understand the common property of different NMF algorithms. Considering that features from multiple views are usually necessary to well characterize an object, a multiview extension of the patch alignment framework was presented in [91] to learn a low-dimensional embedding of the multiview data. In this embedding, distribution of each view is sufficiently smooth and the complementary property of different views is explored. Si et al. [68] introduced an innovative Bregman divergence-based regularization to the patch alignment framework for learning a subspace, in which the knowledge gained from the training examples can be transferred to the test examples. This regularization boosts the performance when training and test examples are not independent and identically distributed. The patch alignment framework was utilized in [77] to learn a submanifold by transferring the local geometry and the discriminative information from the labeled images to the whole (global) image database in active reranking. By the use of the learned submanifold, we can localize the user's intention in the visual feature space. Zhou et al. [103] proposed manifold elastic net (MEN) to find the optimal sparse solution of the patch alignment framework. MEN is equivalent to the lasso penalized least square problem, and thus the popular least angle regression (LARS) algorithm can be directly applied to obtain the optimal sparse solution. Besides, the patch alignment framework can be applied to the understanding of parametric dimensionality reduction learning algorithms [8, 75] and combined with tensor learning methods for developing innovative multilinear dimensionality reduction tools [73, 74].

In this chapter, we first present some related work and then introduce the patch alignment framework. Subsequently, we study several extensions of this framework for nonnegative matrix factorization (NMF), multiview learning, transfer learning, active learning, and sparse learning.

Notations: We present some notations that will be used throughout this chapter if there are no special illustrations. We use $X = [x_1, \ldots, x_N]$ to denote a dataset with N examples. Each column vector $x_i \in \mathbb{R}^m, i = 1, \ldots, N$ is in the original feature space. The corresponding low-dimensional representations of X are $Y = [y_1, \ldots, y_N]$ with each $y_i \in \mathbb{R}^r, i = 1, \ldots, N$ and $r < m$ is the reduced dimensionality. For the linear dimensionality reduction, we use $U \in \mathbb{R}^{m \times r}$ to denote a projection matrix such that $Y = U^{\mathrm{T}} X$.

2 Related Work

This section reviews some recent related work on dimensionality reduction and manifold learning.

2.1 *Dimensionality Reduction*

The high-dimensionality problem often occurs in data analysis tasks [23, 24, 45]. The variable (or feature) selection [25,40,63,100] and subspace learning [14,28,43–46,72,75,86] techniques can be utilized to tackle this problem. For example, a new feature selection algorithm was presented in [100] by constructing localized graphs and considering label information. An unsupervised feature selection approach was proposed in [40] based on parameter covariance matrix minimization, in which the Laplacian regularization was utilized. In [14], the popular linear discriminative analysis (LDA) algorithm was reformulated as a regression problem, which needs no eigenvector computation and thus can be applied to large-scale datasets. By the use of correlation metric, a general formulation was proposed in [28] for feature extraction. Jiang [45] gave an in-depth understanding of the linear dimensionality reduction approaches and suggested to remove the relatively harmful dimensions for robust classification. To reduce the "class separation" problem in LDA, a subspace selection method was proposed in [75] by maximizing the geometric mean of the Kullback–Leibler (KL) divergences between different classes. In [86], a large margin based discriminative dimensionality reduction method was proposed to separate different clusters. Sun et al. [72] presented a two-stage approach for dimensionality reduction, and it can be applied to large-size problem. This approach is proved to be equivalent to solving the generalized eigenvalue problem, which is formulated in many dimensionality reduction algorithms, without the linear independent assumption of the data. In addition, the traditional subspace learning algorithms can be extended to handle multidimensional data by incorporating tensor analysis [27, 60, 61, 74, 84]. We refer to [62] for a literature survey of this kind of work, i.e., multilinear subspace learning.

2.2 *Manifold Learning*

Manifold learning is one of the most important subspace learning techniques. It usually assumes that data are intrinsically low-dimensional and are artificially embedded in a high-dimensional ambient space [22,26,56,69] and much efforts have been made to find such a low-dimensional embedding [89, 90]. Manifold learning has numerous applications, e.g., text categorization [11], data clustering [38, 85], feature analysis [69, 101], image retrieval [7, 88], image processing [29, 70, 71], image registration [50,51], cartoon animation [94,95], etc. The manifold structure of the data was explored in [11] to help selecting data points for labeling in active learning. The graph Laplacian was introduced in [38] to regularize the Gaussian mixture model (GMM) for clustering. Considering that biologically inspired features (BIFs) are sampled from a low-dimensional manifold and embedded in a high-dimensional space, a new dimensionality reduction algorithm was formed in [69] to find a low-dimensional embedding of the BIF. In [7], an Euclidean embedding method was presented to explore the manifold structure of the image low-level visual features. A sparse neighbor selection scheme was proposed in [29] for super-resolution construction based on neighbor embedding. In [50], 3-D images are represented by 4-D manifolds for registration and a diffusion process is utilized to match the embedded maps. Yu et al. [94] introduced a semi-supervised patch alignment framework for complex object correspondence construction by constructing local patches for each point on an object and aligning these patches in a new feature space.

3 The Patch Alignment Framework

In the patch alignment framework, N patches are built in the dataset. Each patch consists of an example and its related ones, which depend on both the characteristics of the dataset and the objective of an algorithm. As shown in Fig. 1, examples are on the S-curve manifold embedded in a three-dimensional space. Local patches should be built based on a given example and its nearest neighbors to capture the local geometry (locality). With these built patches, optimization can be imposed on them based on an objective function, and then alignment trick [102] can be utilized to form a global coordinate.

3.1 *A Summarization of the Patch Alignment Framework*

Part optimization Given example x_i and its k nearest neighbors $\left[x_{i_1}, x_{i_2}, \ldots, x_{i_k}\right]$, the part optimization at x_i is defined by

$$\arg\min_{Y_i} \operatorname{tr}\left(Y_i L_i Y_i^{\mathrm{T}}\right), \tag{1}$$

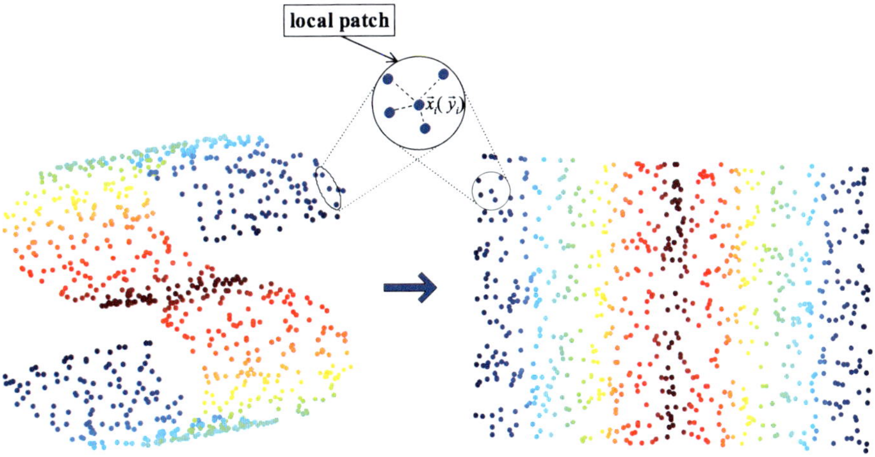

Fig. 1 Patch alignment framework illustration

where $\mathrm{tr}(\cdot)$ is the trace operator, $Y_i = [y_i, y_{i_1}, y_{i_2}, \ldots, y_{i_k}]$ is the projection of the local patch $X_i = [x_i, x_{i_1}, x_{i_2}, \ldots, x_{i_k}]$ onto the low-dimensional subspace, and L_i encodes the local geometric information at example x_i and is chosen algorithm-specifically.

Whole alignment By summarizing part optimizations over all examples, we get

$$\arg\min_{Y_1, Y_2, \ldots, Y_N} \sum_{i=1}^{N} \mathrm{tr}\left(Y_i L_i Y_i^{\mathrm{T}}\right). \tag{2}$$

Let $Y = [y_1, y_2, \ldots, y_N]$ be the projection of all examples $X = [x_1, x_2, \ldots, x_N]$. As for each local patch Y_i should be a subset of the whole alignment Y, the relationship between them can be expressed by

$$Y_i = Y S_i, \tag{3}$$

where S_i is a proper 0-1 matrix called the selection matrix. Thus

$$\begin{aligned} \arg\min_{Y} \sum_{i=1}^{N} \mathrm{tr}\left(Y_i L_i Y_i^{\mathrm{T}}\right) &= \arg\min_{Y} \sum_{i=1}^{N} \mathrm{tr}\left(Y S_i L_i S_i^{\mathrm{T}} Y^{\mathrm{T}}\right) \\ &= \arg\min_{Y} \mathrm{tr}\left(Y L Y^{\mathrm{T}}\right), \end{aligned} \tag{4}$$

with $L = \left(\sum_{i=1}^{N} S_i L_i S_i^{\mathrm{T}}\right)$ called the alignment matrix. By letting $Y = U^{\mathrm{T}} X$ for linearization, (4) can be rewritten as

$$\arg\min_{U} \mathrm{tr}\left(U^{\mathrm{T}} X L X^{\mathrm{T}} U\right). \tag{5}$$

Further, we can impose the orthogonal constraint $U^{\mathrm{T}} U = I$ on the projection matrix U, or the constraint $Y^{\mathrm{T}} Y = I$ on the Y, which leads to $U^{\mathrm{T}} X X^{\mathrm{T}} U = I$. In both cases, (5) is solved by eigen- or generalized eigen-decomposition.

Table 1 Manifold learning algorithms filled in the patch alignment framework

Algorithms	Patch: X_i	Representation of part optimization: L_i	Objective function
LLE/NPE/ONPP	Given example and its neighbors	$\begin{bmatrix} 1 & -c_i^{\mathrm{T}} \\ -c_i & c_i c_i^{\mathrm{T}} \end{bmatrix}$	Nonlinear/ linear/ orthogonal linear
ISOMAP	Given example and the rest ones	$\frac{1}{N} \cdot \tau\left(D_G^i\right)$	Nonlinear
LE/LPP	Given example and its connected ones in the undirected graph	$\begin{bmatrix} \sum_{j=1}^{l}(w_i)_j & -w_i^{\mathrm{T}} \\ -w_i & diag(w_i) \end{bmatrix}$	Nonlinear/ linear
LTSA/LLTSA	Given example and its neighbors	$R_{k+1} - V_i V_i^{\mathrm{T}}$	Nonlinear/ linear
HLLE	Given example and its neighbors	$H_i H_i^{\mathrm{T}}$	Nonlinear

c_i is the reconstruction coefficient in LLE; $\tau(D_G^i)$ is the inner product matrix, where $D_G^i = [d_G(F_i\{m\}, F_i\{n\})]$ and $d_G(F_i\{m\}, F_i\{n\})$ are the approximated geodestic distance between the $F_i\{m\}$th example and the $F_i\{n\}$th example, both of which are on the ith patch; w_i is weighting vector in LE; R_{k+1} is the centralization matrix and V_i denotes d largest right singular vectors of $X_i R_{k+1}$; H_i is the Hessian matrix. One can refer to [98] for more detailed descriptions

Different spectral analysis-based dimensionality reduction algorithms can be unified in the patch alignment framework. Only the way to build the patch X_i and the part optimization L_i varies across different algorithms. Table 1 summarizes these algorithms in the patch alignment framework.

3.2 *Discriminative Locality Alignment*

One representative subspace selection method based on the patch alignment framework above is the discriminative locality alignment (DLA) [98]. In DLA, the discriminative information, i.e., labels of examples, is imposed on the part optimization stage and then the whole alignment stage constructs the global coordinate in the projected low-dimensional subspace.

Given example x_i and its k nearest neighbors $\{x_{i_1}, x_{i_2}, \ldots, x_{i_k}\}$, we divide the k neighbors into two groups according to the label information, i.e., belonging to the same class with x_i or not. Without losing generality, we can assume the first k_1 neighbors $\{x_{i_j}\}_{j=1}^{k_1}$ having the same class label with x_i and the rest $k_2 = k - k_1$ neighbors $\{x_{i_j}\}_{j=k_1+1}^{k_1+k_2}$ having different class labels (otherwise, we just have to resort the indexes properly). And their low-dimensional representations are y_i, $\{y_{i_j}\}_{j=1}^{k_1}$ and $\{y_{i_j}\}_{j=k_1+1}^{k_1+k_2}$, respectively. The key idea of DLA is enforcing y_i close to $\{y_{i_j}\}_{j=1}^{k_1}$ while pushing it apart from $\{y_{i_j}\}_{j=k_1+1}^{k_1+k_2}$. Figure 2 illustrates such motivation.

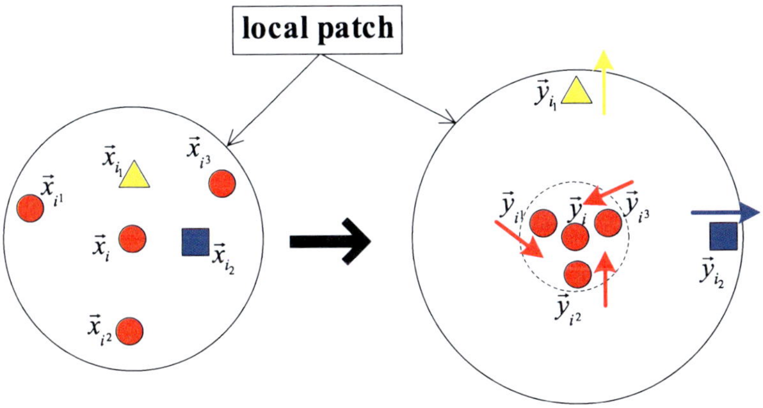

Fig. 2 The motivation of DLA. The examples with the same shape and color come from the same class

For x_i and its same class neighbors, we expect the summation of squared distance in the low-dimensional subspace to be as small as possible, i.e.,

$$\arg\min_{y_i} \sum_{j=1}^{k_1} \left\| y_i - y_{i_j} \right\|^2. \tag{6}$$

However, for x_i and its different class neighbors, we want the corresponding result to be large, i.e.,

$$\arg\max_{y_i} \sum_{j=k_1+1}^{k_1+k_2} \left\| y_i - y_{i_j} \right\|^2. \tag{7}$$

A convenient trade-off between (6) and (7) is

$$\arg\min_{Y_i} \left(\sum_{j=1}^{k_1} \left\| y_i - y_{i_j} \right\|^2 - \gamma \sum_{j=k_1+1}^{k_1+k_2} \left\| y_i - y_{i_j} \right\|^2 \right), \tag{8}$$

where γ is a scaling factor between 0 and 1 to balance the importance between measures of the within-class distance and the between-class distance. Define the coefficients vector

$$\omega_i = \left[\overbrace{1, 1, \ldots, 1}^{k_1}, \overbrace{-\gamma, -\gamma, \ldots, -\gamma}^{k_2} \right]^{\mathrm{T}}, \tag{9}$$

then (8) is readily rewritten as

$$\arg\min_{Y_i} \operatorname{tr}\left(Y_i L_i Y_i^{\mathrm{T}} \right), \tag{10}$$

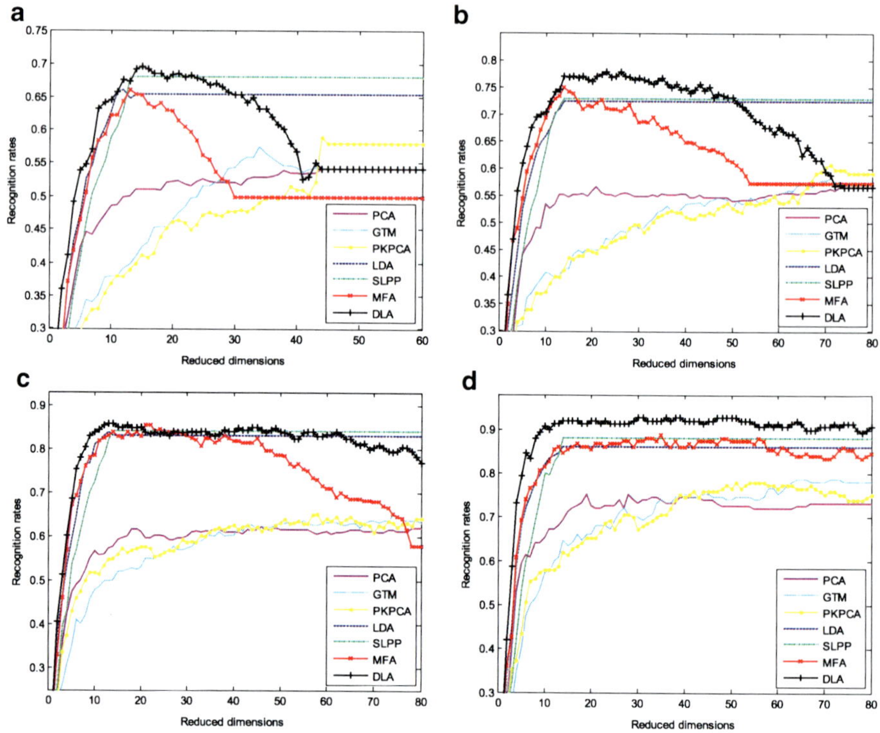

Fig. 3 Recognition rate vs. subspace dimension on Yale dataset. (**a**) Three images per subject for training; (**b**) five images per subject for training; (**c**) seven images per subject for training; (**d**) nine images per subject for training

where

$$L_i = \begin{bmatrix} \sum_{j=1}^{k} \omega_i & -\omega_i^{\mathrm{T}} \\ -\omega_i & \operatorname{diag}(\omega_i) \end{bmatrix}. \tag{11}$$

The low-dimensional embedding $y = U^{\mathrm{T}}x$ can be obtained by just substituting (11) into the whole alignment formula (4) and solving the eigen-decomposition problem with constraint $U^{\mathrm{T}}U = I$. It is worth emphasizing some merits of DLA here: (1) it exploits local geometric information of data distribution; (2) it is ready to deal with the case of nonlinear boundaries for class separation; (3) it avoids the matrix singularity problem.

Finally, we present some performance comparisons of DLA with six representative algorithms, i.e., principal component analysis (PCA) [82], generative topographic mapping (GTM) [9], probabilistic kernel principal components analysis (PKPCA) [81], linear discriminative analysis (LDA) [3], SLPP (LPP1 in [12]), and marginal fisher analysis (MFA) [93], on YALE face image dataset [3].

All face images from the dataset were cropped with reference to the eyes and cropped images were normalized to the 40×40 pixel arrays with 256 gray levels per pixel. Each image was reshaped to one long vector by arranging its pixel values in a fixed order. The YALE dataset contains face images collected from 15 individuals, 11 images for each individual and showing varying facial expressions and configurations. The experimental results are shown in Fig. 3. It can be seen that DLA outperforms the other algorithms.

4 Nonnegative Patch Alignment

In this section, we present the nonnegative extension of the patch alignment framework, i.e., NPAF proposed in [35]. It builds patches for each example, forms one local coordinate for such patch, and aligns all the local coordinates to form the global coordinate of all the examples in the nonnegative subspace. Nonnegative matrix factorization (NMF) was a newly proposed dimensionality reduction technique [36, 37, 48] and NPAF can unify various NMF-related dimensionality reduction algorithms.

4.1 *A Summarization of the Nonnegative Patch Alignment Framework*

Given N nonnegative examples in $\mathbb{R}^m$ that are arranged in matrix $X \in \mathbb{R}^{m\times N}$, NPAF projects them to $\mathbb{R}^r$, wherein r is the reduced dimensionality. Figure 4 gives an example of NPAF. The three-dimensional nonnegative examples (see Fig. 4a) are projected onto the two-dimensional space (see Fig. 4b). The local coordinates are

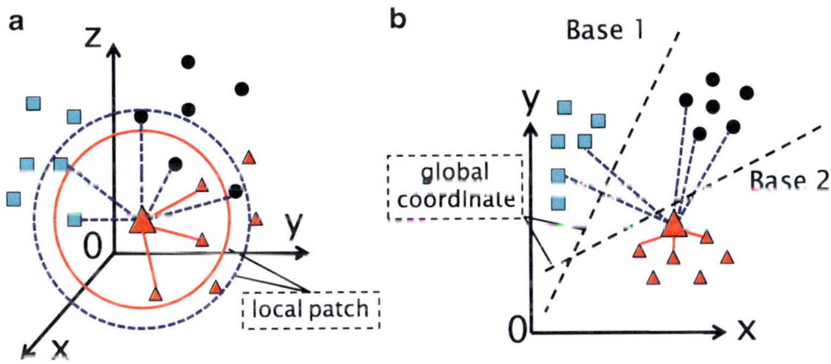

Fig. 4 An example of NPAF: (**a**) nonnegative examples in the three-dimensional space and (**b**) embedded two-dimensional representation

aligned with the global coordinate that is spanned by Base 1 and Base 2. By incorporating the nonnegativity constraint, we obtain the objective of NPAF

$$\arg\min_{W\geq 0, H\geq 0} \frac{\lambda}{2}\mathrm{tr}(HLH^{\mathrm{T}}) + D(X, WH), \tag{12}$$

where λ is a trade-off parameter, $W \in \mathbb{R}^{m\times r}$ signifies the bases vectors, $H \in \mathbb{R}^{r\times N}$ refers to the coordinate, and $D(X, WH)$ is the error between examples X and its approximation WH. The error can be measured by the Kullback–Leibler divergence (KLD), i.e.,

$$\mathrm{KL}(X, WH) = \sum_{ij}\left(x_{ij}\log\frac{x_{ij}}{(WH)_{ij}} - x_{ij} + (WH)_{ij}\right).$$

It can be replaced by the Frobenius matrix norm [33].

A multiplicate update rule (MUR) is developed in [35] to solve NPAF. However, the MUR converges slowly and thus it is difficult to apply the algorithm in practice. Therefore, based on a previous work of using fast gradient descent (FGD) method [34] to accelerate MUR, a new efficient FGD method is then proposed to optimize NPAF. The new FGD method assigns a step size for each column of the matrix factor and searches the optimal step size vector in each iteration round. Since the objective of FGD is convex, the multivariate Newton method can be applied to optimize its objective function. Although the inverse of Hessian matrix in the multivariate Newton method is time-consuming, the special structure of the Hessian matrix can be utilized to efficiently approximate its inverse, and thus FGD efficiently searches the optimal step size vector without increasing the time cost compared with the "old FGD" [34]. Therefore, FGD rapidly reduce the objective function at each iteration round. Various NMF-related dimensionality reduction algorithms, i.e., original NMF [48], local NMF (LNMF) [49], graph regularized NMF (GNMF) [15], and discriminant NMF (DNMF) [96], can be unified in NPAF and we summarize them in Table 2.

4.2 Nonnegative Discriminative Locality Alignment

A new NMF-related dimensionality reduction algorithm, termed "nonnegative discriminative locality alignment" (NDLA), can be obtained by introducing the underlying strategy used in DLA [98] to NPAF. Given an example x_i and its k nearest neighbors, we assume the first k_1 neighbors of a given example x_i having the same class label as x_i and the rest $k_2 = k - k_1$ neighbors having different class labels. To preserve the data local geometric structure, we expect the examples in the same class to be as close as possible in the low-dimensional space and thus obtain the part optimization on the within-class patch $\min_{H_i^{\mathrm{w}}} \mathrm{tr}(H_i^{\mathrm{w}} L_i^{\mathrm{w}} H_i^{\mathrm{w}\mathrm{T}})$, where

Table 2 NMF-related dimensionality reduction algorithms filled in NPAF

Algorithms	How to derive from NPAF (12)	MUR update rule
NMF	Setting $\lambda = 0$	$H \leftarrow H \odot \frac{W^{\mathrm{T}}X}{W^{\mathrm{T}}WH}$, $W \leftarrow W \odot \frac{XH^{\mathrm{T}}}{WHH^{\mathrm{T}}}$
LNMF	Replace L by $-I$ with fixed W Replace L by E_s with fixed H	$H \leftarrow H \odot \sqrt{\frac{\beta H + (W^{\mathrm{T}}X)/(WH)}{W^{\mathrm{T}}E}}$ $W \leftarrow W \odot \sqrt{\frac{(XH^{\mathrm{T}})/(WH)}{\alpha W E_s + EH^{\mathrm{T}}}}$
GNMF	Construct alignment matrix L_g using the part optimization representation L_i, then L_g is equivalent to the graph Laplacian L in GNMF	The same as in NPAF
DNMF	Construct alignment matrices L^W and L^B for $\min_H S_W = \min_H \mathrm{tr}(HL^WH^{\mathrm{T}})$ and $\max_H S_B = \max_H \mathrm{tr}(HL^BH^{\mathrm{T}})$	The same as in NPAF

$E \in \mathbb{R}^{m\times n}$ is a matrix whose entries are all 1, E_s is an all 1 square matrix, and $\mathbf{1}_k = [1,\ldots,1]^{\mathrm{T}} \in \mathbb{R}^k$ denotes a k-dimensional vector whose entries are all 1; $\odot$ is the elementwise product operator; α and β are the trade-off parameters used in LNMF; $L_i = \begin{bmatrix} k & -\mathbf{1}_k^{\mathrm{T}} \\ -\mathbf{1}_k & \mathrm{diag}(\mathbf{1}_k) \end{bmatrix}$; the part optimization representation to construct L^W is $L_i^W = \frac{1}{N_i^2}\begin{bmatrix} (N_i-1)^2 & -(N_i-1)\mathbf{1}_{N_i-1}^{\mathrm{T}} \\ -(N_i-1)\mathbf{1}_{N_i-1} & \mathbf{1}_{N_i-1}\mathbf{1}_{N_i-1}^{\mathrm{T}} \end{bmatrix}$, the part optimization representation to construct L^R is $L_i^B = \frac{N_i}{C^2}\begin{bmatrix} (C-1)^2 & -(C-1)\mathbf{1}_{C-1}^{\mathrm{T}} \\ -(C-1)\mathbf{1}_{C-1} & \mathbf{1}_{C-1}\mathbf{1}_{C-1}^{\mathrm{T}} \end{bmatrix}$, where N_i is the number of examples that have the same class label as x_i and C is the number of classes. One can refer to [35] for detailed descriptions

$$L_i^{\mathrm{w}} = \begin{bmatrix} \Sigma_{j=1}^{k_1}(\mathbf{1}_{k_1})_j & -(\mathbf{1}_{k_1})^{\mathrm{T}} \\ -\mathbf{1}_{k_1} & \mathrm{diag}(\mathbf{1}_{k_1}) \end{bmatrix}.$$

The set of indices, for x_i, on the within-class patch is $F_i^{\mathrm{w}} = \{i, i_1, \ldots, i_{k_1}\}$. To make examples in different classes separable, we obtain the part optimization on the between-class patch $\min_{H_i^{\mathrm{b}}} \mathrm{tr}(H_i^{\mathrm{b}} L_i^{\mathrm{b}} {H_i^{\mathrm{b}}}^{\mathrm{T}})$, where

$$L_i^{\mathrm{b}} = \begin{bmatrix} \Sigma_{j=k_1}^{k_1+k_2}(\mathbf{1}_{k_2})_j & -(\mathbf{1}_{k_2})^{\mathrm{T}} \\ -\mathbf{1}_{k_2} & \mathrm{diag}(\mathbf{1}_{k_2}) \end{bmatrix}.$$

The set of indices, for x_i, on the between-class patch is $F_i^{\mathrm{b}} = \{i, i_{k_1+1}, \ldots, i_k\}$.

By using the whole alignment, we come up with the following two objective functions:

$$\min_H \sum_{i=1}^{n} \mathrm{tr}(H_i L_i^{\mathrm{w}} H_i^{\mathrm{T}}) = \min_H \mathrm{tr}(HL^{\mathrm{w}}H^{\mathrm{T}}), \tag{13}$$

$$\max_H \sum_{i=1}^{n} \mathrm{tr}(H_i L_i^{\mathrm{b}} H_i^{\mathrm{T}}) = \max_H \mathrm{tr}(HL^{\mathrm{b}}H^{\mathrm{T}}), \tag{14}$$

where $L^{\mathrm{w}} = \sum_{i=1}^{N} S_i^{\mathrm{w}} L_i^{\mathrm{w}} S_i^{\mathrm{w}\,\mathrm{T}}$ and $L^{\mathrm{b}} = \sum_{i=1}^{N} S_i^{\mathrm{b}} L_i^{\mathrm{b}} S_i^{\mathrm{b}\,\mathrm{T}}$ are the alignment matrices of within-class patch and between-class patch, respectively, and $S_i^{\mathrm{w}} \in \mathbb{R}^{N\times(k_1+1)}$ and $S_i^{\mathrm{b}} \in \mathbb{R}^{N\times(k_2+1)}$ are selection matrices for the within-class patch and the between-class patch of the example x_i.

By combining (13) and (14), we arrive at

$$\min_{H} \operatorname{tr}\left(H\left((L^{\mathrm{b}})^{-1/2}\right)^{\mathrm{T}} L^{\mathrm{w}} (L^{\mathrm{b}})^{-1/2} H^{\mathrm{T}}\right).$$

Then the optimization problem for NDLA is given by

$$\min_{W\geq 0, H\geq 0} \frac{\lambda}{2} \operatorname{tr}(HLH^{\mathrm{T}}) + \mathrm{KL}(X, WH), \tag{15}$$

where $L = \left((L^{\mathrm{b}})^{-1/2}\right)^{\mathrm{T}} L^{\mathrm{w}} (L^{\mathrm{b}})^{-1/2}$.

4.3 Experimental Evaluation of NDLA

The effectiveness and robustness of NDLA is evaluated by comparing it with six representative algorithms, which are PCA [82], FLDA [3], DLA [98], NMF [48], LNMF [49], and DNMF [96], under different partial occlusions on the popular ORL [66] face image dataset.

In order to make statistical comparisons between classification performances of different algorithms, Dietterich [19] proposed an empirical method which uses five twofold cross-validations followed by a t-test. Alpaydm [2] subsequently proposed to modify Dietterich's method by removing the unsatisfactory aspect of the result depending on the ordering of the folds, which was called 5×2 cv F-test by the author. In particular, five replications of twofold cross-validation were performed. Assuming p_i^j is the difference between the classification error rates of two algorithms on fold $j = 1, 2$ of replication $i = 1, \ldots, 5$, the average on replication i was denoted by $\bar{p}_i = (p_i^1 + p_i^2)/2$, and the estimated variance was $s_i^2 = (p_i^1 - \bar{p}_i)^2 + (p_i^2 - \bar{p}_i)^2$. According to [2], the statistic $F = \sum_{i=1}^{5} \sum_{j=1}^{2} (p_i^j)^2 / 2 \sum_{i=1}^{5} s_i^2$ was approximately **F**-distributed with 10 and 5 degrees of freedom. Throughout this chapter, we rejected the hypothesis that the algorithms have statistically identical error rate with 95% confidence if the F-statistic is greater than 4.74.

The F-statistic defined above is used to statistically compare classification performances of NDLA and the other algorithms on the dataset. All face images were aligned according to the eye position. Each pixel of images was linearly rescaled to the gray level of 256, and each image was rearranged to a long vector.

According to [2], we randomly selected equal number of images from each individual to constitute twofolds, signified as training set and test set, and the rest images make up the validation set. The training set was used to learn bases for

Fig. 5 Image examples of the ORL dataset under different occlusions

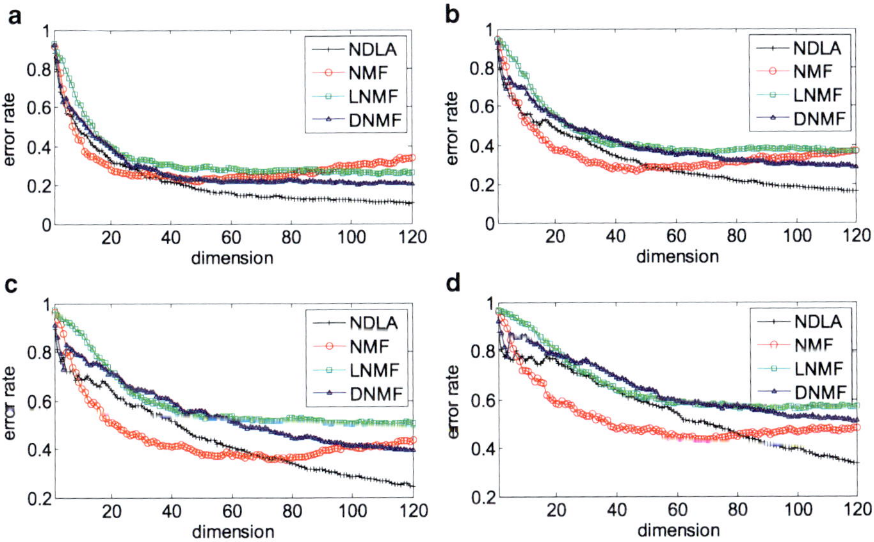

Fig. 6 Average error rate on test set when the partial occlusions size are (**a**) 20×20, (**b**) 25×25, (**c**) 30×30, and (**d**) 35×35 on ORL dataset

the low-dimensional space and the validation set was used to select the best model parameters, then the error rate was calculated as the percentage of examples in the test set which were improperly classified using the nearest neighbor (NN) rule. To evaluate NDLA's robustness to image occlusion, a randomly positioned square partial occlusion of different size $x \times x$, wherein x is the side length, was added to each image in the test set during the classification phase. Figure 5 shows the examples of image and the occluded images of the ORL dataset.

The Cambridge ORL [66] dataset consists of 400 images collected from 40 individuals. There are 10 images for each individual with varying lighting, facial expressions and facial details (with-glasses or no-glasses). All images were taken in the same dark background, and each image was normalized to 112×92 pixel array and reshaped to a long vector. We randomly selected eight images from each individual to constitute the twofolds (training set and test set) and the rest makes up the validation set. Figure 6 shows the average error rate vs. the dimension of the subspace on test set when the side length of partial occlusion $x = 20, 25, 30, 35$. Table 3 gives the average error rates on the twofolds and the dimension corresponding to the best performance for all the algorithms under different partial occlusions.

Table 3 Average error rate (%) followed by F-statistic value of NDLA vs. representative algorithms on ORL dataset under different partial occlusions

Occlusion	20×20		25×25		30×30		35×35	
PCA [82]	12.8(119)	2.29	20.4(117)	11.23∗	33.7(111)	14.93∗	45.8(113)	22.28∗
FLDA [3]	16.6(39)	48.00∗	24.8(39)	6.60∗	33.4(39)	24.14∗	39.9(39)	2.48
DLA [98]	11.7(43)	2.16	18.1(68)	4.24	31.3(120)	5.59∗	45.6(106)	19.66∗
NMF [48]	22.1(49)	10.83∗	28.3(73)	39.22∗	36.1(82)	14.10∗	42.9(75)	6.08∗
LNMF [49]	25.3(116)	31.32∗	37.1(111)	33.56∗	26.3(120)	3.50	57.3(90)	40.63∗
DNMF [96]	20.0(120)	35.51∗	28.2(120)	27.49∗	38.8(120)	28.90∗	50.2(118)	18.92∗
NDLA	**10.8**(120)	–	**14.1**(119)	–	**23.3**(120)	–	**33.3**(120)	–

Tick ∗ indicates that NDLA is statistically superior to the comparator algorithms

Figure 6 shows that NDLA outperforms all the representative NMF-related algorithms on the test set under different partial occlusions. Table 3 shows that the average error rates of NDLA are superior to all the comparator algorithms on the training set and test set. It also shows that NDLA is statistically superior to all the comparator algorithms.

5 Multiview Spectral Embedding

We present a multiview extension of the patch alignment framework in this section. Multimedia data generally have multiple modalities [92, 97], and each modality is usually represented in a high-dimensional feature space which frequently leads to the "curse of dimensionality" problem. In this case, multiview dimensionality reduction provides an effective solution to solve or at least reduce this problem. Existing spectral embedding algorithms assume that examples are drawn from a vector space and thus cannot deal with multiview data directly. A possible solution is to concatenate vectors from different views together as a new vector and then apply spectral embedding algorithms directly on the concatenated vector. However, this concatenation is not physically meaningful because each view has a specific statistical property. Besides, the concatenation ignores the diversity of multiple views and thus cannot efficiently explore the complementary nature of different views. Another solution is the distributed spectral embedding (DSE) proposed in [59]. DSE performs a spectral embedding algorithm on each view independently, and then based on the learned low-dimensional representations, it learns a common low-dimensional embedding which is "close" to each representation as much as possible. Although DSE allows selecting different spectral embedding algorithms for different views, respectively, the original multiple view data are invisible to the final learning process, and thus it cannot well explore the complementary nature of different views.

To effectively and efficiently learn the complementary nature of different views, multiview spectral embedding (MSE) [91] is proposed to learn a low-dimensional

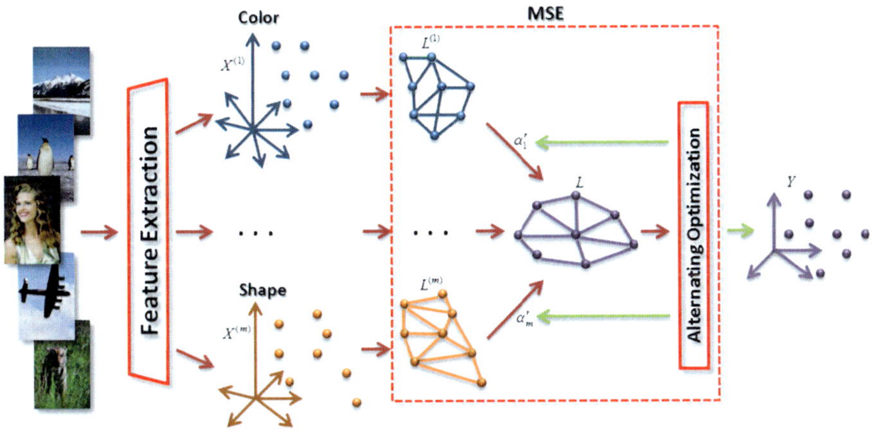

Fig. 7 The working flow of MSE: MSE first builds a patch for an example on a view. Based on patches from different views, the part optimization can be performed to get the optimal low-dimensional embedding for each view. Then all low-dimensional embeddings from different patches are unified together as a whole one by the global coordinate alignment. Finally, the solution of MSE is derived by using the alternating optimization

and sufficiently smooth embedding over all views of a dataset. Figure 7 shows the working principle of MSE. MSE first builds a patch for an example on a view. Based on patches from different views, the part optimization can be performed to get the optimal low-dimensional embedding for each view. Afterward, all low-dimensional embeddings from different patches are unified as a whole one by global coordinate alignment. Finally, the solution of MSE is derived by using the alternating optimization.

5.1 A Summarization of the MSE Formulation

Given a multiple view datum with N objects having V views, i.e., a set of matrices $X = \{X^v \in \mathbb{R}^{m_v \times N}\}_{v=1}^{V}$, each representation $X^v = [x_1^v, \ldots, x_N^v]$ is a feature matrix from view v. Consider an arbitrary point x_i^v and its k related ones in the same view (e.g., nearest neighbors) $x_{i_1}^v, \ldots, x_{i_k}^v$, the patch of x_i^v is defined as $X_i^v = [x_i^v, x_{i_1}^v, \ldots, x_{i_k}^v] \in \mathbb{R}^{m_v \times (k+1)}$. For X_i^v, there is a part mapping $f_i^v : X_i^v \to Y_i^v$, wherein $Y_i^v = [y_i^v, y_{i_1}^v, \ldots, y_{i_k}^v] \in \mathbb{R}^{r \times (k+1)}$. To preserve the locality in the projected low-dimensional space, the part optimization for the ith patch on the vth view is

$$\arg\min_{Y_i^v} \sum_{j=1}^{k} \|y_i^v - y_{i_j}^v\|^2 (\omega_i^v)_p, \tag{16}$$

where ω_i^v is a k-dimensional column vector weighted by $(\omega_i^v)_j = \exp(-\|x_i^v - x_{i_j}^v\|^2/\sigma)$ and σ is a scaling factor. Therefore, (16) can be reformulated as

$$\arg\min_{Y_i^v} \operatorname{tr}\left(Y_i^v L_i^v (Y_i^v)^{\mathrm{T}}\right), \tag{17}$$

where $L_i^v \in \mathbb{R}^{(k+1)\times(k+1)}$ encodes the objective function for the ith patch on the vth view and is given by

$$L_i^v = \begin{bmatrix} \sum_{j=1}^k (\omega_i^v)_j & -(\omega_i^v)^{\mathrm{T}} \\ -\omega_i^v & \operatorname{diag}(\omega_i^v) \end{bmatrix}. \tag{18}$$

Based on the locality information encoded in L_i^v, (17) finds a sufficiently smooth low-dimensional embedding Y_i^v by preserving the intrinsic structure of the ith patch on the vth view.

Because of the complementary property of multiple views to each other, different views definitely have different contributions to the final low-dimensional embedding. In order to well explore the complementary property of different views, a set of nonnegative weights $\alpha = [\alpha_1, \ldots, \alpha_V]$ are imposed on part optimizations of different views independently. The larger α_v is, the more important role the view X_i^v plays in learning to obtain the low-dimensional embedding Y_i^v. By summing over all views, the multiview part optimization for the ith patch is

$$\arg\min_{Y=\{Y_i^v\}_{v=1}^V, \alpha} \sum_{v=1}^V \alpha_v \operatorname{tr}\left(Y_i^v L_i^v (Y_i^v)^{\mathrm{T}}\right). \tag{19}$$

Following (3), we have $Y_i^v = Y S_i^v$, wherein $Y = [y_1, \ldots, y_N]$ is the global coordinate and $S_i^v \in \mathbb{R}^{N\times(k+1)}$ is the selection matrix to encode the spatial relationship of examples in a patch in the original high-dimensional space of the vth view. That is, low-dimensional embeddings in different views are consistent with each other globally. Therefore, (19) can be equivalently rewritten as

$$\arg\min_{Y,\alpha} \sum_{v=1}^V \alpha_v \operatorname{tr}\left(Y S_i^v L_i^v (S_i^v)^{\mathrm{T}} Y^{\mathrm{T}}\right). \tag{20}$$

By summing over all part optimizations defined by (20), the global coordinate alignment is given by

$$\arg\min_{Y,\alpha} \sum_{v=1}^V \alpha_v \operatorname{tr}\left(Y L^v Y^{\mathrm{T}}\right), \tag{21}$$

where L^v is the alignment matrix for the vth view, and it is defined as

$$L^v = \sum_{i=1}^N S_i^v L_i^v (S_i^v)^{\mathrm{T}}. \tag{22}$$

Putting (18) into (22), we have

$$L^v = D^v - W^v, \tag{23}$$

where $W^v \in \mathbb{R}^{N\times N}$ and $[W^v]_{pq} = \exp(-\|x_p^v - x_q^v\|^2/\sigma)$ if x_p^v is among the k-nearest neighbors of x_q^v or vice versa; $[W^v]_{pq} = 0$ otherwise. In addition, D^v is diagonal and $[D^v]_{ii} = \sum_j [W^v]_{ij}$. Therefore, L^v is an unnormalized graph Laplacian matrix [83]. The normalized graph Laplacian matrix $\tilde{L}^v$ is adopted in MSE,

$$\tilde{L}^v = I - (D^v)^{-1/2} W^v (D^v)^{-1/2}. \tag{24}$$

The constraint $YY^{\mathrm{T}} = I$ is imposed on (21) to uniquely determine the low-dimensional embedding Y, i.e.,

$$\begin{aligned} &\arg\min_{Y,\alpha} \sum_{v=1}^{V} \alpha_v \mathrm{tr}\left(Y\tilde{L}^v Y^{\mathrm{T}}\right), \\ &\text{s.t. } YY^{\mathrm{T}} = I; \sum_{v=1}^{V} \alpha_v = 1, \alpha_v \geq 0. \end{aligned} \tag{25}$$

The solution to α in (25) is $\alpha_v = 1$ corresponding to the minimum $\mathrm{tr}(Y\tilde{L}^v Y^{\mathrm{T}})$ over different views and $\alpha_v = 0$ otherwise. This solution means that only one view is finally selected by this method. Then the performance of this method is equivalent to the one from the best view. This solution does not meet the objective on exploring the complementary property of multiple views to get a better embedding than based on a single view. A trick utilized in [87] is adopted to avoid this phenomena, i.e., α_v is replaced by α_v^s with $s > 1$. In this condition, $\sum_{v=1}^{V} \alpha_v^s$ achieves its minimum when $\alpha_v = 1/V$ with respect to $\sum_{v=1}^{V} = 1$, $\alpha_v > 0$. Similar α_v for different views will be obtained by setting $s > 1$, so each view has a particular contribution to the final low-dimensional embedding Y. Therefore, the new objective function is defined as

$$\begin{aligned} &\arg\min_{Y,\alpha} \sum_{v=1}^{V} \alpha_v^s \mathrm{tr}\left(Y\tilde{L}^v Y^{\mathrm{T}}\right) \\ &\text{s.t. } YY^{\mathrm{T}} = I, \sum_{v=1}^{V} \alpha_v = 1, \alpha_v \geq 0, \end{aligned} \tag{26}$$

where $s > 1$. According to (26) and related discussions, MSE finds a low-dimensional sufficiently smooth embedding Y by preserving the locality of each view simultaneously.

An alternating optimization procedure is utilized in [91] to solve (26) by iteratively updating α with fixed $Y\tilde{L}^v Y^{\mathrm{T}}$ and computing Y with fixed $L = \sum_{v=1}^{V} \alpha_v^s \tilde{L}^v$ until convergence.

Fig. 8 Example images in the toy dataset

5.2 *Experimental Evaluation of MSE*

We compare the effectiveness of the proposed MSE with the conventional feature concatenation based spectral embedding (CSE), the distributed spectral embedding (DSE) [59], the average performance of single view based spectral embedding (ASE), and the best performance of single view based spectral embedding (BSE) in both image retrieval and video annotation. In ASE, BSE, CSE, and DSE, the Laplacian eigenmaps (LE) are adopted. The CSE is performed based on the Gaussian normalized concatenated vector from different views. For all experiments, the value of k for patch construction in MSE and that of k-nearest neighbor construction in LE are fixed at 30. In LE and MSE, unweighted graph as defined in [4] is adopted.

Firstly, a toy dataset is used to illustrate the effectiveness of MSE in comparing with CSE-LE and DSE-LE. The toy dataset, a subset of COREL image gallery, consists of three semantic categories, i.e., bus, ship, and train. Each category includes 100 images. Figure 8 shows some example images. For each image, two kinds of low-level visual features are extracted, i.e., 64-dimensional HSV color histogram (HSVCH) and 75-dimensional edge directional histogram (EDH) to represent two different views. Because the two views for an image are generally complementary to each other, s in MSE is empirically set to be five.

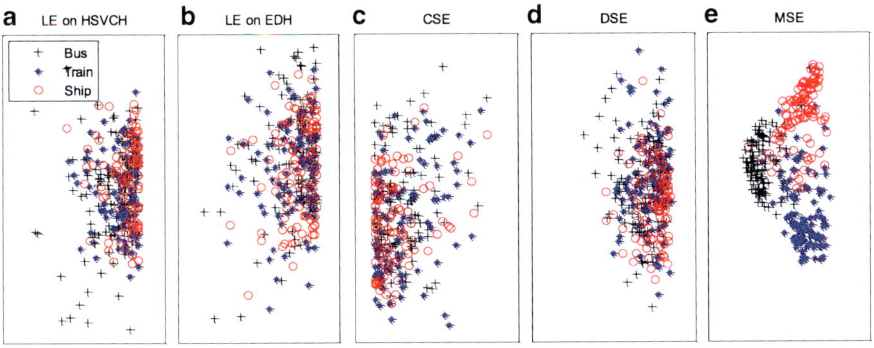

Fig. 9 Low-dimensional embeddings of different spectral embedding algorithms

Figure 9a, b shows the low-dimensional embeddings obtained by LE performed on HSVCH and EDH independently, and (c), (d) and (e) show embeddings obtained by CSE, DSE, and MSE, respectively. The results shown in (a)–(d) demonstrate that existing algorithms merge different categories in the low-dimensional space. On the contrary, the proposed MSE can well separate different categories, because MSE takes the complementary property of different views into consideration for embedding.

In image retrieval, the procedure for performance evaluation is as following: (1) the low-dimensional embedding of an image retrieval dataset is learned by an embedding algorithm, e.g., MSE; and (2) based on a low-dimensional embedding, a standard image retrieval procedure is conducted for all images in the dataset. In detail, for each category, one image is selected as a query, and then all the other images in the dataset (including other categories) are ranked according to the Euclidean distance to the query computed in the low-dimensional embedding. The retrieval performance is evaluated through the average precision (AP) based on the top n images. Mean average precision (MAP) is computed by averaging all APs for different categories. Corel-2000 and Caltech256-2045 are utilized independently for image retrieval test. Corel-2000 is a subset of the COREL photo gallery. Caltech256-2045 is a subset of the Caltech256 dataset [32]. In video annotation, the TRECVID 2008 training set [1] is adopted. The video dataset contains 39,674 shots from 20 concepts. A key frame is extracted from each shot for representation and ten concepts are selected for performance evaluation.

For each image or key frame, five kinds of low-level visual features are extracted to represent five different views. These five features are color moment, color correlogram, HSV color histogram, edge directional histogram, and wavelet texture. Because different views for an image are generally complementary to each other, s in MSE is empirically set as five. The dimensionality of the low-dimensional embedding r is set as 30. The results for the performance comparison on the three datasets are shown in Fig. 10. MSE achieves the best performance on all the three datasets.

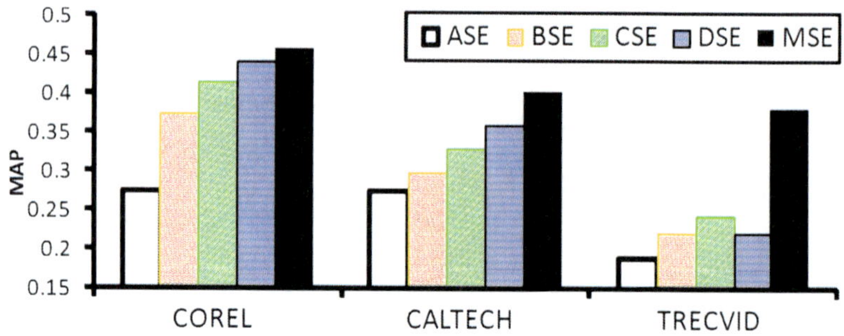

Fig. 10 Performance comparison on the three datasets measured by MAP of top 100 examples

6 Transfer Subspace Learning

In this section, we introduce a new regularization to the patch alignment framework for transfer subspace learning. Conventional algorithms including subspace selection methods are built under the assumption that training and test examples are independent and identically distributed (i.i.d.). For practical applications, however, this assumption is always violated. Particular, in face recognition, faces of the test subjects may not appear in the training set, and thus the distributions of the training set and test set are different. Transfer learning is an effective tool to address this problem and has many practical applications with the cross-domain setting [30, 67]. However, there is little effort of transfer learning made for the subspace learning. To this end, a transfer subspace learning (TSL) framework is proposed [68]. TSL extends conventional subspace learning methods by using a Bregman divergence [10, 57, 58] based regularization. Compared with the popular Tikhonov regularization [80] and manifold regularization [5, 31], the new regularization encourages the difference between the training and test examples in the selected subspace to be minimized. Thus, we can approximately assume the examples of training and test are almost i.i.d. in the learnt subspace.

6.1 A Summarization of the Transfer Subspace Learning Framework

The TSL framework [68] is presented by the following unified form:

$$\arg\min_{U} F(U) + \lambda D_U(P_l||P_u), \tag{27}$$

where $F(U)$ is the objective function of a subspace selection method, e.g., FLDA or PCA, etc., and $D_U(P_l||P_u)$ is the Bregman divergence between the training data distribution P_l and the test data distribution P_u in the low-dimension subspace

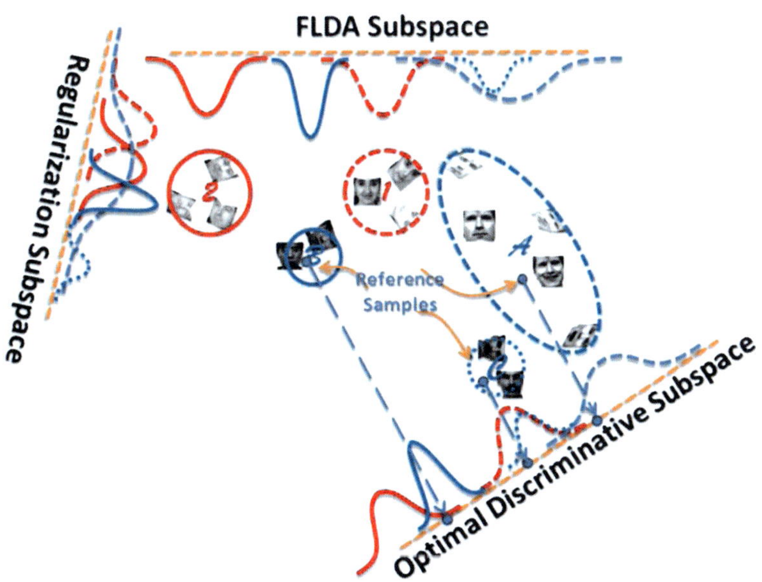

Fig. 11 Two classes of training examples are marked as 1 and 2, while three classes of test examples are marked as A, B, and C. *Blue circles* A and C are merged together in the FLDA subspace, where discrimination of the training examples can be well preserved. Blue circles A and B are mixed in the regularization subspace, where exists the smallest divergence between training domain (1, 2) and test domain (A, B, and C). Blue circles A, B, and C can be well separated in the discriminative subspace, which is obtained by optimizing the combination of the proposed regularization (the divergence between training sets 1, 2 and test sets A, B, C) and FLDA

induced by U, and parameter λ controls the balance between the objective function and the regularization. It should be noted that generally the objective function $F(U)$ only depends on the training data.

For example, when $F(U)$ is chosen to be FLDA's objective, (27) will give a subspace in which the training and test data distributions are close to each other and the discriminative information in the training data is partially preserved. In particular, suppose we have two classes of training examples, represented by two red circles (1 and 2, e.g., face images in the FERET dataset) and three classes of test examples, represented by three blue circles (A, B, and C, e.g., face images in the YALE dataset), as shown in Fig. 11. FLDA finds a subspace that fails to separate the test circle A from the test circle C, but the subspace is helpful to to distinct different subjects in the training set. The minimization of the Bregman divergence would give a subspace that makes the training and test examples almost i.i.d but merges A and B. Apparently, neither of them individually can find a best discriminative subspace for test. However, as shown in the figure, a combination of FLDA and the Bregman regularization does find the optimal subspace for discrimination, wherein A, B and C can be well separated and examples in them can be correctly classified with given references. It is worth emphasizing that the combination works well because the

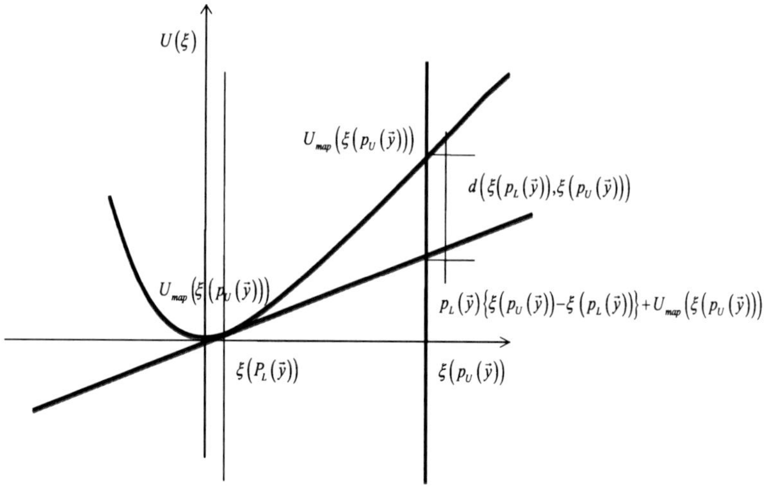

Fig. 12 The Bregman divergence-based regularization

training and test examples are coming from different domains but both domains share some common properties.

In (27), $F(U)$ is usually known and $D_U(P_l||P_u)$ is defined as follows.

Definition 1 (Bregman Divergence Regularization). Let $f : \mathbb{S} \to \mathbb{R}$ be a convex function defined on a closed convex set $\mathbb{S} \subseteq \mathbb{R}^+$. We denote the first-order derivative of f as f', whose inverse function as $\xi = (f')^{-1}$. The probability density for the training and test examples in the projected subspace U are $p_l(y)$ and $p_u(y)$, respectively, wherein $y = U^{\mathrm{T}}x$ is the low-dimensional representation of the example x. The difference at $\xi(p_l(y))$ between the function f and the tangent line to f at $(\xi(p_l(y)), f(\xi(p_l(y))))$ is given by (Fig. 12):

$$\begin{aligned} &d(\xi(p_l(y)), \xi(p_u(y))) \\ &\quad = \{f(\xi(p_u(y))) - f(\xi(p_l(y)))\} - p_l(y)\{\xi(p_u(y)) - \xi(p_l(y))\}. \end{aligned} \tag{28}$$

Based on (28), the Bregman divergence regularization, which measures the distance between $p_l(y)$ and $p_u(y)$, is a convex function given by

$$D_U(P_l||P_u) = \int d(\xi(p_l(y)), \xi(p_u(y)))\,\mathrm{d}\mu, \tag{29}$$

where $\mathrm{d}\mu$ is the Lebesgue measure.

By taking a special form $f(y) = y^2$, $D_U(P_l||P_u)$ can be expressed as

$$D_U(P_l||P_u) = \int (p_l(y) - p_u(y))^2 \, dy = \int \left(p_l(y)^2 - 2p_l(y)\,p_u(y) + p_u(y)^2\right) dy. \quad (30)$$

Further, the kernel density estimation (KDE) technique is used to estimate $p_l(y)$ and $p_u(y)$. Suppose there are N_l training examples $\{x_1, x_2, \ldots, x_{N_l}\}$ and N_u test examples $\{x_1, x_2, \ldots, x_{N_u}\}$, then through projection $y_i = U^{\mathrm{T}} x_i$, we have the estimates [68]

$$p_l(y) = (1/N_l) \sum_{i=1}^{N_l} G_{\Sigma_1}(y - y_i)$$

and

$$p_u(y) = (1/N_u) \sum_{i=N_l+1}^{N_l+N_u} G_{\Sigma_2}(y - y_i),$$

where $G_{\Sigma_1}(y)$ is a Gaussian kernel with covariance Σ_1, so is $G_{\Sigma_2}(y)$. With these estimates, the quadratic divergence (30) is rewritten as

$$D_U(P_l||P_u) = \frac{1}{N_l^2} \sum_{s=1}^{N_l} \sum_{t=1}^{N_l} G_{\Sigma_{11}}(y_t - y_s) + \frac{1}{N_u^2} \sum_{s=N_l+1}^{N_l+N_u} \sum_{t=N_l+1}^{N_l+N_u} G_{\Sigma_{22}}(y_t - y_s) - \frac{2}{N_l N_u} \sum_{s=1}^{N_l} \sum_{t=N_l+1}^{N_l+N_u} G_{\Sigma_{12}}(y_t - y_s), \quad (31)$$

where $\Sigma_{11} = \Sigma_1 + \Sigma_1$, $\Sigma_{12} = \Sigma_1 + \Sigma_2$ and $\Sigma_{22} = \Sigma_2 + \Sigma_2$. Further, by basis matrix calculus, we obtain the derivative of $D_U(P_l||P_u)$ with respect to U as follows:

$$\begin{aligned}
\frac{\partial D_U(P_l||P_u)}{\partial U} = {} & \frac{2}{N_l^2} \sum_{i=1}^{N_l} \sum_{t=1}^{N_l} G_{\Sigma_{11}}(y_i - y_t)(\Sigma_{11})^{-1}(y_t - y_i)x_i^{\mathrm{T}} \\
& - \frac{2}{N_l N_u} \sum_{l=1}^{N_l} \sum_{t=N_l+1}^{N_l+N_u} G_{\Sigma_{12}}(y_t - y_i)(\Sigma_{12})^{-1}(y_t - y_i)x_i^{\mathrm{T}} \\
& + \frac{2}{N_u^2} \sum_{i=N_l+1}^{N_l+N_u} \sum_{t=N_l+1}^{N_l+N_u} G_{\Sigma_{22}}(y_i - y_t)(\Sigma_{22})^{-1}(y_t - y_i)x_i^{\mathrm{T}} \\
& - \frac{2}{N_l N_u} \sum_{i=N_l+1}^{N_l+N_u} \sum_{t=1}^{N_l} G_{\Sigma_{12}}(y_t - y_i)(\Sigma_{12})^{-1}(y_t - y_i)x_i^{\mathrm{T}}.
\end{aligned} \quad (32)$$

We now give an example of TSL, the transferred DLA (TDLA). As presented in Sect. 3, the DLA subspace is obtained by minimizing the function

$$\begin{aligned} F(U) &= \mathrm{tr}(U^{\mathrm{T}} X L X^{\mathrm{T}} U) \\ \text{s.t. } U^{\mathrm{T}} U &= I. \end{aligned} \tag{33}$$

The derivative of $F(U)$ with respect to W is

$$\frac{\partial F(U)}{\partial U} = \left(XLX^{\mathrm{T}} + (XLX^{\mathrm{T}})^{\mathrm{T}}\right)U. \tag{34}$$

Based on (32) and (34), we can solve TDLA iteratively subject to $U^{\mathrm{T}}U = I$.

We refer to [68] for detailed descriptions of some other examples of TSL, which are the transferred principal component analysis (TPCA), the transferred Fisher's linear discriminant analysis (TFLDA), the transferred locality preserving projections (TLPP) with supervised setting and the transferred marginal Fisher's analysis (TMFA).

6.2 Experimental Evaluation of TSL

Based on the YALE, UMIST, and a subset of FERET datasets, cross-domain face recognition is performed by applying the TSL framework. In detail, we have (1) Y2F: the training set is on YALE and the test set is on FERET; (2) F2Y: the training set is on FERET and the test set is on YALE; and (3) YU2F: the training set is on the combination of YALE and UMIST and the test set is on FERET. In the training stage, the labeling information of test images is blind to all subspace learning algorithms. However, one reference image for each test class is preserved so that the classification can be done in the test stage. The nearest neighbor classifier is adopted for classification, i.e., we calculate the distance between a test image and every reference image and predict the label of the test image as that of the nearest reference image.

To evaluate the effectiveness of the TSL framework, the TSL algorithms, e.g., TPCA, TFLDA, TLPP, TMFA, and TDLA, are compared with conventional subspace learning algorithms, e.g., PCA [82], FLDA [3], LPP [41], MFA [93], DLA [98], and the semi-supervised discriminant analysis (SDA) [13]. In addition, an unsupervised transfer learning algorithm MMDE [64] is introduced for comparison. Table 4 shows the recognition rate of each algorithm with the corresponding optimal subspace dimension. In detail, conventional subspace learning algorithms, e.g., FLDA, LPP, and MFA, perform poorly because they assume training and test examples are i.i.d. variables and this assumption is unsuitable for cross-domain tasks. Although SDA learns a subspace by taking test examples into account, it assumes examples in a same class are drawn from an identical underlying manifold. Therefore, SDA is not designed for the cross-domain tasks. Although MMDE

Table 4 Recognition rates of different algorithms under three experimental settings

	Y2F	F2Y	YU2F
LDA	39.71(70)	36.36(30)	29.57(30)
LPP	44.57(65)	44.24(15)	45.00(35)
MFA	40.57(65)	34.54(60)	27.85(70)
DLA	50.43(80)	50.73(15)	50.86(65)
SDA	44.42(65)	41.81(40)	32.00(35)
MMDE	45.60(60)	42.00(75)	49.75(80)
TLDA	57.28(15)	50.51(20)	55.57(45)
TLPP	58.28(30)	53.93(25)	58.42(30)
TMFA	63.14(70)	56.96(35)	65.42(70)
TDLA	63.12(60)	61.82(30)	65.57(70)

The number in the parenthesis is the corresponding subspace dimensionality

considers the distribution bias between the training and the test examples, it ignores the discriminative information contained in the training examples. We have given an example in the synthetic data test to show that the training discriminative information is helpful to separate test classes. Example TSL algorithms perform consistently and significantly better than others, because the training discriminative information can be transferred to test examples by minimizing the distribution distance between the training and the test examples. In particular, TDLA performs best among all TSL examples because it inherits the merits of DLA in preserving both the discriminative information of different classes and the local geometry of examples in an identical class.

7 Active Reranking

The patch alignment framework is also utilized in an active learning based reranking algorithm to learn a submanifold, which can encode the user's intention. The textual information is usually insufficient for semantic image retrieval, a natural solution is to exploit the usage of the visual information for refining the text-based search result. However, image search reranking usually fails to capture the user's intention when the query term is ambiguous. Therefore, reranking with user interactions, or active reranking [77], is highly demanded to effectively improve the search performance.

7.1 A Summarization of Active Reranking

In active reranking, the essential problem is how to capture the user's intention, i.e., to distinguish query relevant images from irrelevant ones. An image may be relevant for one user but irrelevant for another. In other words, the semantic

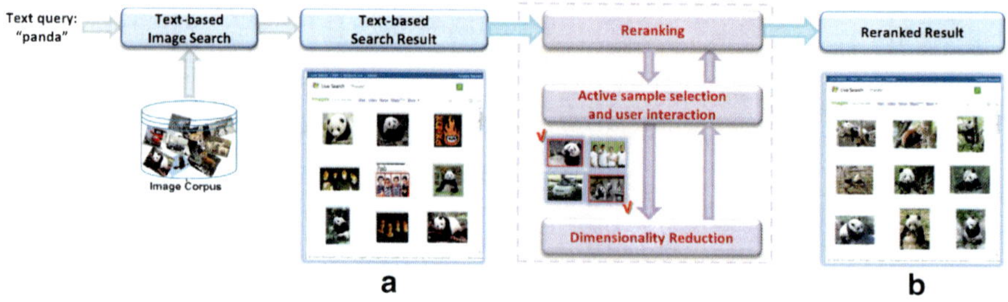

Fig. 13 Framework for active reranking illustrated with the query "panda." When the query is submitted, the text-based image search engine returns a coarse result (**a**). Then the active reranking process is adopted to obtain a more satisfactory result (**b**), by learning the user's intention

space is user-driven, according to their different intentions but with identical query keywords. Therefore, two aspects are proposed in [77] to target the user-driven intention: (1) collecting labeling information from users to obtain the specified semantic space, and (2) localizing the visual characteristics of the user's intention in this specific semantic space.

To collect the labeling information from users efficiently, a new structural information (SInfo) based strategy is proposed to actively select the most informative query images. Then a novel local–global discriminative (LGD) dimensionality reduction algorithm is developed to localize the visual characteristics of the user's intention. It is assumed in [77] that the query relevant images, which represent the user's intention, are lying on a low-dimensional submanifold of the original ambient (visual feature) space. LGD learns the submanifold by transferring both the local geometry and the discriminative information from labeled images to unlabeled ones. The learned submanifold preserves both the local geometry of labeled relevant images and the discriminative information to separate relevant from irrelevant images. As a consequence, the well-known semantic gap between low-level visual features and high-level semantics is narrowed to further enhance the reranking performance on this submanifold.

Figure 13 shows the general framework for active reranking in web image search. Take the query term "panda" as an example. When "panda" is submitted to the web image search engine, an initial text-based search result is returned to the user, as shown in Fig. 13a (only the top nine images are given for illustration). This result is unsatisfactory because both person and animal images are retrieved as top results. This is caused by the ambiguity of the query term. To solve this problem, active reranking, i.e., reranking with user interactions, is proposed. As shown in Fig. 13, four images are first selected according to an active example selection strategy, and then the user is required to label them. If the user labels the animal pandas as query relevant (indicated by "✓" in Fig. 13) and other two images (person, car) as query irrelevant, then we can learn that the animal panda is the user's intention. To represent this intention, i.e., the animal panda, a discriminative submanifold should be exploited to separate query relevant images from irrelevant ones. A dimensionality reduction step is thus introduced to localize the visual characteristics of the user's intention. With the knowledge of the user's intention, including both the labeling information and the learned discriminative submanifold, the reranking process is conducted and different kinds of animal pandas are returned, as shown in Fig. 13b. Sometimes, several interaction rounds are preferred to achieve a more satisfactory performance.

7.2 SInfo Active Example Selection

In active reranking, it is direct and reasonable to measure the ambiguity with the ranking scores obtained in the reranking process. For an image I_i, $0 \leq s_i \leq 1$ is its ranking score, where $s_i = 1$ means I_i is definitely query relevant, while $s_i = 0$ means I_i is totally irrelevant. s_i and $(1 - s_i)$ can be regarded as the probability of I_i to be relevant and irrelevant, respectively. Then the ambiguity can be measured via the information entropy, which is a widely used example in the information theory. The ambiguity of I_i is

$$H_r(I_i) = -s_i \log s_i - (1 - s_i) \log (1 - s_i). \tag{35}$$

Because the reranking is conducted based on the initial text-based search result [78], the ambiguity in the initial text-based search result should also be taken into account, i.e.,

$$H_{\bar{s}}(I_i) = -\bar{s}_i \log \bar{s}_i - (1 - \bar{s}_i) \log (1 - \bar{s}_i), \tag{36}$$

where $0 \leq \bar{s}_i \leq 1$ is the initial text-based search ranking score for I_i.

By combining (35) and (36), the total ambiguity for I_i is

$$H(I_i) = \alpha H_s(I_i) + (1 - \alpha) H_{\bar{s}}(I_i), \tag{37}$$

where $\alpha \in [0, 1]$ is a trade-off parameter to control the influence of the two ambiguity terms.

To avoid the small sample size problem in active example selection, the representativeness can be estimated in an unsupervised manner. Intuitively, labeling an image in a dense area will be more helpful than labeling an isolated one because the labeling information of the image can be shared with other surrounding images. As a consequence, we can measure the representativeness of image I_i via the probability density $p(I_i)$, which can be estimated by using the kernel density estimation (KDE),

$$p(I_i) = \frac{1}{|\mathcal{N}_i|} \sum_{I_j \subset \mathcal{N}_i} k(x_i - x_j), \tag{38}$$

where $\mathcal{N}_i$ is the set of neighbors of I_i. x_i is the visual feature for image I_i. $k(x)$ is a kernel function that satisfies both $k(x) > 0$ and $\int k(x)\mathrm{d}x = 1$. The Gaussian kernel is adopted in [77].

Since the most informative images should meet both ambiguity and representativeness simultaneously, the structural information of image I_i, $\mathrm{SI}(I_i)$, can be measured by the product of the two terms, i.e.,

$$\mathrm{SI}(I_i) = p(I_i)H(I_i).$$

Then the most informative image I^* is selected from the unlabeled image set $\mathcal{U}$ according to

$$I^* = \arg\max_{I_i \in \mathcal{U}} \mathrm{SI}(I_i). \tag{39}$$

In practical applications, to provide a good user experience, it would be better to ask users to label a small number of images than only one image in each round. This is because users will lose their patience after a few rounds. Thus, the batch mode is utilized to select several images in each round. A simple method is to select the top-n most informative images. The disadvantage of this method is that the selected n images may be redundant and cluster in a small area in the high-dimensional feature space. Thus, we seek to select a batch of most informative images and maintain their diversity at the same time.

The angle-diversity criterion [16] is a good choice to achieve this purpose. This criterion iteratively selects images which are most informative and also be diverse to the already selected image set $\mathcal{S}$. For an unlabeled image I_i, the diversity between I_i and $\mathcal{S}$ is measured by the minimal angle between I_i and each image $I_j \in \mathcal{S}$. Then, the images are selected iteratively according to

$$\arg\max_{I_i \in \mathcal{U}} \left(\eta \mathrm{SI}(I_i) + (1-\eta) \min_{I_j \in \mathcal{S}} \frac{-x_i \cdot x_j}{\|x_i\| \|x_j\|} \right), \tag{40}$$

where $\eta \in [0,1]$ is a trade-off parameter which is introduced to balance the effects of the two components: the structural information and the angle-diversity.

7.3 LGD Dimensionality Reduction

By mining user's labeling information, we can learn a submanifold to encode the user's intention. In [77], a linear subspace U is used to approximate this submanifold. Recall that $X = [x_1, \ldots, x_N] \in \mathbb{R}^{m \times N}$ is the feature matrix of of an image set $\mathcal{I} = \{I_1, \ldots, I_N\}$, then the images in the subspace can be represented as $Y = U^{\mathrm{T}}X = [y_1, \ldots, y_N] \in \mathbb{R}^{r \times N}$ $(r < m)$ with $y_i \in \mathbb{R}^r$ for image I_i.

The LGD dimensionality reduction algorithm considers both the local information contained in the labeled images and the global information of the whole image database simultaneously. In detail, LGD transfers the local information, including both the local geometry of the labeled relevant images and the discriminative information in the labeled images, to the global domain (the whole image database). This cross-domain transfer process is completed by building different local and global patches for each image and then aligning those patches together to learn a consistent coordinate. One patch is a local area formed by a set of neighboring images. We have three types of images: labeled relevant, labeled irrelevant, and unlabeled. Therefore, we build three types of patches, which are: (1) *local patches for labeled relevant images* to represent the local geometry of them and the discriminative information to separate relevant images from irrelevant ones, (2) *local patches for labeled irrelevant images* to represent the discriminative information to separate irrelevant images from relevant ones, and (3) *global patches for both labeled and unlabeled images* for transferring both the local geometry and the discriminative information from all labeled images to the unlabeled ones.

For convenience, we use superscript "+" to denote the labeled relevant images and "−" to denote the labeled irrelevant ones. If there is no superscript, it refers to an arbitrary image which may be labeled relevant, labeled irrelevant or unlabeled.

The query relevant examples may vary in appearance and corresponding visual features. For this reason, instead of requiring relevant images to be close to each other in the projected subspace, it is more proper to remain the local geometry of the relevant images while separating relevant images from all irrelevant ones. Therefore, the local patch for a labeled relevant image I_i^+ should preserve both the local geometry of relevant images and the discriminative information between the relevant images and all irrelevant images. We model the local patch for the low-dimensional representation y_i^+ of the labeled relevant image I_i^+ as

$$\min \|y_i^+ - \sum_{j=1}^{k_1} (c_i)_j y_{i_j}\|^2 - \gamma \sum_{j=k_1+1}^{k_1+k_2} \|y_i^+ - y_{i_j}\|^2. \tag{41}$$

The $\{I_{i_1}, \ldots, I_{i_{k_1}}\}$ are I_i^+'s k_1 nearest neighbors in the labeled relevant image set "+" and $\{I_{i_{(k_1+1)}}, \cdots, I_{i_{(k_1+k_2)}}\}$ are its k_2 nearest neighbors in the labeled irrelevant image set "−." The combination coefficient γ is a trade-off factor between the two parts.

The first part in (41) is used to preserve the local geometry of labeled relevant images before and after projection, thus the linear combination coefficient vector c_i is required to reconstruct I_i^+ from its neighboring relevant images with minimal error,

$$\begin{aligned} &\arg\min_{c_i} \ \|x_i^+ - \sum_{j=1}^{k_1} (c_i)_j x_{i_j}\|^2 \\ &\text{s.t.} \quad \sum_{j=1}^{k_1} (c_i)_j = 1. \end{aligned} \tag{42}$$

Solving problem (42), we can get $(c_i)_j = \sum_{t=1}^{k_1} G_{jt}^{-1} / (\sum_{p=1}^{k_1} \sum_{q=1}^{k_1} G_{pq}^{-1})$ with the local gram matrix $G_{jt} = (x_i^+ - x_{i_j})^{\mathrm{T}} (x_i^+ - x_{i_t})$.

We can rewrite (41) in a more compact form. For the first part, which models the local geometry of relevant images,

$$\|y_i^+ - \sum_{j=1}^{k_1} (c_i)_j y_{i_j}\|^2 = \mathrm{tr}\left(Y_i^+ L_i^A (Y_i^+)^{\mathrm{T}}\right), \tag{43}$$

where $Y_i^+ = [y_i^+, y_{i_1}, \ldots, y_{i_{k_1}}, y_{i_{k_1+1}}, \ldots, y_{i_{(k_1+k_2)}}]$ and $L_i^A = \begin{bmatrix} 1 & -\bar{c}_i^{\mathrm{T}} \\ \bar{c}_i & \bar{c}_i \bar{c}_i^{\mathrm{T}} \end{bmatrix}$ with $\bar{c}_i = [c_i^{\mathrm{T}}, \underbrace{0, \cdots, 0}_{k_2}]^{\mathrm{T}}$.

The second part models the discriminative information for separating relevant image I_i^+ from all irrelevant ones, i.e.,

$$-\gamma \sum_{j=k_1+1}^{k_1+k_2} \|y_i^+ - y_{i_j}\|^2 = \mathrm{tr}\left(Y_i^+ L_i^B (Y_i^+)^{\mathrm{T}}\right), \tag{44}$$

where $L_i^B = \begin{bmatrix} \sum_{j=1}^{k_1+k_2} (\omega_i)_j & -\omega_i^{\mathrm{T}} \\ -\omega_i & \mathrm{diag}(\omega_i) \end{bmatrix}$ and $\omega_i = [\underbrace{0, \ldots, 0}_{k_1}, \underbrace{-\gamma, \ldots, -\gamma}_{k_2}]^{\mathrm{T}}$. By combining (43) and (44) together into (41), we have

$$\min \|y_i^+ - \sum_{j=1}^{k_1} (c_i)_j y_{i_j}\|^2 - \gamma \sum_{j=k_1+1}^{k_1+k_2} \|y_i^+ - y_{i_j}\|^2 = \min \mathrm{tr}\left(Y_i^+ L_i^+ (Y_i^+)^{\mathrm{T}}\right),$$

where $L_i^+ = L_i^A + L_i^B$.

Discriminative information is also partially encoded in all irrelevant images, so we construct local patches for labeled irrelevant images by separating each irrelevant image from all relevant images. Because each irrelevant image is irrelevant in its own way, it could be unreasonable to keep the local geometry of the irrelevant images. In [77], the local patch for the low-dimensional representation y_i^- of labeled irrelevant image I_i^- is modeled as

$$\min - \sum_{j=1}^{k} \|y_i^- - y_{i_j}\|^2 = \min \mathrm{tr}\left(Y_i^- L_i^- (Y_i^-)^{\mathrm{T}}\right). \tag{45}$$

The $\{I_{i_1}, \cdots, I_{i_k}\}$ is I_i^-'s k nearest neighbors in the labeled relevant image set "+." The matrix L_i^- can be calculated in the way similar to that of computing L_i^B in (44) by setting $k_1 = 0$ and $k_2 = k$.

In active reranking, users would like to label only a small number of images. With only the labeled images, the learned subspace will bias to that spanned by these labeled images and cannot generalize well to the large amount of unlabeled data. Therefore some semi-supervised methods have been proposed which also

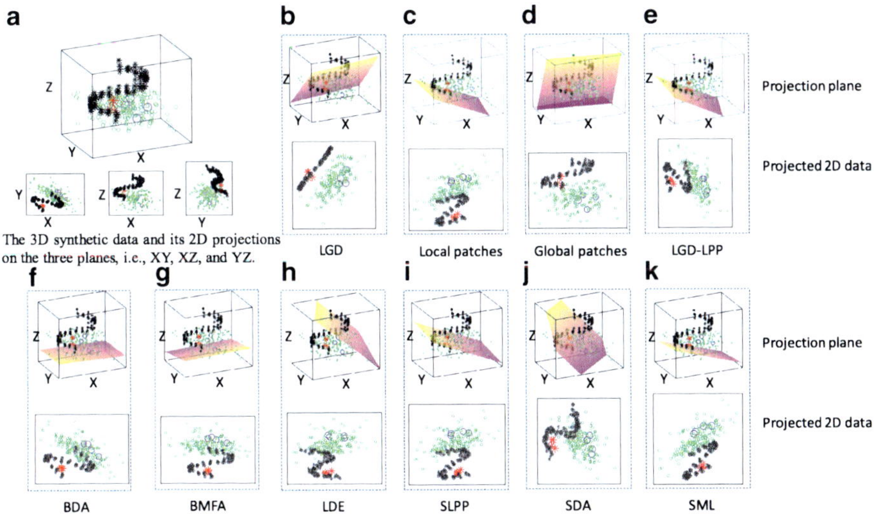

Fig. 14 A 3D synthetic dataset for dimension reduction illustration. In this dataset, *big red* "*" and *big blue* "o" denote labeled relevant and irrelevant examples, respectively. *Small black* "*" and *small green* "o" are unlabeled relevant and irrelevant examples, respectively. As given in (**b**), LGD reveals the submanifold of the relevant examples and separates the relevant examples from the irrelevant ones in the projected 2D subspace. When other dimension reduction algorithms are adopted, the relevant and irrelevant examples are overlapped in the projected subspace, as shown in (**c**)–(**k**)

take the unlabeled images into utilization. However, because only relevant images are lying on an unknown manifold and the distribution of irrelevant images is nearly flat, conventional manifold regularizations which assume both relevant and irrelevant examples are drawn from unknown manifolds prone to over-fit to unlabeled examples. As a consequence, another method is considered in [77] to model unlabeled images in active reranking.

Global patches are introduced in [77] to make use of both the labeled and unlabeled images. The global patches transfer the local geometry and the discriminative information, which is exploited in the domain of labeled images, to the domain of unlabeled images. With the global patches, we aim to preserve the principal subspace to keep the submanifold of relevant images. The noise information contained in the ambient space should be eliminated. The principal component analysis (PCA) is a suitable choice, which maximizes the mutual information between the ambient space and the corresponding projected subspace.

A synthetic example is shown in Fig. 14 to illustrate the advantage of global patches for dimensionality reduction. From the results of different conventional dimensionality reduction algorithms presented in Fig. 4c–k, we can see that the relevant and irrelevant examples are overlapped in the projected subspace and the submanifold of the relevant examples is not well preserved. This is caused by the problems existing in these algorithms as aforementioned.

To avoid these problems, LGD learns the submanifold by transferring both the local geometry and the discriminative information from labeled examples to all unlabeled examples. Global patches are built for each example (including both labeled and unlabeled) to complete the cross-domain knowledge transferring process. According to the alignment scheme in [102], the global patch for the low-dimensional representation y_i of the image I_i is modeled in a similar way to local patches,

$$\max \operatorname{tr}\left((y_i - y^m)(y_i - y^m)^{\mathrm{T}}\right), \tag{46}$$

where y^m is the centroid of the projected low-dimensional feature. A variant version of the original definition of PCA is used here to achieve a formula-level consistency for both local and global patches. Equation (46) can be rewritten as

$$\max \operatorname{tr}\left((y_i - y^m)(y_i - y^m)^{\mathrm{T}}\right) = \max \operatorname{tr}(Y_i L_i^{\mathrm{PCA}} Y_i^{\mathrm{T}}),$$

where $Y_i = [y_i, y_{i_1}, \ldots, y_{i_{N-1}}]$ with $\{I_{i_1}, \ldots, I_{i_{N-1}}\}$ are the rest $N-1$ images beyond I_i, and

$$L_i^{\mathrm{PCA}} = \frac{1}{N^2}\begin{bmatrix} (N-1)^2 & -(N-1)\mathbf{1}_{N-1}^{\mathrm{T}} \\ -(N-1)\mathbf{1}_{N-1} & \mathbf{1}_{N-1}\mathbf{1}_{N-1}^{\mathrm{T}} \end{bmatrix}.$$

Here, the vector $\mathbf{1}_{N-1} = [1, \ldots, 1]^{\mathrm{T}} \in \mathbb{R}^{N-1}$.

By combining both local and global patches, LGD approximates the intrinsic submanifold of relevant examples, as shown in Fig. 14b. Relevant examples can be separated from irrelevant ones in the projected 2D subspace. Besides, we show results of only local patches and only global patches for dimension reduction in Fig. 14c, d, respectively. Neither of them can perform well.

The effectiveness of the PCA-based global patches is investigated by replacing them with LPP-based patches, which are built in a similar way for each example. This LPP-based LGD is named as LGD-LPP and its performance is shown in Fig. 14e. This result is unsatisfactory because LPP assumes there is a manifold for both labeled and unlabeled examples which violates the true distribution of irrelevant examples. On the other hand, by using PCA-based global patches, the subspace with maximum variance is preserved, so manifold structure of relevant examples can also be preserved. By integrating global patches and local patches, we can discover the intrinsic submanifold of relevant examples and separate relevant examples from irrelevant examples.

Finally, we can align the calculated local and global patches together into a consistent coordinate. For each image I_i, $Y_i = [y_i, y_{i_1}, \ldots, y_{i_k}]$ can be rewritten as $Y_i = YS_i$, where $Y = [y_1, \ldots, y_N]$ and still $S_i \in \mathbb{R}^{N \times (k+1)}$ is the selection matrix. Then, we can combine all the patches defined in (41), (45) and (46) together

$$\begin{aligned} &\sum_{I_i^+} \min \operatorname{tr}\left(Y_i^+ L_i^+ (Y_i^+)^{\mathrm{T}}\right) + \sum_{I_i^-} \min \operatorname{tr}\left(Y_i^- L_i^- (Y_i^-)^{\mathrm{T}}\right) \\ &\quad + \lambda \sum_{i=1}^{N} \max \operatorname{tr}\left(Y_i L_i^{\mathrm{PCA}} Y_i^{\mathrm{T}}\right) = \max \operatorname{tr}\left(U^{\mathrm{T}} X L X^{\mathrm{T}} U\right), \end{aligned} \tag{47}$$

where $L = \lambda \sum_{i=1}^{N} S_i L_i^{\mathrm{PCA}} S_i^{\mathrm{T}} - \sum_{I_i^+} S_i^+ L_i^+ (S_i^+)^{\mathrm{T}} - \sum_{I_i^-} S_i^- L_i^- (S_i^-)^{\mathrm{T}}$ and $\lambda \geq 0$ is a control parameter. By imposing $U^{\mathrm{T}}U = I$, the projection matrix $U = [u_1, \ldots, u_r]$ can be obtained by solving the standard eigen-decomposition problem $XLX^{\mathrm{T}}u = \lambda u$, where U is consisting of the eigenvectors corresponding to the r largest eigenvalues.

7.4 Experimental Evaluation of Active Reranking on Web Image Search Dataset

Experiments are conducted on a real web image search dataset. In this dataset, there are 105 queries selected seriously from a commercial image search engine query log as well as popular tags of Flickr. These queries cover a large range of topics, including named person, named object, general object, and scene. For each query, a maximum of 1,000 images returned by commercial image search engines, i.e., Google, Live, and Yahoo, were collected as the initial text-based search results. This dataset contains 94,341 images in total. For each query, three participants were asked to judge whether the returned images are query relevant or irrelevant. An image is labeled as query relevant if at least two of the three participants judged it as relevant and vice versa.

Images are represented by 428-dimensional low-level visual features, including 225-dimensional color moment in LAB color space, 128-dimensional wavelet texture as well as 75-dimensional edge distribution histogram. For the initial text search score list $\bar{s}$, because images are all downloaded from web search engines (e.g., Google, Live, and Yahoo), we only know ranks of images in the text-based search and their scores are not available. According to [42], the normalized rank is adopted as the pseudo score, $\bar{s}_i = 1 - \frac{i}{N}$ for the ith ranked image, where $i = 1, \ldots, N$ and N is the number of images returned by the web search engine for a query term.

For active example selection, five images were selected to interact with the user in each interaction round and four rounds were considered. Therefore, for each query, there were 20 images labeled by the user totally. The performance is also measured by average precision (AP). We calculated the APs at different positions from top-1 to top-100 to obtain the AP curve. We averaged the APs over all the 105 queries to get the mean average precision (MAP) for overall performance evaluation.

The effectiveness of SInfo is investigated by comparing it with other three methods. "Error Reduction" [105], "Most Uncertain" [16], and "Random." To be noted, here both the reranking and the active example selection were conducted in the original feature space.

Figure 15 summarizes the comparison results. The "Baseline" curve gives the performance of the text-based search results and the "RerankInitial" curve is the performance of the unsupervised reranking without user interactions. The "SInfo," "Error Reduction," "Most Uncertain," and "Random" curves denote the performances of the reranked results with query images selected according to these four strategies, respectively.

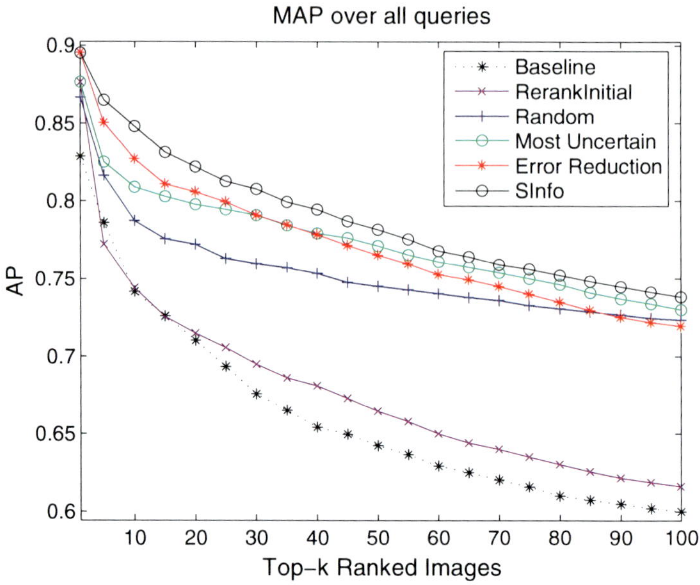

Fig. 15 MAP over all queries with different example selection strategies

Figure 15 shows the effectiveness of the active reranking framework as well as the superiority of the SInfo example selection strategy. Curves in this figure show that user's labeling information helps enhance the reranking performance. User interactions can improve the average performance, no matter which example selection strategy is adopted. Moreover, among these four strategies, SInfo performs best and achieves a significant performance improvement. This is because SInfo considers both the ambiguity and the representativeness while the "Most Uncertain" and "Random" only take one side of them into account. For "Error Reduction" and "Most Uncertain," they both suffer from the small example size problem while SInfo alleviates this influence by taking representativeness into account in an unsupervised manner.

To test the effectiveness of LGD, the active reranking is conducted in the projected subspace by using different dimension reduction algorithms. The SInfo example selection strategy was adopted in this experiment.

LGD is compared with several representative algorithms, including unsupervised algorithm, i.e., PCA, supervised ones, i.e., BDA [104], LDE [17], and SLPP [12], as well as semi-supervised ones, i.e., SML [52], SDA [13], and LGD-LPP. The subspace dimension was set to 100 for all algorithms empirically. Figure 16 shows the results. The "SInfo" curve denotes the reranked results of active reranking which is conducted in the original feature space without dimension reduction with the examples selected via SInfo. This curve is identical to the "SInfo" curve in Fig. 15.

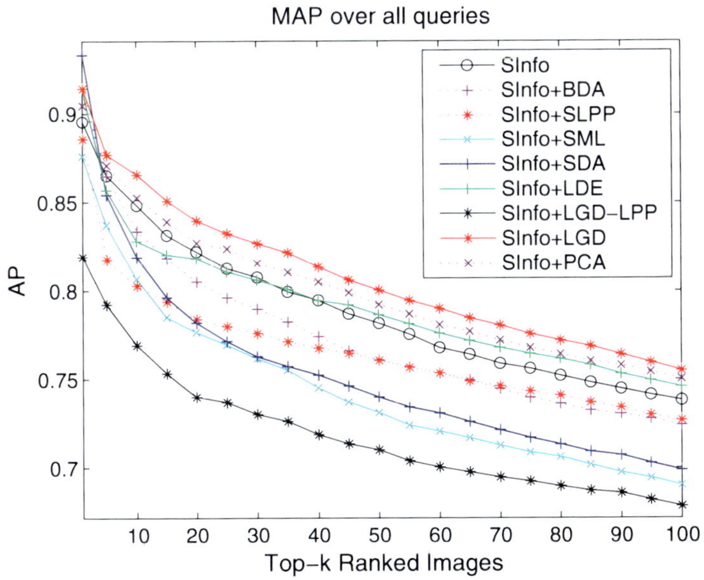

Fig. 16 MAP over all queries with different dimension reduction algorithms

The performance of reranking via different dimension reduction algorithms is denoted as SInfo+DR algorithm name, e.g., "SInfo+LGD" for performance of LGD.

Figure 16 shows that LGD performs best among these algorithms and achieves a more satisfactory performance than "SInfo." It reflects the effectiveness of LGD in localizing the visual characteristics of the user intention. For the other dimension reduction algorithms, reranked performances are either slightly improved or dramatically decreased. PCA fails to capture the user-driven intention since it ignores the labeling information. BDA, LDE, and SLPP, which are all supervised dimension reduction algorithms, only utilize a few labeled images. Thus, the subspace learned by them is biased to that spanned by several labeled images and cannot generalize well to the large amount of unlabeled ones.

For semi-supervised algorithms, SDA is unsuitable for the reranking task because it assumes that images in an identical class are sampled from a Gaussian. However, in web image search, each irrelevant image is irrelevant in its own way and thus images in the irrelevant class are not similar to each other, i.e., it is inconvenient to assume that irrelevant images are from an identical Gaussian. Therefore, SDA performed poorly. SML assumes that all images are sampled from a nonlinear manifold. In image search, irrelevant images usually scatter in the whole space, i.e., they may be distributed uniformly. SML is prone to over-fit to unlabeled images because of the improper manifold regularization assumption.

8 Manifold Elastic Net

Finally, we present manifold elastic net (MEN) [103], which is a sparse learning [53–55] method built upon the patch alignment framework. The key feature of MEN is that it is able to achieve sparse basis (projection matrix) by imposing the popular elastic net penalty (i.e., the combination of the lasso penalty and the $L2$ norm penalty). As sparse basis is more interpretable both psychologically and physiologically, MEN is expected to give more meaningful results on face recognition, which will be shown in experiments later.

For derivation convenience, we use the transpose of the original defined X, Y, which means that $X \in \mathbb{R}^{N \times m}$, $Y \in \mathbb{R}^{N \times r}$, and thus $Y = XU$.

8.1 A Summarization of MEN

First, MEN uses the same part optimization and whole alignment as in DLA, i.e., the following minimization is considered:

$$\arg\min_{Y} \operatorname{tr}\left(Y^{\mathrm{T}} L Y\right). \tag{48}$$

However, rather than substitute $Y = XU$ directly, (48) is reformed equivalent as below

$$\arg\min_{Y,U} \operatorname{tr}\left(Y^{\mathrm{T}} L Y\right) + \beta \left\| Y - XU \right\|^2. \tag{49}$$

Note that (49) indeed will lead to $Y = XU$. Given the equivalence between the two formulations, the latter is more convenient to incorporate the minimization of classification error. To enhance the performance of MEN for classification problems, the classification error minimization is considered, i.e.,

$$\arg\min_{U} \left\| Z - XU \right\|^2, \tag{50}$$

where $Z \in \mathbb{R}^{N \times C}$ is an indicator matrix carefully designed in [103] and C is the number of classes. By combing (49) and (50), we get the main objective of MEN

$$\arg\min_{Y,U} \left\| Z - XU \right\|^2 + \alpha \operatorname{tr}\left(Y^{\mathrm{T}} L Y\right) + \beta \left\| Y - XU \right\|^2, \tag{51}$$

where α and β are trade-off parameters to control the impacts of different terms.

To obtain a sparse projection matrix U, an ideal approach is to restrict the number of nonzeros entries in it, i.e., using the L_0 norm as a penalty over (51). However, the L_0 norm penalized (51) is an NP-hard problem and thus intractable practically. One attractive way of approximating the L_0 norm is the L_1 norm, i.e., the Lasso penalty [79], which is convex and actually the closet convex relaxation of the L_0 norm. Various efficient algorithms exit for solving Lasso penalized least square regression lasso problem, including the LARS [21]. However, the lasso penalty has the following two disadvantages: (1) the number of selected variables is limited by the number of observations and (2) the lasso penalized model can only select one variable from a group of correlated ones and does not care which one is selected. These limitations of Lasso are well addressed by the so-called elastic net penalty, which combines the L_2 and L_1 norm together. MEN adopts the elastic net penalty [106]. In detail, the L_2 of the projection matrix is helpful to increase the dimension (and the rank) of the combination of the data matrix and the response. In addition, the combination of the L_1 and L_2 of the projection matrix is convex with respect to the projection matrix and thus the obtained projection matrix has the grouping effect property. The final form of MEN is given by

$$\arg\min_{Y,U} \|Z - XU\|^2 + \alpha \mathrm{tr}\left(Y^{\mathrm{T}} L Y\right) + \beta \|Y - XU\|^2 + \lambda_1 \|U\|_1 + \lambda_2 \|U\|_2^2. \tag{52}$$

In the following, we show that (52) is equivalent to the lasso penalized least square problem and thus LARS can be directly applied to solve it.

By setting the differentiate of the objective function (52) with respect to Y as 0, we have

$$Y = \beta(\alpha L + \beta I)^{-1} X U. \tag{53}$$

We can eliminate Y in the objective function defined in (52) according to (53) and obtain

$$\arg\min_{U} U^{\mathrm{T}} X^{\mathrm{T}} A X U - 2U^{\mathrm{T}} X^{\mathrm{T}} Z + \lambda_1 \|U\|_1 + \lambda_2 \|U\|_2^2. \tag{54}$$

where A is an asymmetric matrix computed form L:

$$\begin{aligned} A = {} & \alpha\left(\beta(\alpha L + \beta I)^{-1}\right)^{\mathrm{T}} L(\beta(\alpha L + I)) \\ & + \beta\left(\beta(\alpha L + \beta I)^{-1} - I\right)^{\mathrm{T}} (\beta(\alpha L + \beta I) - I) + I. \end{aligned} \tag{55}$$

Because $2X^{\mathrm{T}}AX = X^{\mathrm{T}}(A + A^{\mathrm{T}})X$ and the eigenvalue decomposition of $(A + A^{\mathrm{T}})/2$ can be written as WDW^{T}, the objective function defined in (54) without the elastic net penalty can be rewritten as

$$U^{\mathrm{T}} X^{\mathrm{T}} A X U - 2U^{\mathrm{T}} X^{\mathrm{T}} Z = \left\| \left((D^{1/2} W^{\mathrm{T}})^{\mathrm{T}} \right)^{-1} Z - (D^{1/2} W^{\mathrm{T}}) X U \right\|_2^2. \tag{56}$$

By further setting

$$X^* = (1 + \lambda_2)^{-1/2} \begin{bmatrix} (D^{1/2W^{\mathrm{T}}})X \\ \sqrt{\lambda_2} I_{m \times m} \end{bmatrix} \in \mathbb{R}^{(N+m) \times m} \text{ and}$$

$$Y^* = \begin{bmatrix} \left((D^{1/2} W^{\mathrm{T}})^{\mathrm{T}}\right)^{-1} Z \\ \mathbf{0}_{m \times 1} \end{bmatrix} \in \mathbb{R}^{(N+m) \times 1}$$

in (54), we get

$$\arg\min_{U^*} = \|Z^* - X^*U^*\|_2^2 + \lambda\|U^*\|_1 \tag{57}$$

where $\lambda = \lambda_1/(1+\lambda_2)$ and $U^* = \sqrt{(}1+\lambda_2)U$.

According to (57), the LARS algorithm can be applied to obtain the optimal solution of MEN. Though some other L_1 least square algorithms, e.g., block coordinate descent and fixed-point algorithms, may have advantages in speed, LARS is chosen in MEN because it satisfies KKT conditions at each step and thus it can obtain the global solutions on different sparse levels in one run.

The LARS for MEN begins with a coefficient vector U^* (a column in the projection matrix with the ith entry $(U^*)_i$) with all zero entries. A variable (a column vector in X, i.e., a particular feature) in $\mathbb{R}^N$, which is most correlated with the objective function, is added to the active set A. Then the corresponding coefficient in U^* increases as large as possible until a second variable (another column vector in X, i.e., another feature) in $\mathbb{R}^N$ has the same correlation as the first variable. Instead of continuously increasing the coefficient vector in the direction of the first variable, LARS proceeds on a direction equiangular over all variables in the active set A until a new variable earns its way into A. To make the coefficient U^* become n-sparse (at most n nonzero entries), the above procedure is conducted for n loops. We refer to [103] for a detailed description of this algorithm.

8.2 Experimental Evaluation of MEN

We report an empirical evaluation of MEN on the FERET dataset. From the total 13,539 face images of 1,565 individuals, 100 individuals with 7 images per subject are randomly selected in the experiment. Four or five images per individual are selected as training set, and the remaining is used for test. We run the experiment five times, the average recognition rates are calculated. Six representative dimension reduction algorithms, i.e., principal component analysis (PCA) [82], Fisher's linear discriminant analysis (FLDA) [3], discriminative locality alignment (DLA) [98], supervised locality preserving projection (SLPP) [12], NPE [39], and sparse principal component analysis (SPCA) [18], are also performed for performance comparison.

The recognition performance is summarized in Fig. 17. Apparently, the seven algorithms are divided into three groups according to their performance. The baseline level methods are PCA and SPCA, which is because they are both unsupervised methods and thus may not give satisfying performance due to the missing of label information. LPP, NPE, and LDA only show moderate performance. In contrast, DLA and MEN give rise to significant improvements. Further, the sparsity of MEN makes it outperform DLA. The best performance of MEN is actually not supervising, since it considers the most aspects on data representation and distribution, including the sparse property, the local geometry information, and classification error minimization.

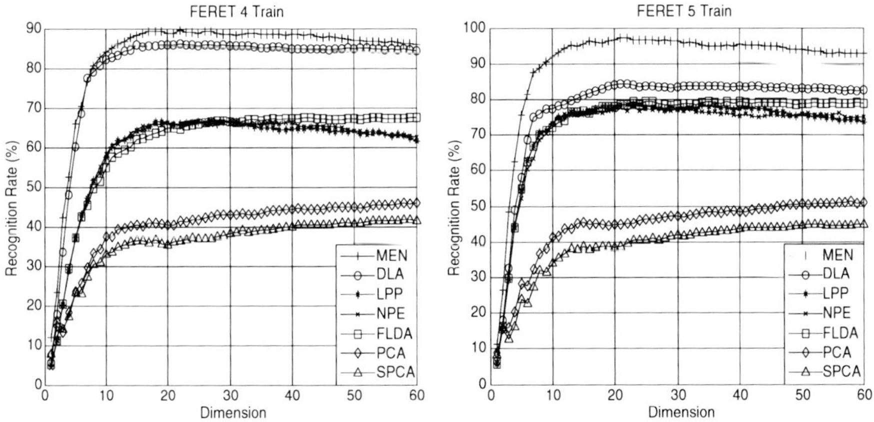

Fig. 17 Performance evaluation on the FERET dataset

Fig. 18 Plots of first 10 bases obtained from 7 dimensionality reduction algorithms on FERET. For each column, from top to bottom: MEN, DLA, LPP, NPE, FLDA, PCA, and SPCA

Figure 18 shows the first ten bases selected by different subspace selection methods. One can see that the bases selected by LPP, NPE, and FLDA are contaminated by considerable noises, which explains why they only give moderate recognition performance. The bases from PCA, i.e., eigenfaces, are smooth but present relatively few discriminative information. In terms of sparsity, SPCA gives the desired bases;

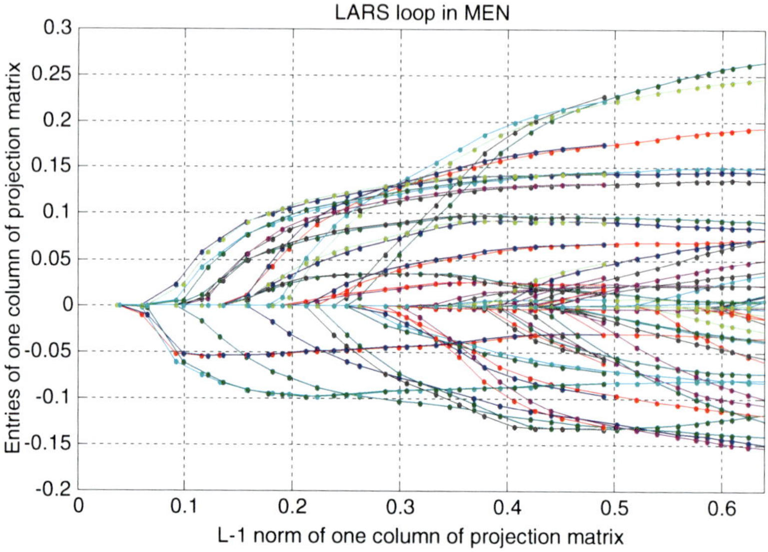

Fig. 19 Entries of one column of projection matrix vs. its L_1 norm in one LARS loop of MEN

however the problem is that the patterns presented in these bases are not grouped so they cannot provide meaningful interpretation. The bases from MEN, which we call "MEN's faces," have a low level of noise and are also reasonably sparse. And more importantly, thanks to the elastic net penalty, the sparse pattern of MEN's bases are satisfyingly grouped, which gives meaningful interpretations, e.g., most discriminative facial features are obtained, including eyebrows, eyes, nose, mouth, ears, and facial contours.

The optimization algorithm of MEN is built upon LARS. In each LARS loop of the MEN algorithm, all entries of one column in the projection matrix are zeros initially. They are sequentially added into the active set according to their importance. The values of active ones are increased with equal altering correlation. In this process, the L_1 norm of the column vector is augmented gradually. Figure 19 shows the altering tracks of some entries of the column vector in one LARS loop. These tracks are called "coefficient paths" in LARS. As shown by these plots, one can observe that every coefficient path starts from zero when the corresponding variable becomes active and then changes its direction when another variable is added into the active set. All the paths keep in the directions which make the correlations of their corresponding variables equally altering. The L_1 norm is increasing along the greedy augment of entries. The coefficient paths proceed along the gradient decent direction of objective function on the subspace, which is spanned by the active variables.

In addition, Fig. 20 shows 10 of the 1,600 coefficient paths from LAPS loop. It can be seen that MEN selects ten important features sequentially. For each feature, its corresponding coefficient path and the "MEN face" when the feature is added into

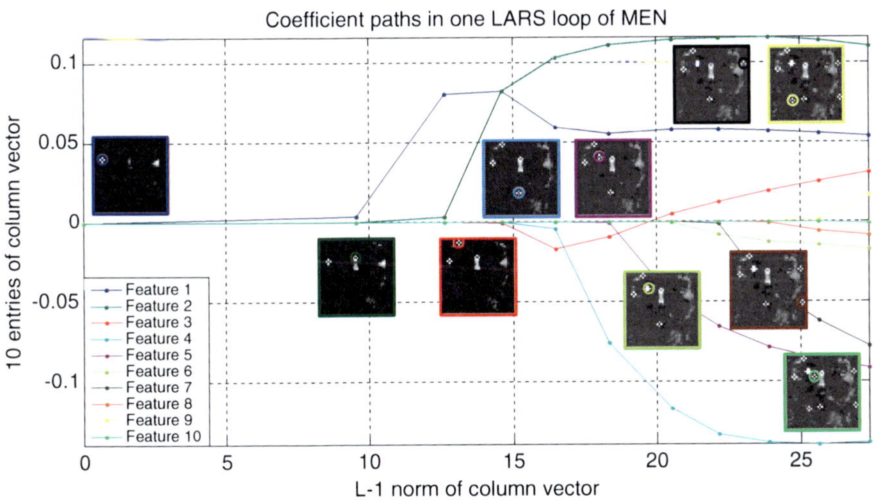

Fig. 20 Coefficient paths of ten features in one column vector

active set are assigned the same color which is different with the other 9 features. In each "MEN face," the new added active feature is marked by a small circle, and all the active features are marked by white crosses. The features selected by MEN can produce explicit interpretation of the relationship between facial features and face recognition: feature 1 is the left ear, feature 2 is the top of nose, feature 3 is on the head contour, feature 4 is the mouth, feature 5 and feature 6 are on the left eye, feature 7 is the right ear, and feature 8 is the left corner of mouth. These features are already verified of great importance in face recognition by many other famous face recognition methods. Moreover, Fig. 20 also shows MEN can group correlated features, e.g., feature 5 and feature 6 are selected sequentially because they are both on the left eye. In addition, features which are not very important, such as feature 9 and feature 10 in Fig. 20, are selected after the selection of the other more significant features and assigned smaller value than those more important ones. Therefore, MEN is a powerful algorithm in feature selection.

9 Summary

In this chapter, we present the patch alignment framework and its several extensions. The patch alignment framework unified various manifold learning based dimensionality reduction algorithm. By introducing the nonnegative constraint, we obtain the nonnegative patch alignment framework. To deal with multimedia data, which usually contain features from different views, a multiview extension of the patch alignment framework is developed. Considering that training and test examples may be not independent and identically distributed, a new regularization is added to the patch alignment to learn a transfer subspace. The patch alignment framework is also utilized in active reranking to learn a submanifold for encoding the user's intention by transferring information from the images labeled by users to the whole image dataset. Finally, the patch alignment framework is extended for sparse dimensionality reduction.

References

1. Trec video retrieval evaluation 2008, 2008.
2. Alpaydm E (1999) Combined 5× 2 cv f test for comparing supervised classification learning algorithms. Neural Comput 11(8):1885–1892
3. Belhumeur P, Hespanha J, Kriegman D (1997) Eigenfaces vs. fisherfaces: Recognition using class specific linear projection. IEEE Trans Pattern Anal Mach Intell 19(7):711–720
4. Belkin M, Niyogi P (2001) Laplacian eigenmaps and spectral techniques for embedding and clustering. MIT Press, In: Advances in neural information processing systems, pp 585–591
5. Belkin M, Niyogi P, Sindhwani V (2006) Manifold regularization: A geometric framework for learning from labeled and unlabeled examples. J Mach Learn Res 7:2399–2434
6. Bengio Y, Paiement J, Vincent P, Delalleau O, Le Roux N, Ouimet M (2004) Out-of-sample extensions for lle, isomap, mds, eigenmaps, and spectral clustering. MIT Press, In: Advances in neural information processing systems, pp 177–184
7. Bian W, Tao D (2010) Biased discriminant euclidean embedding for content-based image retrieval. IEEE Trans Image Process 19(2):545–554
8. Bian W, Tao, D (2011) Max-min distance analysis by using sequential sdp relaxation for dimension reduction. IEEE Trans Pattern Anal Mach Intell 33(5):1037–1050
9. Bishop C, Svensén M, Williams C (1998) Gtm: The generative topographic mapping. Neural Comput 10(1):215–234
10. Bregman L (1967) The relaxation method of finding the common point of convex sets and its application to the solution of problems in convex programming. USSR Comput Math Math Phys 7(3):200–217
11. Cai D, He X (2012) Manifold adaptive experimental design for text categorization. IEEE Trans Knowl Data Eng 24(4):707–719
12. Cai D, He X, Han J (2005) Using graph model for face analysis. Tech. Rep. 2636, Department of Computer Science, University of Illinois at Urbana-Champaign
13. Cai D, He X, Han J (2007) Semi-supervised discriminant analysis. In: IEEE international conference on computer vision, pp 1–7
14. Cai D, He X, Han J (2008) Srda: An efficient algorithm for large-scale discriminant analysis. IEEE Trans Knowl Data Eng 20(1):1–12
15. Cai D, He X, Han J, Huang T (2011) Graph regularized nonnegative matrix factorization for data representation. IEEE Trans Pattern Anal Mach Intell 33(8):1548–1560
16. Chang E, Tong S, Goh K, Chang C (2005) Support vector machine concept-dependent active learning for image retrieval. IEEE Trans Multimed 2:1–35
17. Chen H, Chang H, Liu T (2005) Local discriminant embedding and its variants. In: IEEE conference on computer vision and pattern recognition, vol 2, pp 846–853
18. d'Aspremont A, El Ghaoui L, Jordan M, Lanckriet G (2007) A direct formulation for sparse pca using semidefinite programming. SIAM Rev 49(3):434–448
19. Dietterich T (1998) Approximate statistical tests for comparing supervised classification learning algorithms. Neural Comput 10(7):1895–1923
20. Donoho D, Grimes C (2003) Hessian eigenmaps: Locally linear embedding techniques for high-dimensional data. Proc Nat Acad Sci USA 100(10):5591–5596

21. Efron B, Hastie T, Johnstone I, Tibshirani R (2004) Least angle regression. Ann Stat 32(2):407–499
22. Escolano F, Hancock E, Lozano M (2011) Graph matching through entropic manifold alignment. In: IEEE Conference on Comput Vision Pattern Recogn, pp 2417–2424
23. Fan J, Fan Y (2008) High dimensional classification using features annealed independence rules. Ann Stat 36(6):2605–2637
24. Fan J, Lv J (2010) A selective overview of variable selection in high dimensional feature space. Stat Sinica 20(1):101
25. Fan Y, Li R (2012) Variable selection in linear mixed effects models. Ann Stat
26. Fu Y, Huang T (2008) Human age estimation with regression on discriminative aging manifold. IEEE Trans Multimed 10(4):578–584
27. Fu Y, Huang T (2008) Image classification using correlation tensor analysis. IEEE Trans Image Process 17(2):226–234
28. Fu Y, Yan S, Huang T (2008) Correlation metric for generalized feature extraction. IEEE Trans Pattern Anal Mach Intell 30(12):2229–2235
29. Gao X, Zhang K, Tao D, Li X (2012) Image super-resolution with sparse neighbor embedding. IEEE Trans Image Process 21(7):3194–3205
30. Geng B, Tao D, Xu C (2010) DAML: Domain adaptation metric learning. IEEE Trans Image Process 20(10):2980–2989
31. Geng B, Tao D, Xu C, Yang L, Hua X (2012) Ensemble manifold regularization. IEEE Trans Pattern Anal Mach Intell 34(6):1227–1233
32. Griffin G, Holub A, Perona P (2007) California Institute of Technology, Caltech-256 object category dataset
33. Guan N, Tao D, Luo Z, Yuan B (2010) Fast gradient descent for non-negative patch alignment framework. Tech. Rep., University of Technology, Sydney
34. Guan N, Tao D, Luo Z, Yuan B (2011) Manifold regularized discriminative nonnegative matrix factorization with fast gradient descent. IEEE Trans Image Process 20(7):2030–2048
35. Guan N, Tao D, Luo Z, Yuan B (2011) Non-negative patch alignment framework. IEEE Trans Neural Networks 22(8):1218–1230
36. Guan N, Tao D, Luo Z, Yuan B (2012) Nenmf: An optimal gradient method for nonnegative matrix factorization. IEEE Trans Signal Process 60(6):2882–2898
37. Guan N, Tao D, Luo Z, Yuan B (2012) Online nonnegative matrix factorization with robust stochastic approximation. IEEE Trans Neural Networks Learn Syst 23(7):1087–1099
38. He X, Cai D, Shao Y, Bao H, Han J (2011) Laplacian regularized gaussian mixture model for data clustering. IEEE Trans Knowl Data Eng 23(9):1406–1418
39. He X, Cai D, Yan S, Zhang H (2005) Neighborhood preserving embedding. In: IEEE international conference on computer vision, pp 1208–1213
40. He X, Ji M, Zhang C, Bao H (2011) A variance minimization criterion to feature selection using laplacian regularization. IEEE Trans Pattern Anal Mach Intell 33(10):2013–2025
41. He X, Niyogi P (2004) Locality preserving projections. MIT Press, In: Advances in neural information processing systems, pp 153–160
42. Hsu W, Kennedy L, Chang S (2006) Video search reranking via information bottleneck principle. In: ACM international conference on multimedia, pp 35–44
43. Ji S, Ye J (2009) Linear dimensionality reduction for multi-label classification. Morgan Kaufmann Publishers Inc. In: International joint conference on artifical intelligence, pp 1077–1082
44. Jiang X (2009) Asymmetric principal component and discriminant analyses for pattern classification. IEEE Trans Pattern Anal Mach Intell 31(5):931–937
45. Jiang X (2011) Linear subspace learning-based dimensionality reduction. IEEE Signal Process Mag 28(2):16–26
46. Jiang X, Mandal B, Kot A (2009) Complete discriminant evaluation and feature extraction in kernel space for face recognition. Mach Vision Appl 20(1):35–46
47. Kokiopoulou E, Saad Y (2007) Orthogonal neighborhood preserving projections: A projection-based dimensionality reduction technique. IEEE Trans Pattern Anal Mach Intell 29(12):2143–2156

48. Lee D, Seung H et al (1999) Learning the parts of objects by non-negative matrix factorization. Nature 401(6755):788–791
49. Li S, Hou X, Zhang H, Cheng Q (2001) Learning spatially localized, parts-based representation. In: IEEE conference on computer vision and pattern recognition, pp 1–207
50. Li X, Long X, Laurienti P, Wyatt C (2012) Registration of images with varying topology using embedded maps. IEEE Trans Med Imag 31(3):749–765
51. Li X, Long X, Wyatt C (2011) Registration of images with topological change via riemannian embedding. In: IEEE international symposium on biomedical imaging: from nano to macro, pp 1247–1252
52. Lin Y, Liu T, Chen H (2005) Semantic manifold learning for image retrieval. In: ACM international conference on multimedia, pp 249–258
53. Liu J, Chen J, Ye J (2009) Large-scale sparse logistic regression. In: ACM international conference on knowledge discovery and data mining, pp 547–556
54. Liu J, Ye J (2009) Efficient euclidean projections in linear time. In: International conference on machine learning, pp 657–664
55. Liu J, Yuan L, Ye J (2010) An efficient algorithm for a class of fused lasso problems. In: ACM international conference on knowledge discovery and data mining, pp 323–332
56. Liu M, Vemuri B (2011) Rboost: Riemannian distance based regularized boosting. In: 2011 IEEE international symposium on biomedical imaging: from nano to macro, pp 1831–1834
57. Liu M, Vemuri B (2011) Robust and efficient regularized boosting using total bregman divergence. In: IEEE conference on computer vision and pattern recognition, pp 2897–2902
58. Liu M, Vemuri B, Amari S, Nielsen F (2010) Total bregman divergence and its applications to shape retrieval. In: IEEE conference on computer vision and pattern recognition, pp 3463–3468
59. Long B, Yu P, Zhang Z (2008) A general model for multiple view unsupervised learning. In: SIAM international conference on data mining, pp 822–833
60. Lu H, Plataniotis K, Venetsanopoulos A (2008) Mpca: Multilinear principal component analysis of tensor objects. IEEE Trans Neural Networks 19(1):18–39
61. Lu H, Plataniotis K, Venetsanopoulos A (2009) Uncorrelated multilinear principal component analysis for unsupervised multilinear subspace learning. IEEE Trans Neural Networks 20(11):1820–1836
62. Lu H, Plataniotis K, Venetsanopoulos A (2011) A survey of multilinear subspace learning for tensor data. Pattern Recogn 44(7):1540–1551
63. Lv J, Fan Y (2009) A unified approach to model selection and sparse recovery using regularized least squares. Ann Stat 37(6A):3498–3528
64. Pan S, Kwok J, Yang Q (2008) Transfer learning via dimensionality reduction. AAAI Press, In: National conference on artificial intelligence, pp 677–682
65. Roweis S, Saul L (2000) Nonlinear dimensionality reduction by locally linear embedding. Science 290:2323–2326
66. Samaria F, Harter A (1994) Parameterisation of a stochastic model for human face identification. In: IEEE workshop on applications of computer vision, pp 138–142
67. Si S, Tao D, Chan K (2010) Evolutionary cross-domain discriminative hessian eigenmaps. IEEE Trans Image Process 19(4):1075–1086
68. Si S, Tao D, Geng B (2010) Bregman divergence-based regularization for transfer subspace learning. IEEE Trans Knowl Data Eng 22(7):929–942
69. Song D, Tao D (2010) Biologically inspired feature manifold for scene classification. IEEE Trans Image Process 19(1):174–184
70. Song M, Tao D, Chen C, Bu J, Luo J, Zhang C (2012) Probabilistic exposure fusion. IEEE Trans Image Process 21(1):341–357
71. Song M, Tao D, Chen C, Li X, Chen C (2010) Color to gray: Visual cue preservation. IEEE Trans Pattern Anal Mach Intell 32(9):1537–1552
72. Sun L, Ceran B, Ye J (2010) A scalable two-stage approach for a class of dimensionality reduction techniques. In: ACM international conference on knowledge discovery and data mining, pp 313–322

73. Tao D, Li X, Wu X, Hu W, Maybank S (2007) Supervised tensor learning. Knowl Inform Syst 13(1):1–42
74. Tao D, Li X, Wu X, Maybank S (2007) General tensor discriminant analysis and gabor features for gait recognition. IEEE Trans Pattern Anal Mach Intell 29(10):1700–1715
75. Tao D, Li X, Wu X, Maybank S (2009) Geometric mean for subspace selection. IEEE Trans Pattern Anal Mach Intell 31(2):260–274
76. Tenenbaum J, Silva V, Langford J (2000) A global geometric framework for nonlinear dimensionality reduction. Science 290(5500):2319–2323
77. Tian X, Tao D, Hua X, Wu X (2010) Active reranking for web image search. IEEE Trans Image Process 19(3):805–820
78. Tian X, Yang L, Wang J, Yang Y, Wu X, Hua X (2008) Bayesian video search reranking. In: ACM international conference on multimedia, pp 131–140
79. Tibshirani R (1996) Regression shrinkage and selection via the lasso. J Roy Stat Soc Ser B (Methodol), 58(1):267–288
80. Tikhonov A (1963) Regularization of incorrectly posed problems. Soviet Math Dokl 4:1624–1627
81. Tipping M, Bishop C (1999) Probabilistic principal component analysis. J Roy Stat Soc Ser B (Stat Methodol) 61(3):611–622
82. Turk M, Pentland A (1991) Face recognition using eigenfaces. In: IEEE conference on computer vision and pattern recognition, pp 586–591
83. Von Luxburg U (2007) A tutorial on spectral clustering. Stat Comput 17(4):395–416
84. Wang F, Wang X (2009) Neighborhood discriminant tensor mapping. Neurocomputing 72(7–9):2035–2039
85. Wang F, Wang X, Li T (2009) Maximum margin clustering on data manifolds. In: IEEE international conference on data mining, pp 1028–1033
86. Wang F, Zhao B, Zhang C (2011) Unsupervised large margin discriminative projection. IEEE Trans Neural Networks 22(9):1446–1456
87. Wang M, Hua X, Yuan X, Song Y, Dai L (2007) Optimizing multi-graph learning: towards a unified video annotation scheme. In: ACM international conference on multimedia, pp 862–871
88. Wang X, Tao D, Li Z (2011) Subspaces indexing model on grassmann manifold for image search. IEEE Trans Image Process 20(9):2627–2635
89. Wilson R, Hancock E (2010) Spherical embedding and classification. Struct Syntact Stat Pattern Recogn, 6218:589–599
90. Wilson R, Hancock E, Pekalska E, Duin R (2010) Spherical embeddings for non-euclidean dissimilarities. In: IEEE conference on computer vision and pattern recognition, pp 1903–1910
91. Xia T, Tao D, Mei T, Zhang Y (2010) Multiview spectral embedding. IEEE Trans Syst Man Cybernet Part B: Cybernet 40(6):1438–1446
92. Xie B, Mu Y, Tao D, Huang K (2011) m-sne: Multiview stochastic neighbor embedding. IEEE Trans Syst Man Cybernet Part B: Cybernet 41(4):1088–1096
93. Yan S, Xu D, Zhang B, Zhang H, Yang Q, Lin S (2007) Graph embedding and extensions: A general framework for dimensionality reduction. IEEE Trans Pattern Anal Mach Intell 29(1):40–51
94. Yu J, Liu D, Tao D, Seah H (2011) Complex object correspondence construction in two-dimensional animation. IEEE Trans Image Process 20(11):3257–3269
95. Yu J, Tao D, Wang M, Cheng J (2011) Semi-automatic cartoon generation by motion planning. Multimedia Syst 17(5):409–419
96. Zafeiriou S, Tefas A, Buciu I, Pitas I (2006) Exploiting discriminant information in nonnegative matrix factorization with application to frontal face verification. IEEE Trans Neural Networks 17(3):683–695
97. Zhang L, Tao D, Huang X (2012) On combining multiple features for hyperspectral remote sensing image classification. IEEE Trans Geosci Remote Sensing 50(3):879–893

98. Zhang T, Tao D, Li X, Yang J (2009) Patch alignment for dimensionality reduction. IEEE Trans Knowl Data Eng 21(9):1299–1313
99. Zhang T, Yang J, Zhao D, Ge X (2007) Linear local tangent space alignment and application to face recognition. Neurocomputing 70(7):1547–1553
100. Zhang Z, Hancock E (2012) Localized graph-based feature selection for clustering. Springer, Image Analysis and Recognition, pp 1–10
101. Zhang Z, Tao D (2012) Slow feature analysis for human action recognition. IEEE Trans Pattern Anal Mach Intell 34(3):436–450
102. Zhang Z, Zha H (2005) Principal manifolds and nonlinear dimension reduction via local tangent space alignment. SIAM J Scientif Comput 26(1):313–338
103. Zhou T, Tao D, Wu X (2011) Manifold elastic net: A unified framework for sparse dimension reduction. Data Mining Knowl Discov 22(3):340–371
104. Zhou X, Huang T (2001) Small sample learning during multimedia retrieval using biasmap. In: IEEE conference on computer vision and pattern recognition, vol 1, pp 1–11
105. Zhu X, Lafferty J, Ghahramani Z (2003) Combining active learning and semi-supervised learning using gaussian fields and harmonic functions. Elsevier, In: International conference on machine learning, pp 58–65
106. Zou H, Hastie T (2005) Regularization and variable selection via the elastic net. J Roy Stat Soc Ser B (Stat Methodol) 67(2):301–320

Improving Classifications Through Graph Embeddings

Anirban Chatterjee, Sanjukta Bhowmick, and Padma Raghavan

1 Introduction

Unsupervised classification is used to identify similar entities in a dataset and is extensively used in many application domains such as spam filtering [5], medical diagnosis [15], demographic research [13], etc. Unsupervised classification using K-Means generally clusters data based on (1) distance-based attributes of the dataset [4, 16, 17, 23] or (2) combinatorial properties of a weighted graph representation of the dataset [8].

Classification schemes, such as K-Means [11], that use distance-based attributes view entities of the dataset as existing in an n-dimensional feature space. The value of the i-th feature of an entity determines its coordinate in the i-th dimension of the feature space. The distance between the entities is used as a classification metric. The entities that lie close to each other are assigned to the same cluster.

Combinatorial techniques for clustering, such as Multilevel K-Means (GraClus) [8], represent the dataset as a weighted graph, where the entities are represented by vertices. The edge weights of the graph indicate the degree of similarity between the entities. A highly weighted subgraph forms a class and its vertices (entities) are given the same label.

A. Chatterjee (✉)
Mathworks, 3 Apple Hill Drive, Natick, MA 01760, USA
e-mail: anirban.chatterjee@mathworks.com

S. Bhowmick
University of Nebraska at Omaha, 6001 Dodge Street, Omaha, NE 68182, USA
e-mail: sbhowmick@unomaha.edu

P. Raghavan
Department of Computer Science and Engineering, The Pennsylvania State University, University Park, PA 16802, USA
e-mail: raghavan@cse.psu.edu

Y. Fu and Y. Ma (eds.), *Graph Embedding for Pattern Analysis*,
DOI 10.1007/978-1-4614-4457-2_5,

In this chapter, we present a feature subspace transformation (FST) scheme to transform the dataset before the application of K-Means (or other distance-based clustering schemes). A unique attribute of our FST-K-Means method is that it utilizes both distance-based and combinatorial attributes of the original dataset to seek improvements in the internal and external quality metrics of unsupervised classification. FST-K-Means starts by forming a weighted graph with the entities as vertices that are connected by weighted edges indicating a measure of shared features. The vertices of the graph are initially viewed as being embedded in the high-dimensional feature subspace, i.e., the coordinates of each vertex (entity) are given by the values of its feature vector in the original dataset. This initial layout of the weighted graph is transformed by a special form of a force-directed graph embedding algorithm that attracts similar entities. The nodal coordinates of the embedded graph provide a feature subspace transformation of the original dataset; K-Means is then applied to the transformed dataset.

The remainder of this chapter is organized as follows. In Sect. 2, we provide a brief review of related classification schemes and a graph layout algorithm that we adapt for use in our FST-K-Means. In Sect. 3, we develop FST-K-Means including its feature subspace transformation scheme and efficient implementation. In Sect. 4, we provide an empirical evaluation of the effectiveness of FST-K-Means using a test suite of datasets from a variety of domains. In Sect. 5, we attempt to quantitatively characterize the performance of FST-K-Means by demonstrating that the clustering obtained from our method satisfies optimality constraints for the internal quality. We present brief concluding remarks in Sect. 6. We would like to note that a shorter version of this research appears in [6].

2 Related Work

We now provide a brief overview of unsupervised classification using K-Means [11] and GraClus [8] with a focus on aspects that are relevant for our empirical evaluation in Sect. 4. We also review the Fruchterman-Reingold (FR) [10] graph layout method that we later adapt in Sect. 3 for use as a component in our FST-K-Means.

The most common implementations of the K-Means algorithm are by Steinhaus [23], Lloyd [16], Ball and Hall [4], and MacQueen [17]. The four independent implementations of K-Means in their first step initialize the positions of the centroids (the number of centroids is equal to the number of predetermined classes). Then the entities are assigned to their closest centroid to form the initial clusters. The centroids are then recalculated based on the center of mass of these clusters, and the entities are reassigned to clusters according to the new centroids. These steps are repeated until convergence is achieved.

Graph approaches to classification have also become popular, especially, *Multilevel K-Means (GraClus)* [8] that performs unsupervised classification through graph clustering. The dataset is represented as a weighted graph, where each entity is a vertex and vertices are connected by weighted edges. High edge weights denote greater similarity between corresponding entities. Similar entities are identified by clustering of vertices with heavy edge weights. GraClus uses multilevel graph coarsening algorithms. The initial partition of the graph is done using a spectral algorithm [20] and the subsequent coarsening stages use a weighted kernel K-Means.

Chatterjee et al. [7] developed a similarity graph neighborhood (SGN) based training data transformation scheme, aimed at enhancing the accuracy of a supervised classifier. The SGN transform is primarily based on resultant displacement calculations to update entity positions. This transform uses class information of the entities to decide the direction of the displacement when unified with the learning phase of a supervised classifier. Alternatively, FST calculates attractive forces between neighboring entities to update entity positions. The underlying assumption in FST is that entities connected by an edge in the similarity graph should move closer in the feature space. This is fundamentally different from SGN where connected entities in the similarity graph could potentially displace away from each other if they belong to different classes. Another important difference between FST and SGN is that FST updates the position of the entities iteratively unlike SGN which applies displacements only once to update entity positions.

2.1 *Graph Layouts with the Fruchterman and Reingold Scheme*

Graph layout methods are designed to produce aesthetically pleasing graphs embedded in a two- (or three)-dimensional space. A popular technique is the Fruchterman and Reingold (FR) [10] algorithm that considers a graph as a collection of objects (vertices) connected together by springs (edges). Initially, the vertices in the graph are placed at randomly assigned coordinates. The FR model assumes that there are two kinds of forces acting on the vertices, (1) attraction due to the springs (only between connected vertices) and (2) repulsion due to mutually charged objects (between all vertices). If the Euclidean distance between two vertices u and v is d_{uv} and k is a constant proportional to the square root of the ratio of the embedding area by the number of vertices, then the attractive force $F_{\mathrm{A}}^{\mathrm{FR}}$ and the repulsive force $F_{\mathrm{R}}^{\mathrm{FR}}$ are calculated as:

$$F_{\mathrm{A}}^{\mathrm{FR}}(u,v) = -k^2/d_{uv} \text{ and } F_{\mathrm{R}}^{\mathrm{FR}}(u,v) = d_{uv}^2/k. \tag{1}$$

At each iteration of FR, the vertices are moved in proportion to the calculated attractive or repulsive forces until the desired layout is achieved.

In Sect. 3, we adapt the FR algorithm as a component in our feature subspace transformation in high dimensions. Through this transformation, we essentially seek embeddings that enhance cluster cohesiveness while improving the accuracy of classification.

3 FST-K-Means: Feature Subspace Transformations for Enhanced Classification

We now develop our feature subspace transformation scheme which seeks to produce a transformed dataset to which K-Means can be applied to yield high-quality classifications. Our FST-K-Means seeks to utilize both distance-based and combinatorial measures through a geometric interpretation of the entities of the dataset in its high-dimensional feature space.

Consider a dataset of N entities and R features represented by an $N \times R$ sparse matrix A. The i-th row, $a_{i,*}$ of the matrix A, represents the feature vector of the i-th entity in the dataset. Thus, we can view entity i as being embedded in the high dimensional feature space (dimension $\leq R$) with coordinates given by the feature vector $a_{i,*}$.

The $N \times N$ matrix $B \approx AA^{\mathrm{T}}$ represents the entity-to-entity relationship. B forms the adjacency matrix of the undirected graph $G(B,A)$, where $b_{i,j}$ is the edge weight between vertices i and j in the graph, and $a_{i,k}$ is the coordinate of vertex i in the k-th dimension.

FST is then applied to $G(B,A)$ to transform the coordinates of the vertices producing the graph $G(B,\tilde{A})$. It should be noted that the structure of the adjacency matrix, i.e., the set of edges in the graph, remains unchanged. The transformed coordinates of the vertices are now represented by the $N \times R$ sparse matrix $\tilde{A}$. The transformed dataset or, equivalently, its matrix representation $\tilde{A}$ has exactly the same sparsity structure as the original, i.e., $\tilde{a}_{i,j} \neq 0$ if and only if $a_{i,j} \neq 0$. However, typically $\tilde{a}_{i,j} \neq a_{i,j}$ as a result of the feature subspace transformation (FST), thus changing the embedding of entity i. Now the i-th entity is represented in a transformed feature subspace (dimension $\leq R$) by the feature vector $\tilde{a}_{i,*}$. K-Means is applied to this transformed dataset represented by $\tilde{A}$ to yield FST-K-Means.

FST-K-Means comprises three mains steps: (1) forming an entity-to-entity sparse, weighted, embedded graph $G(B,A)$, (2) feature subspace transformation (FST) of $G(B,A)$ to yield transformed embedded graph $G(B,\tilde{A})$, and (3) applying K-Means to $\tilde{A}$ for classification.

Figure 1 illustrates these steps using a simple example.

3.1 *Forming an Entity-to-Entity Weighted, Embedded Graph* $G(B,A)$

Consider the dataset represented by the $N \times R$ sparse matrix A. Form $B \approx AA^{\mathrm{T}}$. Although B could be computed exactly as AA^{T}, approximations that compute only a subset of representative values may also be used. Observe that $b_{i,j}$, which represents the relationship between entities i and j, is given by $a_{i,*} \cdot a_{j,*}$, the dot product of the feature vectors of the i-th and j-th entities; thus $b_{i,j}$ is proportional to their

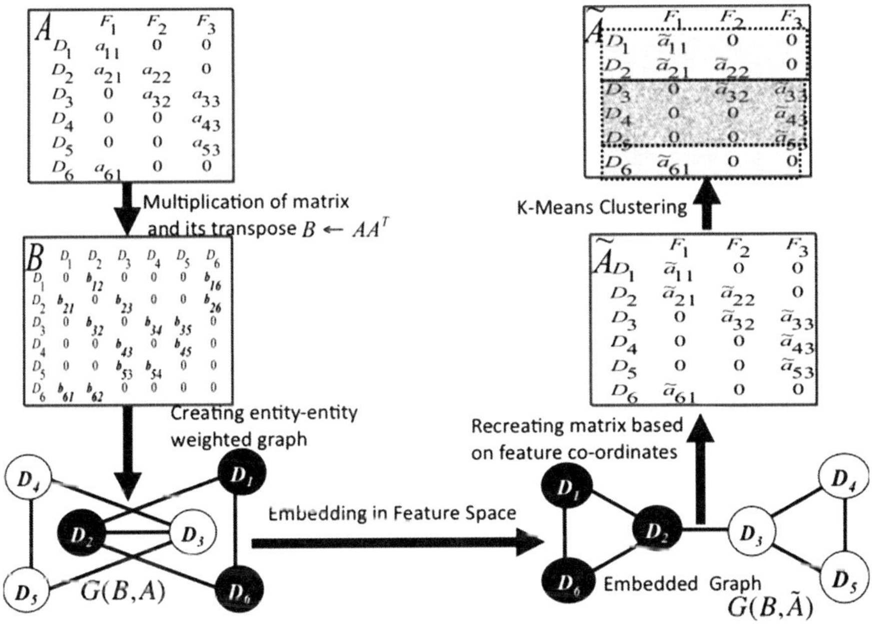

Fig. 1 Illustration of the main steps of FST-K-Means. A is a sparse matrix representing a dataset with 6 entities and 3 features. $B \approx AA^{\mathrm{T}}$ is the adjacency matrix of the weighted graph $G(B,A)$ with 6 vertices and 7 edges. FST is applied on G(B,A) to transform the coordinates of the vertices. Observe that the final embedded graph $G(B,\tilde{A})$ has the same sparsity structure as $G(B,A)$. The sparse matrix $\tilde{A}$ represents the dataset with the transformed feature space. K-Means is applied to the dataset $\tilde{A}$ to produce high-quality clustering

cosine distance in the feature space. Next, view the matrix B as representing the adjacency matrix of the undirected weighted graph $G(B)$, where vertices (entities) v and u are connected by edge (u,v) if $b_{u,v}$ is nonzero; the weight of the edge (u,v) is set to $b_{u,v}$. Finally, consider the weighted graph $G(B)$ of entities as being located in the high-dimensional feature space of A, i.e., vertex v has coordinates $a_{v,*}$. Thus, $G(B,A)$ represents the combinatorial information of the entity to entity relationship similar to graph clustering methods like GraClus. However, $G(B,A)$ uses the distance attributes in A to add geometric information in the form of the coordinates of vertices (entities).

3.2 *Feature Subspace Transformation of* $G(B,A)$ *to* $G(B,\tilde{A})$

We develop FST as a variant of the FR-graph layout algorithm [10], which is described in Sect. 2, to obtain $G(B,A)$ from $G(B,A)$

Algorithm 4: procedure FST(A)

$n \leftarrow entities$
$B \leftarrow AA^T$
for $i = 1$ to MAX_ITER **do**
  Initialize displacement $\Delta \leftarrow (\mathbf{0}_1^T, \ldots, \mathbf{0}_n^T)$

  {Compute displacement}
  for all edges(u, v) in $G(B)$ **do**
    $F_{uv}^A \leftarrow \frac{-k^2 \times w_{uv}}{d_{uv} \times i}$
    $dist = ||(A_v - A_u)||$
    $\Delta_u = \Delta_u + \frac{(A_v - A_u)}{dist} \times F_{uv}^A$
    $\Delta_v = \Delta_v - \frac{(A_v - A_u)}{dist} \times F_{uv}^A$
  end for

  {Update entity positions}
  for $j = 1$ to n **do**
    $A_j \leftarrow A_j + \Delta_j$
  end for
end for

Although FST is motivated from the FR-graph layout algorithm, it is significantly different in the following aspects.

1. FST operates in the high-dimensional feature subspace unlike the FR scheme which seeks layouts in 2 or 3 dimensions. Thus, unlike FR which begins by randomly assigning coordinates in two- or three-dimensional space to the vertices of a graph, we begin with an embedding of the graph in the feature subspace.
2. The original FR scheme assumes that the vertices can be moved freely in any dimension of the embedding space. However, for our purposes, it is important to restrict this to prevent entities from developing spurious relationships to features, i.e., relationships that were not present in the original A. We therefore allow vertices to move only in the dimensions where their original feature values were nonzero.
3. The goal in FST is to bring highly connected vertices closer in the feature space, in contrast to FR objectives that aim to obtain a visually pleasing layout. Therefore, at each iteration of FST, we move the vertices based only on the attractive force (between connected vertices) and eliminate the balancing effect of the repulsive force. Furthermore, we scale the attractive force by the edge weights to reflect higher attraction from greater similarity between the corresponding entities. Together, these modifications can cause heavily connected vertices to converge to the nearly the same position. While this effect might be desirable for classification, it hampers the computation of attractive force for the next iteration of FST. In particular, very small distances between vertices cause an overflow error due to division by zero. We mitigate this problem by scaling the force by the number of iterations to ensure that at higher iterations, the effect of

the displacement is less pronounced. In summary, at FST iteration i, the attractive force between two vertices u and v with edge weight $w_{u,v}$ and Euclidean distance $d_{u,v}$ is given by:

$$F_i^{\mathrm{FST}} = -\frac{k^2 * w_{u,v}}{d_{u,v} * i} \tag{2}$$

In the expression above, k is a constant proportional to the square root of the ratio of the embedding area by the number of vertices as in the original FR scheme.

3.3 Applying K-Means to $\tilde{A}$ for Classification

$\tilde{A}$ forms the matrix representation of the dataset in the transformed feature space, where $\tilde{a}_{i,*}$ represents the new coordinates of vertex (entity) i. The sparsity structure of $\tilde{A}$ is identical to that of A, the original dataset. K-Means is now applied to $\tilde{A}$, the transformed dataset.

3.4 Computational Costs and Stopping Criteria of FST

In FST-K-Means, the cost per iteration of K-Means stays unaffected because it operates on $\tilde{A}$ which has exactly the same nonzero pattern as the original A, i.e., $\tilde{a}_{i,j} \neq 0$ if and only if $a_{i,j} \neq 0$. Additionally, although FST may change the number of K-Means iterations, it should not affect the worst-case complexity which is superpolynomial with a lower bound of $2^{\Omega\sqrt{(N)}}$ iterations for a dataset of N entries [1]. Thus the main overheads in FST-K-Means are those for FST.

The first step, that of forming $G(B,A)$, is similar to the graph setup step in GraClus and other graph clustering methods. Even if B is computed exactly, its costs are no more than $\Sigma_{i=1:n}(nnz_i^2)$ where nnz_i is the number of nonzero feature values of the i-th entity. The cost of FST is given by the number of iterations $\times$ the cost per iteration, which is proportional to the number of edges in the graph of B.

Depending on the degree of similarity between the entities, this graph can be more dense, thereby increasing the cost per iteration of FST. Consequently, an implementation of FST could benefit from using sparse approximations to the graph through sampling, though this could adversely affect the classification results.

3.5 Stopping Criteria for FST

The number of iterations for an ideal embedding using FST varies according to the feature values in the dataset. In this section we describe how we identify convergence criteria that promote improved classification.

An ideal embedding would simultaneously satisfy the two following properties: (1) similar entities are close together and (2) dissimilar entities are far apart. The first property is incorporated into the design of FST. We, therefore, seek to identify a near-ideal embedding based on the second property, inter-cluster distance [22], which is estimated by the distance of the entities from their global mean.

We next show in Lemma 1, how the distance of the entities from their global mean is related to feature variance. We use this relation to determine our convergence test for terminating FST iterations.

Lemma 1. *Consider a dataset of N entities and R features represented as a sparse matrix A; the entity i can be viewed as embedded in the feature space at the coordinates given by the i-th feature vector,* $a_{i,*} = [a_{i,1}, \dots, a_{i,R}]$. *Let* f_q *be the feature variance vector and let* d_q *be the vector of the distance of each entity from the global mean (centroid of all entities) at iteration q of FST. Now the following relation is satisfied by the f and d vectors:*

$$\|f_q\|_1 = \frac{1}{N}\|d_q\|_2$$

Proof. Let $a_{i,j}$ denote the feature j of entity i. Now $a_{i,j}$ is also the j-th coordinate of entity i. Let the mean of the feature vector $a_{*,j}$ be $\tau_j = \frac{1}{N}\Sigma_{k=1}^{N}(a_{k,j})$ and the variance of feature vector $a_{*,j}$ be $\phi_j = \frac{1}{N}\Sigma_{k=1}^{N}(a_{k,j} - \tau_j)^2$.

Let $f_q = [\phi_1, \phi_2, \dots, \phi_R]$ be the vector of feature variances. Now $\|f_q\|_1 = \Sigma_{j=1}^{R}|\phi_j| = \Sigma_{j=1}^{R}\phi_j$, because $\phi_j \geq 0$.

Let the global mean of the entities be represented by μ. The i-th coordinate of μ is calculated as $\mu_i = \frac{1}{N}\Sigma_{k=1}^{N}(a_{k,i}) = \tau_i$.

Let d_i be the distance of entity i from μ. Therefore $\delta_i = \sqrt{\Sigma_{k=1}^{R}(a_{i,k} - \mu_k)^2} = \sqrt{\Sigma_{k=1}^{R}(a_{i,k} - \tau_k)^2}$. Let $d_q = [\delta_1, \delta_2, \dots, \delta_N]$ be the vector representing the distance of the entities from the global mean μ. The 2-norm of d_q is given by:

$$\begin{aligned}
\|d_q\|_2 &= \Sigma_{i=1}^{N}(\delta_i)^2 \\
&= \Sigma_{i=1}^{N}(\Sigma_{k=1}^{R}(a_{i,k} - \tau_k)^2) \\
&= \Sigma_{k=1}^{R}(\Sigma_{i=1}^{N}(a_{i,k} - \tau_k)^2) \\
&= \Sigma_{k=1}^{R}(N\phi_k) \\
&= N\|f_q\|_1
\end{aligned}$$

A high value of $\|f_q\|_1$ implies that there is a high variance among the features of the entities relative to the transformed space. High variance among features indicates easily distinguishable entities. Lemma 1 shows that $\|f_q\|_1 = \frac{1}{N}\|d_q\|_2$, i.e., high feature variance is proportional to the average distance of entities from their global mean. The value of d_q can be easily computed and we base our stopping criteria on this value.

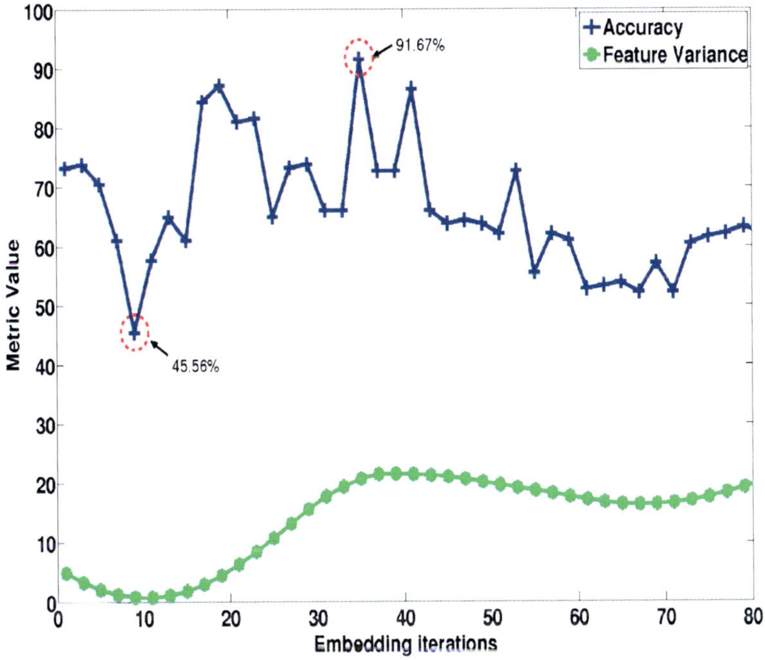

Fig. 2 Plots of classification accuracy and the 1-norm of feature variance vector across FST iterations. FST iterations are continued until feature variance decreases relative to the previous iteration

Our heuristic for determining the termination of FST is as follows: continue layout iterations until $\|d_{i+1}\|_2 \leq \|d_i\|_2$. That is, we terminate FST when successive iterations fail to increase the global mean of the distance of the entities. As illustrated in Fig. 2, this heuristic produces an embedding that can be classified with high accuracy for a sample dataset.

3.6 *Impact of FST on Clustering*

We now consider the impact of FST on a sample dataset, namely *splice* [19], with two clusters. Figure 3 shows the layout of the entities of *splice* in the original and the transformed feature space. For ease of viewing, the entities are projected to the first three principal components. Observe that the clusters are not apparent in the original dataset. However, FST repositions the entities in the feature space into two distinct clusters, thereby potentially facilitating classification.

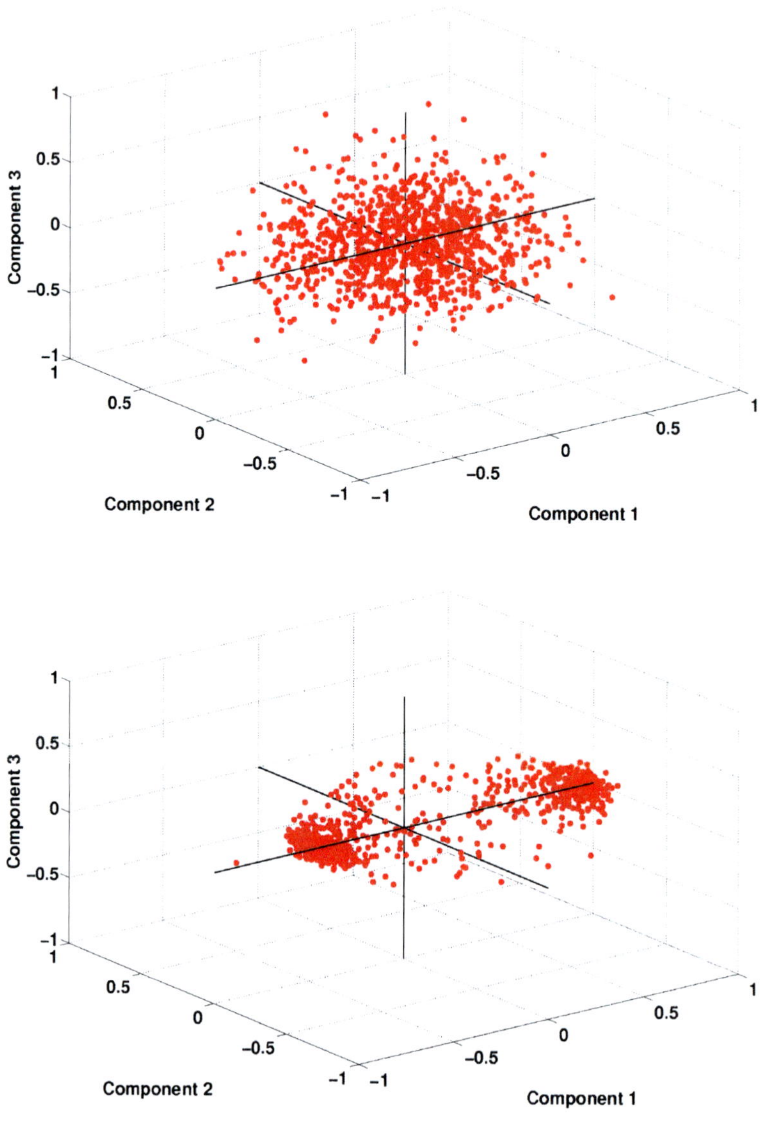

Fig. 3 Layout of entities in *splice* in the original (*top*) and transformed (*bottom*) feature space, projected onto the first three principal components. Observe that two clusters are more distinct after FST

4 Evaluation and Discussion

We now provide an empirical evaluation of the quality of classification using FST-K-Means. We define external and internal quality metrics for evaluating classification results. We use these metrics for a comparative evaluation of the performance of

FST-K-Means, K-Means, and GraClus on a test suite of eight datasets from a variety of applications.

4.1 Metrics of Classification Quality

The quality of classification is typically evaluated using external and internal metrics.

The external metric evaluates the *accuracy* of classification (the higher the value, the better) as the ratio of correctly classified entities to the total number of entities. This metric is dependent on a subjective predetermined labeling. This measure, though an indicator of correctness, cannot be incorporated into the design of classification algorithms. We henceforth refer to this metric as "accuracy" denoted by P defined as:

$$P = \frac{\text{Number of correctly classified entities}}{\text{Total number of entities}}. \tag{3}$$

The internal metric measures the cohesiveness of the clusters, i.e., the sum of the square of the distance of the clustered points from their centroids. We henceforth refer to this metric as "cohesiveness" denoted by J. If $M_1, \ldots, M_k$ are the k clusters and μ_i represents the centroid of cluster M_i, the cohesiveness J is defined as:

$$J = \Sigma_{h=1}^{k} \Sigma_{x \in M_h} ||x - \mu_h||^2. \tag{4}$$

If similar entities are clustered closely, then they are generally easier to classify and so a lower value of J is preferred. The cohesiveness, J, is an internal metric because it can be incorporated into the design of the classification algorithm and several classification methods seek to minimize the objective function given by J.

4.2 Experimental Setup

Our experiments compare the accuracy of our feature subspace transformation method coupled with K-Means clustering versus a commercial implementation of K-Means. We first apply K-Means clustering to the unmodified collection of feature values as obtained from the dataset. We refer to experiments based on this dataset as *K-Means*. We then apply K-Means clustering to the dataset transformed using FST as described in Sect. 3. We refer to this scheme as *FST-K-Means* in our experimental evaluation.

In both set of experiments, we use K-Means based on Llyod's algorithm (as implemented in the MATLAB [12] Statistical Toolbox), where the centroids are initialized randomly and K-Means is limited to a maximum of 25 iterations.

The quality of clustering using K-Means varies depending on the initial choice of centroids. To mitigate this effect, we execute 100 runs of K-Means, each time with a different set of initial centroids. To ensure fair comparison, the initial centroids for each execution of FST-K-Means are selected to be exactly the same as those for the corresponding K-Means execution.

The value of the cohesiveness metric is dependent on the feature vector of the entities. Therefore, for the set FST-K-Means where the values of the features have been modified, we compute cohesiveness by first identifying the elements in the cluster and then transforming them back to the original space. We thereby ensure that the modified datasets are used *exclusively* for cluster identification and the values of cohesiveness (J) are not affected by the transformed subspace.

The GraClus scheme is executed on the weighted graph representation. The results of GraClus do not significantly change across runs. Consequently, one execution of this scheme is sufficient for our empirical evaluation.

The computational cost of FST is only the number of nonzeros in matrix B, which is less than $O(N^2)$, per iteration where N is the number of entities.

4.2.1 Evaluation of FST-K-Means on a Synthetic Dataset

We first evaluate our model using artificially generated data. Our goal is to easily demonstrate and explain the steps of our FST-K-Means algorithm using this data, before we test our method on benchmark datasets from different repositories.

Figure 4 represents our test case, in which three gaussian clusters are formed in 2 dimensions using three different cluster means and covariance matrices. Figure 4a represents the original synthetic data. Figure 4b represents the same data points but after the FST transformation has been applied to it. Even before we proceed to formally clustering the two datasets (original and transformed), a visual inspection of the layout suggests that clusters obtained using the transformed dataset are likely to be distinct. Finally, on clustering the two datasets using K-Means, we observe that K-Means on the transformed dataset is 3% more accurate than K-Means on the original data. Matrices $Z_{\text{K-Means}}$ and $Z_{\text{FST-K-Means}}$ in (5) and (6), respectively, indicate the confusion matrices obtained using the two datasets. In our example, the cluster label of each entity is known a priori; therefore, after clustering the original or transformed data we compare it to the true cluster label. The diagonal element $Z^{(ii)}_{\text{K-Means}}$ of matrix $Z_{\text{K-Means}}$ indicates the number of entities that were assigned correctly to cluster i, and the off-diagonal element $Z^{(ij)}_{\text{K-Means}}$ where $i \neq j$ indicates the number of elements of cluster i that were mapped to cluster j ($Z_{\text{FST-K-Means}}$ is formed in the same way as $Z_{\text{K-Means}}$). We observe that our FST transformation scheme clearly obtains higher precision and accuracy:

$$Z_{\text{K-Means}} = \begin{pmatrix} \mathbf{59} & 16 & 25 \\ 6 & \mathbf{78} & 16 \\ 6 & 3 & \mathbf{91} \end{pmatrix} \tag{5}$$

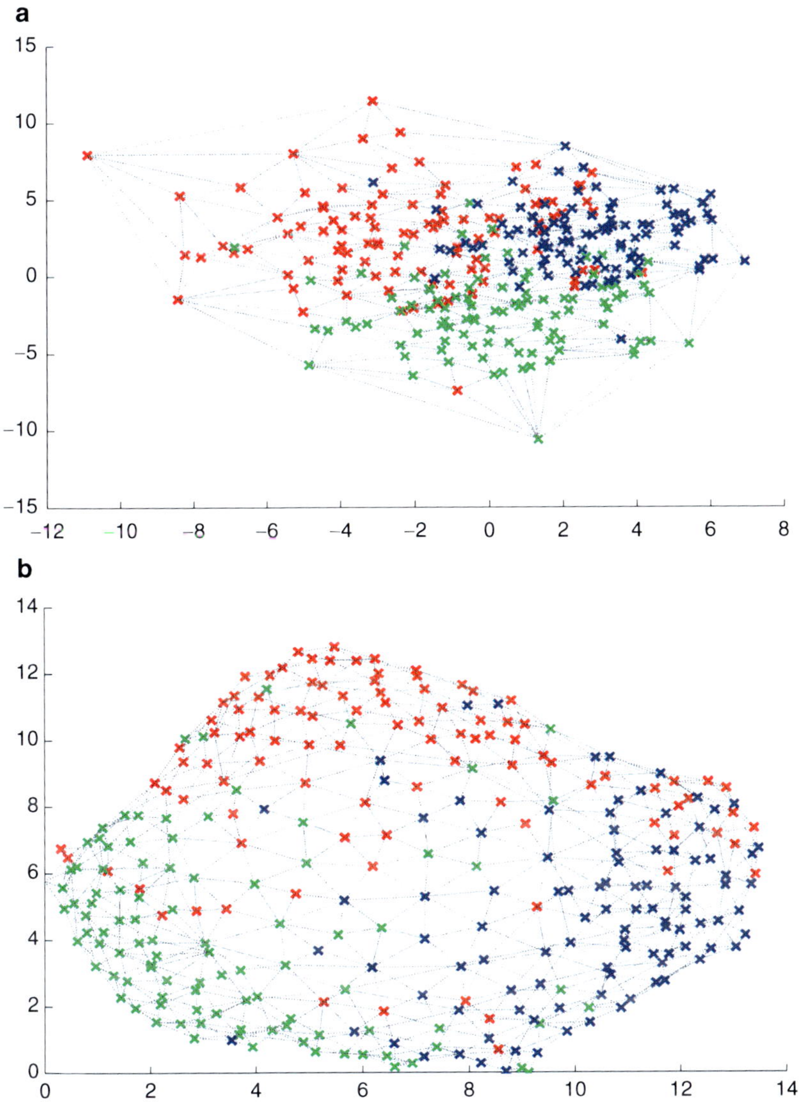

Fig. 4 Layout of entities in two dimensional synthetic dataset in the original (*top*) and transformed (*bottom*) feature space

$$Z_{\text{FST-K-Means}} = \begin{pmatrix} \mathbf{68} & 11 & 21 \\ 10 & \mathbf{82} & 8 \\ 7 & 8 & \mathbf{85} \end{pmatrix} \tag{6}$$

Table 1 Test suite of datasets

Name	Samples	Features	Source (Type)
adult_a2a	2,265	123	UCI (Census)
australian	690	14	UCI (Credit Card)
breast-cancer	683	10	UCI (Census)
dna	2,000	180	Statlog (Medical)
splice	1,000	60	Delve (Medical)
180txt	180	19,698	SMART(Text)
300txt	300	53,914	SMART (Text)
20news	1,061	16,127	Yahoo Newsgroup (Text)

4.2.2 Evaluation of FST-K-Means on Benchmark Datasets

We test K-Means and FST-K-Means classification algorithms on datasets from the UCI [2], Delve [19], Statlog [18] , SMART [21], and Yahoo 20Newsgroup [14] repositories. The parameters of each dataset along with their application domain are shown in Table 1. Datasets *dna*, *180txt*, and *300txt* have three classes. The rest of the datasets represent binary classification problems. We select two classes, *comp.graphics* and *alt.atheism*, from the Yahoo 20Newsgroup dataset and form the *20news* binary classification set.

4.3 *Comparative Evaluation of K-Means, GraClus, and FST-K-Means*

We now compare the quality of classification using the accuracy and cohesiveness metrics for the datasets in Table 1 using K-Means, GraClus, and FST-K-Means.

4.4 *Accuracy*

Table 2 reports the accuracy (P) of classification of the three schemes. The values for K-Means and FST-K-Means are the mode (most frequently occurring value) over 100 executions. These results indicate that with the exception of the *breast-cancer* dataset, the accuracy of FST-K-Means is either the highest value (5 out of 8 datasets) or is comparable to the highest value. In general, GraClus has lower accuracy than either FST-K-Means or K-Means with comparable values for only two of the datasets, namely, *dna* and *180txt*.

Table 4 shows the improvement in accuracy of FST-K-Means relative to K-Means and GraClus. The improvement in accuracy as a percentage of method A relative to method B is defined as $\frac{(P(A)-P(B))\times 100}{P(B)}$, where $P(A)$ and $P(B)$ denote accuracies

Table 2 Accuracy of classification of K-Means, GraClus, and FST-K-Means

Datasets	Classification accuracy (P) K-Means	GraClus	FST-K-Means
adult_a2a	70.60	52.49	*74.17*
australian	*85.51*	74.20	85.36
breast-cancer	*93.70*	69.69	83.16
dna	*72.68*	70.75	70.75
splice	55.80	53.20	*69.90*
180txt	73.33	91.67	*91.67*
300txt	78.67	64.33	*95.00*
20news	46.74	54.85	*73.70*

Table 3 Accuracy of classification for PCA (top three principal components) and FST-K-Means

Datasets	Classification Accuracy (P) PCA	FST-K-Means	Relative Improvement
adult_a2a	65.81	*74.17*	+12.70
australian	71.17	*85.36*	+19.93
breast-cancer	*96.05*	83.16	−13.42
dna	60.41	*70.75*	+17.12
splice	65.10	*69.90*	+7.37
180txt	61.78	*91.67*	+48.38
300txt	62.00	*95.00*	+53.22
20news	49.50	*73.70*	+48.49

of methods A and B, respectively; positive values represent improvements while negative values indicate degradation. We see that the improvement in accuracy for FST-K-Means can be as high as 57.6% relative to K-Means (for *20news*) and as high as 47.7% relative to GraClus (for *300txt*). On average, the accuracy metric obtained from FST-K-Means shows an improvement of 14.9% relative to K-Means and 23.6% relative to GraClus.

4.4.1 Comparison of FST-K-Means with K-Means After PCA Dimension Reduction

Table 3 compares the accuracy obtained by K-Means using top three principal components (PCA-K-Means) of the dataset with FST-K-Means. We observe that in 7 out of 8 datasets FST-K-Means performs better than PCA-K-Means and on average achieves 24.27% better accuracy relative to PCA-K-Means. This is an empirical proof that K-Means benefits from the high-dimensional embedding of FST. We also observe that the relative improvement in accuracy between PCA-K-Means and FST-K-Means is particularly high for text datasets (*180txt, 300txt, and 20news*). Typically, feature selection in noisy text data is a difficult problem and the right set of principal components that can provide high accuracy is often unknown [3]. Our feature subspace transformation brings related entities closer, in effect, common features gain more importance than unrelated features.

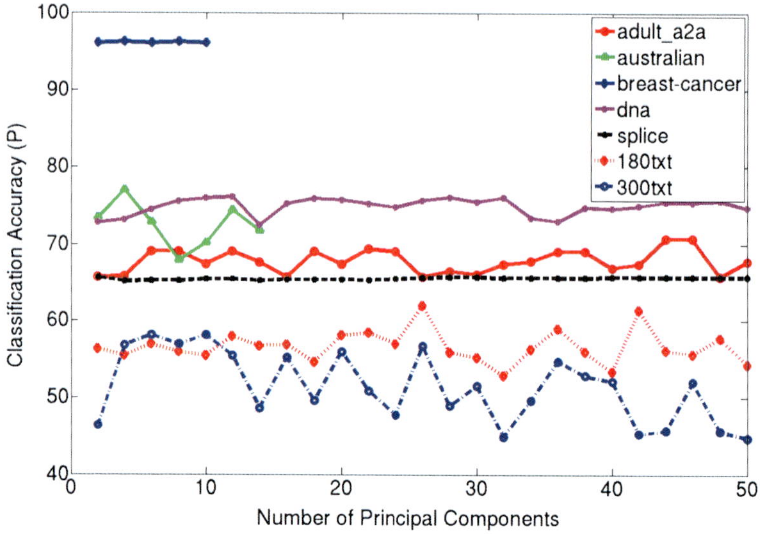

Fig. 5 Sensitivity of classification accuracy (P) of K-Means to number of principal components

Table 4 Improvement or degradation (negative values) percentage of accuracy of FST-K-Means relative to K-Means and GraClus

Datasets	Relative improvement (accuracy) K-Means (percentage)	GraClus
adult_a2a	5	41.3
australian	−1	15
breast-cancer	−11	19
dna	−2	0
splice	25.2	31.4
180txt	25.0	0
300txt	20.75	*47.7*
20news	*57.6*	34.3
Average	14.9	23.6

Additionally, in Fig. 5 we present a sensitivity study of using up to 50 principal components on K-Means accuracy. Table 3 reports results for top 3 principal components, while Fig. 5 presents the accuracy of PCA-K-Means upto 50 principal components. It is clear that FST-K-Means performs better than PCA-K-Means.

4.5 Cohesiveness

Table 5 compares cluster cohesiveness (J), the internal quality metric across all three schemes. Once again, the values for K-Means and FST-K-Means are the mode

Table 5 Cluster cohesiveness of K-Means, GraClus, PCA(top three principal components), and FST

Datasets	Cluster cohesiveness (J) K-Means	GraClus	PCA	FST
Adult_a2a	24,013	16,665	*15,522*	16,721
australian	4,266	3,034	2,791	*2,638*
breast-cancer	2,475	2,203	*980*	1,366
dna	84,063	65,035	*65,029*	65,545
splice	31,883	31,618	*30,830*	31,205
180txt	25,681	*23,776*	23,806	24,131
300txt	47,235	44,667	*44,539*	45,052
20news	3,851,900	3,483,591	*3,029,614*	3,341,400

Table 6 Improvement or degradation (negative values) percentage of cohesiveness of FST-K-Means relative to K-Means and GraClus

Datasets	Relative improvement (cohesiveness) K-Means (percentage)	GraClus
adult_a2a	30.3	−0.33
australian	38.1	13.0
breast-cancer	*44.8*	*37.9*
dna	22.0	−0.78
splice	2.1	1.3
180txt	6.0	−1.4
300txt	4.6	−0.86
20news	13.2	4.0
Average	20.2	6.6

(most frequently occurring value) over 100 executions. Recall that a lower value of cohesiveness is better than a higher value. The cohesiveness measure of FST-K-Means is the lowest in 4 out of 8 datasets, and it is comparable to the lowest values (obtained by GraClus) for the remaining 4 datasets.

Table 6 shows the improvement of the cohesiveness metric of FST-K-Means relative to K-Means and GraClus. The improvement in cohesiveness as a percentage of method A relative to method B is defined as $\frac{(J(B)-J(A))\times 100}{J(B)}$, where $J(A)$ and $J(B)$ denote cohesiveness of methods A and B, respectively; positive values represent improvements while negative values indicate degradation. FST-K-Means achieves as high as 44.8% improvement in cohesiveness relative to K-Means and 37.9% improvement relative to GraClus. On average FST-K-Means realizes an improvement in cohesiveness of 20.2% relative to K-Means and 6.6% relative to (GraClus).

In summary, the results in Tables 2–6 clearly demonstrate that FST-K-Means is successful in improving accuracy beyond K-Means at cohesiveness measures that are significantly better than K-Means and comparable to GraClus. The superior accuracy of K-Means is related to the effectiveness of the distance-based measure for clustering. Likewise, the superior cohesiveness of GraClus derives from using the combinatorial connectivity measure. We conjecture that FST-K-Means shows superior accuracy *and* cohesiveness from a successful combination of both distance and combinatorial measures through FST.

5 Toward Optimal Classification

Unsupervised classification ideally seeks 100% accuracy in labeling entities. However, as discussed in Sect. 4, accuracy is a subjective measure based on external user specifications. It is therefore difficult to analyze algorithms based on this metric. On the other hand, the internal metric, i.e., the cluster cohesiveness J, provides an objective function dependent only on the coordinates (features) of the entities and centroids. Thus, this metric is more amenable for use in analysis of clustering algorithms.

Let the dataset be represented by the $N \times R$ sparse matrix A with N entities and R features. Let the i-th row vector be denoted by $a_{i,*} = [a_{i,1}, a_{i,2}, \ldots, a_{i,R}]$. Now the global mean of the entities (i.e., the centroid over all entities) is given by the R-dimensional vector $\bar{a} = \frac{1}{N}\Sigma_{i=1}^{N} a_i$, where the j-th component of $\bar{a}$ is denoted by $\bar{a}_j = \frac{1}{N}\Sigma_{i=1}^{N} a_{i,j}$ for $j = 1, \ldots, R$. Let Y be the centered data matrix, where each column $y_{i,*} = (a_{i,*} - \bar{a})^{\mathrm{T}}$ or equivalently $y_{i,j} = a_{i,j} - \bar{a}_j$ for $i = 1, \ldots, N$ and $j = 1, \ldots, R$.

It has been shown in [9] that for a classification with k clusters, the optimal value of the cohesiveness J can be bounded as shown below, where $N \times \bar{y}^2$ is the trace of $Y^{\mathrm{T}}Y$ and λ_i is the i-th principal eigenvalue of $Y^{rmT}Y$:

$$N\bar{y}^2 - \Sigma_{i=1}^{k-1}(\lambda_i) \leq J \leq N\bar{y}^2.$$

Table 7 shows the lower and upper bounds for the cohesiveness measure in the original feature space along with the range observed for this measure (minimum and maximum values) over 100 trials of FST-K-Means and K-Means. The bounds of the 20news group were excluded due to computational complexity in calculating eigenvalues, posed by the size of the data. These results indicate that unlike K-Means, FST-K-Means always achieves near-optimal values for J and it consistently satisfies the bounds for the optimal cohesiveness. Thus FST-K-Means moves K-Means toward cohesiveness optimality and consequently toward enhanced classification accuracy.

6 Summary

We have developed and evaluated a feature subspace transformation scheme to combine the advantages of distance-based and graph-based clustering methods to enhance K-Means clustering. However, our transformation is general and as part of

Table 7 The lower bound, range, and upper bound of cohesiveness across 100 runs of FST-K-Means (top half) and K-Means in the original feature space

Cluster cohesiveness: FST-K-Means				
Datasets	Lower *bound*	Min	Max	Upper bound
Adult_a2a	15,250	15,866	17,208	17,380
australian	2,442	2,638	3,008	3,493
breast-cancer	747	1,366	1,366	5,169
dna	63,890	65,525	65,865	67,190
splice	30,389	31,205	31,205	31,942
180txt	23,188	23,765	24,178	25,290
300txt	43,708	44,800	45,194	46,512
Cluster cohesiveness: K-Means				
Datasets	*Lower bound*	Min	Max	Upper bound
Adult_a2a	15,250	*24,013*	24,409	17,380
australian	2,442	*4,266*	4,458	3,493
breast-cancer	747	2,475	2,475	5,169
dna	63,890	*84,062*	84,123	67,190
splice	30,389	31,882	31,884	31,942
180txt	23,188	*25,651*	25,730	25,290
300txt	43,708	*47,220*	47,288	46,512

Observe that FST-K-Means consistently satisfies the optimality bounds while K-Means fails to do so for most datasets. The highlighted numbers in the lower table indicate datasets for which the minimum value of cohesiveness exceeds the upper bound on optimality

our plans for future work we propose to adapt FST to improve other unsupervised clustering methods.

We empirically demonstrate that our method, FST-K-Means, improves both the accuracy and cohesiveness relative to popular classification methods like K-Means and GraClus.

We also show that the values of the cohesiveness metric of FST-K-Means consistently satisfy the optimality criterion (i.e., lie within the theoretical bounds for the optimal value) for datasets in our test suite. Thus, FST-K-Means can be viewed as moving K-Means toward cohesiveness optimality to achieve improved classification.

Acknowledgments This research was supported in part by the National Science Foundation through grants CNS 0720749, OCI 0821527, and CCF 0830679. Additionally, Dr. Sanjukta Bhowmick would like to acknowledge the support of the College of Information Science and Technology at the University of Nebraska at Omaha.

References

1. Arthur D, Vassilvitskii S (2006) How slow is the k-means method? In: SCG '06: Proceedings of the twenty-second annual symposium on Computational geometry. ACM, New York, pp 144–153, http://doi.acm.org/10.1145/1137856.1137880
2. Asuncion A, Newman D (2007) UCI machine learning repository. http://www.ics.uci.edu/~mlearn/MLRepository.html
3. Baker LD, McCallum AK (1998) Distributional clustering of words for text classification. In: SIGIR '98: Proceedings of the 21st annual international ACM SIGIR conference on Research and development in information retrieval. ACM, New York, pp 96–103, http://doi.acm.org/10.1145/290941.290970
4. Ball G, Hall D (1965) Isodata, a novel method of data analysis and pattern classification. Tech report NTIS AD 699616, Stanford Research Institute, Stanford, CA
5. Bíró I, Szabó J, Benczúr A (2008) Latent dirichlet allocation in web spam filtering. In: AIRWeb '08: Proceedings of the 4th international workshop on Adversarial information retrieval on the web. ACM, New York, pp 29–32, http://doi.acm.org/10.1145/1451983.1451991
6. Chatterjee A, Bhowmick S, Raghavan R (2010) Feature subspace transformations for enhancing k-means clustering. In: Proceedings of the 19th ACM international conference on information and knowledge management, ACM
7. Chatterjee A, Raghavan P, Bhowmick S (2012) Similarity graph neighborhoods for enhanced supervised classification. In: Procedia computer science, Elsiever, pp 577–586, 10.1016/j.procs.2012.04.062
8. Dhillon I, Guan Y, Kulis B (2007) Weighted graph cuts without eigenvectors a multilevel approach. IEEE Trans Pattern Anal Mach Intell 29(11):1944–1957, 10.1109/TPAMI.2007.1115
9. Ding C, He X (2004) K-means clustering via principal component analysis. ACM, New York, pp 225–232
10. Fruchterman TMJ, Reingold EM (1991) Graph drawing by force-directed placement. Software Pract Exp 21(11):1129–1164
11. Hartigan JA, Wong MA (1979) A k-means clustering algorithm. Appl Stat 28(1)
12. Inc TM (2007) Matlab and simulink for technical computing. http://www.mathworks.com
13. Kim K, Ahn H (2008) A recommender system using ga k-means clustering in an online shopping market. Expert Syst Appl 34(2):1200-1209, 10.1016/j.eswa.2006.12.025
14. Lang K (1995) Newsweeder: Learning to filter netnews. In: Proceedings of the twelfth international conference on machine learning, pp 331–339
15. Li X (2008) A volume segmentation algorithm for medical image based on k-means clustering. In: IIH-MSP '08: Proceedings of the 2008 international conference on intelligent information hiding and multimedia signal processing. IEEE Computer Society, Washington, pp 881–884, http://dx.doi.org/10.1109/IIH-MSP.2008.161
16. Lloyd S (1982) Least squares quantization in pcm. IEEE Trans Inform Theory 28:129–137
17. MacQueen J (1967) Some methods for classification and analysis of multivariate observations. In: Fifth Berkeley symposium on mathematics, statistics and probability. University of California Press, California
18. Michie D, Spiegelhalter DJ, Taylor C (1994) Machine learning, neural and statistical classification, Prentice hall. Edn. 1.
19. Neal R (1998) Assessing relevance determination methods using delve. In: Neural networks and machine learning. Springer, Berlin, pp 97–129, URL http://www.cs.toronto.edu/~delve/
20. Ng A, Jordan M, Weiss Y (2001) On spectral clustering: Analysis and an algorithm In: T. Dietterich, S. Becker, Z. Ghahramani (eds). In Advances in Neural Information Processing Systems, pp. 849–856
21. Salton G (1971) Smart data set. ftp://ftp.cs.cornell.edu/pub/smart
22. Savaresi SM, Boley DL, Bittanti S, Gazzaniga G (2002) Cluster selection in divisive clustering algorithms. In: SIAM datamining conference, Arlington, VA
23. Steinhaus H (1956) Sur la division des corp materials en parties. Bull Acad Polon Sci IV (C1. III):801–804

Learning with ℓ^1-Graph for High Dimensional Data Analysis

Jianchao Yang, Bin Cheng, Shuicheng Yan, Yun Fu, and Thomas Huang

1 Introduction

An informative graph, directed or undirected, is critical for those graph-orientated algorithms designed for data analysis, such as clustering, subspace learning, and semi-supervised learning. Data clustering often starts with a pairwise similarity graph and then translates into a graph partition problem [19], and thus the quality of the graph essentially determines the clustering quality. The pioneering works on manifold learning, e.g., ISOMAP [20], locally linear embedding (LLE) [17], and Laplacian eigenmaps [3], all rely on different graphs reflecting the data similarities. Most popular subspace learning algorithms, e.g., principal component analysis (PCA) [13], linear discriminant analysis (LDA) [1], and locality preserving projections (LPP) [11], can all be explained within the graph embedding framework in [22]. Also, most semi-supervised learning algorithms are driven by certain graphs constructed over both labeled and unlabeled data. Zhu et al. [25] explored the harmonic property of Gaussian random field over the graph for semi-supervised learning. Belkin and Niyogi [2] instead learned a regression function that fits the labels on the labeled data while preserving the smoothness of data manifold modeled by a graph.

J. Yang (✉)
Advanced Technology Laboratory, Adobe Systems Inc.,
321 Park Avenue, San Jose, CA 95110

B. Cheng • S. Yan
ECE Department, National University of Singapore

Y. Fu
ECE Department, Northeastern University

T. Huang
ECE Department, University of Illinois at Urbana-Champaign

Y. Fu and Y. Ma (eds.), *Graph Embedding for Pattern Analysis*,
DOI 10.1007/978-1-4614-4457-2_6,

There exist two popular ways for graph construction: the k-nearest-neighbor (k-nn) method and the ϵ-ball method. In k-nn graph construction, for each datum, its k-nearest-neighbor samples are connected with it, while in ϵ-ball graph construction, data samples falling into its ϵ-ball neighborhood are connected with it. Then various approaches, e.g, binary, Gaussian kernel [3], or ℓ^2-reconstruction [17], are used to determine the graph edge weights. Since the ultimate goals of the constructed graphs are for data analysis, the following graph characteristics are desired:

1. *Similarity fidelity*. The desired graph should reflect the underlying relationships between data samples, i.e., data samples from the same class or cluster should be connected and not be connected otherwise.
2. *Robustness to noise*. Noise and outliers are inevitable in practical applications, especially for visual data, and therefore, the graph construction procedure should be robust to such corruptions. The graph constructed based on k-nn or ϵ-ball method is founded on the pair-wise Euclidean distance, which is very sensitive to noise, especially in high-dimensional spaces. That is, the graph structure is not stable with the presence of noise.
3. *Adaptive sparsity*. Studies on manifold learning [3] show that sparse graph can convey valuable information for classification purposes. Also for large-scale applications, sparse graphs are required due to the storage limitation. In real-world scenarios, it is likely that data samples are not evenly sampled from the space. Therefore, we want to adaptively select the most likely kindred neighbors for each datum.

In this work, we are particularly interested in building an informative graph for data analysis in high-dimensional spaces, where the data samples are sparsely distributed in the space and the local neighborhood becomes less and less useful for k-nn and ϵ-ball methods [9]. Empirically, many types of high-dimensional data tend to lie in a union of low-dimensional linear subspaces [21], where our ℓ^1-graph is of particular interest. In Sect. 2, we present the procedure to construct our robust ℓ^1-graph by taking advantage of the overall contextual information instead of only pairwise Euclidean distance as conventionally. The neighboring samples of a datum and their corresponding edges weights are simultaneously derived by solving an ℓ^1-norm minimization problem, where each datum is reconstructed by a sparse linear combination of the remaining samples and a noise term. Compared with graphs constructed by k-nn and ϵ-ball methods, the ℓ^1-graph can better recover the subspace structures in high-dimensional spaces, is robust to noise, and can adaptively and sparsely choose the kindred neighbors. Figure 1 demonstrates the graph robustness comparison between our ℓ^1-graph and the k-nn graph. The first row illustrates some data samples from the USPS dataset [12]. The left column shows the edges weights in ℓ^1-graph; the middle row shows the current datum (noise added from the third row on); and the right column shows the edges wights in the k-nn graph. Here, the horizontal axes indicate the index number of the remaining data samples. The number in parenthesis of each row and column shows the number of neighbors changed compared with the noiseless results in the second row. As shown, our

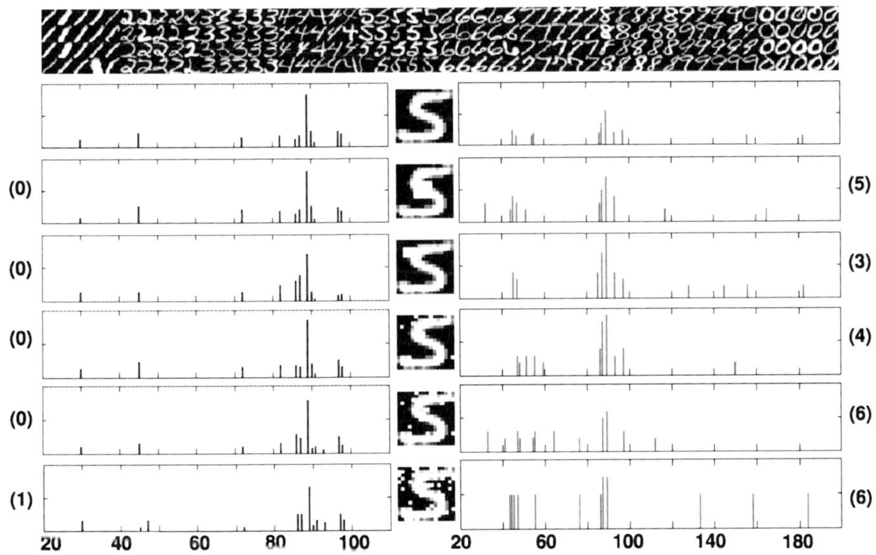

Fig. 1 Robustness comparison for neighbors selected by ℓ^1-graph and k-nn graph, where different levels of noise are added to the middle testing sample

ℓ^1-graph shows much greater robustness to noise, compared with k-nn graph.[1] On the same dataset, Fig. 2 shows the kindred neighbor selection comparison between the ℓ^1-graph and k-nn graph on 19 testing samples. As shown, the ℓ^1-graph can adaptively and sparsely select the neighboring samples. The red bars indicate the numbers of the neighbors selected by ℓ^1-graph (weights are not considered here). The green bars indicates the numbers of kindred samples selected by ℓ^1-graph, while the blue bars indicate those selected by k-nn graph. Note that for the k-nn graph we adjust k based on the red bar number for each testing sample. Clearly, the ℓ^1-graph selects more kindred data samples than the k-nn graph, implying that our ℓ^1-graph better recovers the underlying data relationships.

This ℓ^1-graph is then utilized in Sect. 3 to instantiate a series of graph-oriented algorithms for various machine learning tasks, e.g., data clustering, subspace learning, and semi-supervised learning. Owing to the above advantages over classical graphs, the ℓ^1-graph brings consistent performance gain in all these tasks as detailed in Sect. 4.

[1] We compare with k-nn graph only because ϵ-ball graph is generally inferior to k-nn graph in practice.

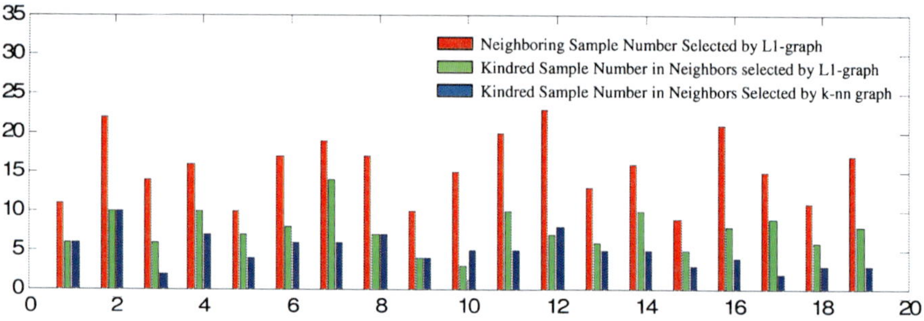

Fig. 2 The numbers of kindred neighbors selected by ℓ^1-graph and k-nn graph on 19 testing samples

2 Rationales on ℓ^1-graph

We denote the training samples by a matrix $X = [x_1, x_2, \ldots, x_N], x_i \in \mathbb{R}^m$, where N is the sample number and m is the feature dimension. For each x_i, we want to find its connections and weights with the remaining samples X/x_i, which will be represented as a graph. In this section, we discuss the motivation and construction procedure for our ℓ^1-graph. The graph construction includes both neighbor selection and graph edge weight assignment, which are assumed to be unsupervised in this work, without harnessing any data label information.

2.1 Motivation

The k-nn and ϵ-ball [3, 17] approaches are the two most popular ones for graph construction in the literature. Both of them determine the neighboring samples based on the *pairwise* Euclidean distance, which is, however, very sensitive to data noises, especially in high-dimensional spaces, i.e., the graph structure may dramatically change with the presence of noise. Also, both methods are based on the local neighborhood on the data manifold, which becomes less and less useful in high-dimensional feature spaces due to the curse of dimensionality [9]. Moreover, both k-nn and ϵ-ball methods are prone to overfitting or underfitting when data samples are not evenly distributed, i.e., the global parameter k and ϵ may be too large for some data samples while too small for other samples. To avoid the above limitations of the conventional graph construction methods, we propose a novel way of graph construction based on seeking the sparse representations of the data samples by ℓ^1-norm minimization, and thus the new graph is termed ℓ^1-graph.

Our ℓ^1-graph is motivated by the recent advances in sparse representation [7, 15, 21] for high-dimensional data analysis. The quest for sparse representation has a long history in signal processing [18]. Recent studies show that sparse representation [16] appears to be biologically plausible as well as essential for high-dimensional signal recovery [5, 8] and effective for pattern recognition [21]. In particular, Wright et al. [21] proposed to perform face recognition by seeking a robust sparse representation of the test sample with respect to the training samples from all classes. Inspecting which classes those nonzero coefficients fall into reveals the identity of the test sample. The rationale behind this work is to assume that human faces lie in a union of low-dimensional linear subspaces for different subjects, and the sparse representation through ℓ^1-norm minimization can correctly identify which linear subspace a testing sample belongs to. In this work, we generalize this observation from faces [21] to more general high-dimensional data types, where the data approximately distributes as a union of low-dimensional linear subspaces, and we show that building a graph based on such sparse representations is very effective for many graph-orientated machine learning algorithms.

2.2 Robust Sparse Representation

Much interest has been shown in seeking sparse representations with respect to an overcomplete dictionary of the basis elements in signal processing [5, 8, 18] and pattern recognition [21]. Suppose we have an underdetermined system of linear equations: $x = D\alpha$, where $x \in \mathbb{R}^m$ is the vector to be represented, $\alpha \in \mathbb{R}^n$ is the unknown reconstruction coefficients, and $D \in \mathbb{R}^{m\times n}(m < n)$ is the overcomplete dictionary with n bases. Among the infinite many solutions to the above underdetermined linear system, we are interested in the sparsest solution, i.e., the one with the fewest non-zero coefficients.

$$\min_{\alpha} \|\alpha\|_0, \quad s.t.\ x = D\alpha. \tag{1}$$

where the ℓ^0-norm $\|\cdot\|_0$ counts the number of nonzero entries in a vector. It is well known that the above optimization is NP-hard in general case. However, recent results [7] show that if the solution is sparse enough, the sparse representation can be approximated by the following convex ℓ^1-norm relaxation,

$$\min_{\alpha} \|\alpha\|_1, \quad s.t.\ x = D\alpha, \tag{2}$$

which can be solved efficiently by linear programming [6]. In practice, there may exist noise on certain elements of x. In order to recover these elements and provide a robust estimation of α, we can reformulate

$$x = D\alpha + \zeta + \xi = [D\ I]\begin{bmatrix}\alpha\\ \zeta\end{bmatrix} + \xi, \tag{3}$$

where $\zeta \in \mathbb{R}^m$ is a sparse error term and $\xi \in \mathbb{R}^m$ denotes a Gaussian noise term. Then by setting $B = [D\ I] \in \mathbb{R}^{m\times(m+n)}$ and $\alpha' = \begin{bmatrix}\alpha\\ \zeta\end{bmatrix}$, we can solve the following

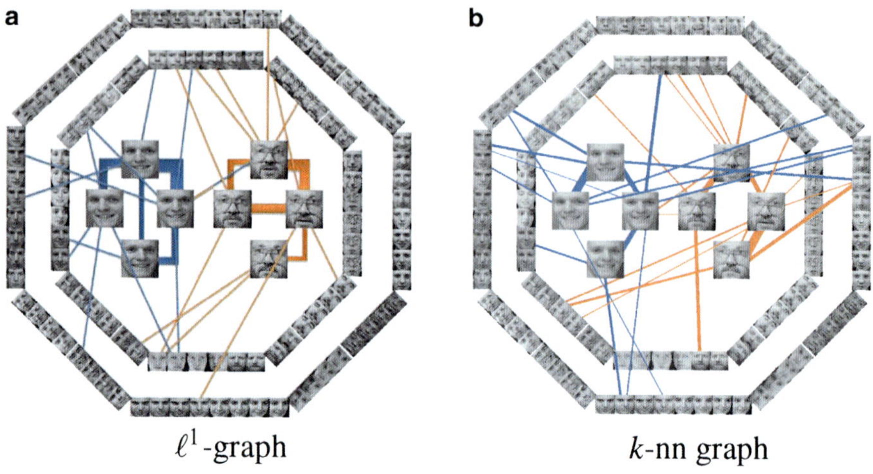

Fig. 3 Visualization comparison of (**a**) the ℓ^1-graph and (**b**) the k-nn graph. The thickness of the edge lines indicates the edge weights. Gaussian kernel is used for the k-nn graph. For ease of display, we only show the graph edges related to the samples from two classes, where in total 30 classes from the YALE-B database are used for graph construction

ℓ^1-norm minimization to recover the sparse representation,

$$\min_{\alpha'} \|\alpha'\|_1, \quad s.t. \quad \|x - B\alpha'\|_2^2 \le \epsilon, \tag{4}$$

This optimization problem is convex and can be readily solved by many optimization toolboxes (http://sparselab.stanford.edu). It has been shown that even with the presence of noise on x, (4) can correctly recover the underlying sparse representation (and thus can denoise x).

2.3 ℓ^1-Graph Construction

The ℓ^1-graph characterizes the overall behavior of the entire sample set by their sparse representations. The construction procedure is formally summarized in Algorithm 1. Figure 3 illustrates the ℓ^1-graph constructed on the data samples from YALE-B face database [14]. As shown, the ℓ^1-graph indeed better connects those kindred samples for faces, and thus it conveys more discriminative information later processing, which aligns well with [21]. Figure 4 depicts another example of ℓ^1-graph on the USPS digit database [12]. Top ten neighbors selected by ℓ^1-graph and k-nn graph are shown, respectively, for the same testing sample. It is clearly that the ℓ^1-graph can select many more kindred examples than the k-nn graph, again demonstrating that the ℓ^1-graph better discovers the underlying data relationships.

Algorithm 1: ℓ^1-graph construction

Inputs: A sample data set denoted by the matrix $X = [x_1, x_2, \ldots, x_N]$, where $x_i \in \mathbb{R}^m$, and parameter ϵ based on the noise level.
for $i = 1 \to N$ **do**
  compute the sparse representation for x_i by

$$\alpha_i = \arg\min_{\alpha} \|\alpha\|_1, \quad \text{s.t.} \quad \|x - B\alpha\|_2^2 \leq \epsilon,$$

  where $B = [x_1, \ldots, x_{i-1}, 0, x_{i+1}, \ldots, x_N, I]$.
end for
Construct an affinity matrix for X by setting $W_{ij} = W_{ji} = \alpha_i(j)$.
Outputs: ℓ^1-graph G $= \{X, W\}$ with the samples set X as graph vertices and W the graph weight matrix.

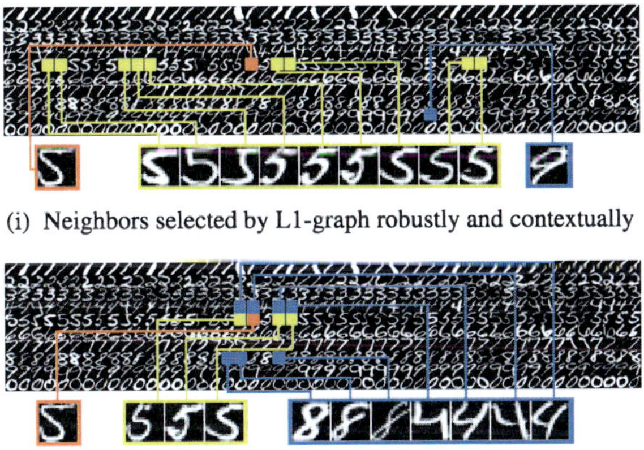

(i) Neighbors selected by L1-graph robustly and contextually

(ii) Pair-distance based k nearest neighbors with many outliers

Fig. 4 Illustration on the positions of a testing sample (*red*), its kindred neighbors (*yellow*), and its inhomogeneous neighbors (*blue*) selected by **(i)** ℓ^1-graph and **(ii)** k-nn graph on samples from the USPS digit database [12]

2.3.1 Discussions

1. Our ℓ^1-graph constructed in (4) is based on the assumption that the feature dimension, m, is high and the data distribution can be well approximated as a union of low-dimensional linear subspaces. Therefore, our ℓ^1-graph would not be applicable for low dimensional, say 2 or 3 dimensions, feature spaces.
2. In our implementation, we found that data normalization, i.e., $\|x_i\|_2 = 1$, is needed for obtaining semantically reasonable coefficients.
3. The k-nn graph is flexible in terms of the selection metric, whose optimality is heavily data dependent. In this work, we simply use the most common Euclidean distance for searching the k nearest neighbors.
4. In many practical applications, it may happen that some data samples in X are highly correlated, which may cause unstable solution for (4). In such cases, we could add a small ℓ^2 norm regularization on the solution as in elastic net [26],

$$\min_{\alpha'} \|\alpha'\|_1 + \beta\|\alpha'\|_2^2, \quad s.t. \quad \|x - B\alpha'\|_2^2 \leq \epsilon. \tag{5}$$

3 Learning with ℓ^1-Graph

An informative graph is critical for those graph-oriented learning algorithms. Similar to classical graphs constructed by k-nn or ϵ-ball method, ℓ^1-graph can be applied with different learning algorithms for various tasks, e.g., data clustering, subspace learning, and semi-supervised learning. In this section, we briefly introduce how to use ℓ^1-graph for these tasks.

3.1 Spectral Clustering

Data clustering is the unsupervised classification of samples into different groups, or more precisely, the partition of samples into subsets, such that the data within each subset are similar to each other. Spectral clustering [19] is among the most popular algorithms for this task, which is summarized as follows.

1. Find the symmetrical similarity matrix by $W = (|W| + |W^{\mathrm{T}}|)/2$.
2. Compute the graph Laplacian matrix $L = D^{-1/2}WD^{-1/2}$, where D is a diagonal matrix with $D_{ii} = \sum_j w_{ij}$.
3. Find eigenvectors $c_1, c_2, \ldots, c_K$ of L corresponding to the K largest eigenvalues, and form the matrix $C = [c_1, c_2, \ldots, c_K]$ by stacking the eigenvectors in columns.
4. Treat each row of C as a point in $\mathbb{R}^K$, and cluster them into K clusters via K-means.
5. Assign x_i to the cluster j if the i-th row of the matrix C is assigned to the cluster j.

3.2 Subspace Learning

In many computer vision and pattern recognition problems, the signal or feature involved is often in a very high-dimensional space, and thus, it is necessary and beneficial to transform the data from the original high-dimensional space to a low-dimensional one for alleviating the curse of dimensionality. A popular dimensionality reduction and feature extraction technique is subspace learning, which has been successfully applied in various recognition applications. Representative subspace learning methods include principal component analysis (PCA) [13], fisher linear discriminant analysis (LDA) [1], locality preserving projection (LPP) [11], and neighborhood preserving embedding (NPE) [10]. As shown by Yan et al. [22], all these works can be formulated into a unified graph embedding framework based on an adjacency graph.

Similar to the graph construction process in locally linear embedding (LLE), our ℓ^1-graph characterizes the neighborhood reconstruction relationships. In LLE, the graph is constructed by reconstructing each datum with its k-nearest neighbors or samples within the ϵ-ball neighborhood. However, both LLE and its linear extension, neighborhood preserving embedding (NPE) [10], rely on a global graph parameter (k or ϵ), which may lead to over-fitting or under-fitting for different data samples due to their uneven distribution in the signal space. In comparison, our ℓ^1-graph can adaptively select the the most relevant neighbors based on the trade-off between reconstruction quality and representation sparsity. Following the same idea of the NPE algorithm, we apply our ℓ^1-graph for subspace learning.

The purpose of subspace learning is to search for a transformation matrix $P \in \mathbb{R}^{m \times d}$ (usually $d \ll m$), where m is the dimension of the original high-dimensional space and d is the dimension of the projected low dimensional space. The transformation matrix P is used to project the original high-dimensional data in into a low-dimensional space while preserving the original data relationships. The ℓ^1-graph discovers the underlying sparse reconstruction relationships among the data samples, and it is desirable to preserve these reconstruction relationships in the low-dimensional feature space. The pursue of such a transformation matrix can be formulated as the following optimization problem

$$\min_{P^{\mathrm{T}}XX^{\mathrm{T}}P=I} \sum_{i=1}^{N} \|P^{\mathrm{T}}x_i - \sum_{j=1}^{N} W_{ij}P^{\mathrm{T}}x_j\|^2, \tag{6}$$

where W_{ij} is determined by our ℓ^1-graph, and the constraints $P^{\mathrm{T}}XX^{\mathrm{T}}P = I$ is to remove the arbitrary scaling factor in the projection. After several easy mathematical manipulations, the above optimization reduces to finding

$$\min_{P^{\mathrm{T}}XX^{\mathrm{T}}P=I} \operatorname{trace}(P^{\mathrm{T}}XMX^{\mathrm{T}}P), \tag{7}$$

where $M = (I - W)^{\mathrm{T}}(I - W)$. This minimization problem is readily to be solved with the generalized eigenvalue decomposition as

$$XMX^{\mathrm{T}}p_{m+1-j} = \lambda_j XX^{\mathrm{T}}p_{m+1-j}, \tag{8}$$

where p_{m+1-j} is the $(m + 1 - j)$-th column vector of the matrix P as well as the eigenvector corresponding to the j-th largest eigenvalue λ_j. The derived transformation matrix is then used for dimensionality reduction, $y_i = P^{\mathrm{T}}x_i$, where y_i is the corresponding low dimensional representation of the sample x_i.

3.3 *Semi-supervised Learning*

As shown in Figs. 1 and 2, our ℓ^1-graph is robust to data noise and better discovers the underlying data relationships compared with conventional graphs based on k-nearest neighbors. These properties recommend ℓ^1-graph as a good candidate for label propagation over the graph. Semi-supervised learning has attracted much attention recently and has been widely used for both regression and classification tasks. The main idea of semi-supervised learning is to utilize unlabeled data for improving the classification and generalization capability on the testing data. Commonly, the unlabeled data is used as an extract regularization term in the traditional supervised learning objective function. In this work, we build our unsupervised ℓ^1-graph over both labeled and unlabeled data to better explore the data relationships in the entire data sampling space. Then this ℓ^1-graph can serve as a regularization for many supervised learning algorithms. In this work, we take Marginal Fisher Analysis (MFA) [22] as an example for the supervised learning part. Similar to the philosophy in [4, 23], the objective for ℓ^1-graph-based semi-supervised learning is defined as

$$\min_P \frac{\gamma S_c(P) + (1-\gamma)\sum_{i=1}^N \|P^\mathrm{T} x_i - \sum_{j=1}^N W_{ij} P^\mathrm{T} x_j\|^2}{S_p(P)},$$

where $\gamma \in (0,1)$ is a threshold for balancing the supervised term and ℓ^1-graph regularization term, and the supervised part is defined as

$$S_c(P) = \sum_i \sum_{j \in N_{k_1}^+(i)} \|P^\mathrm{T} x_i - P^\mathrm{T} x_j\|^2, \tag{9}$$

$$S_p(P) = \sum_i \sum_{(i,j) \in P_{k_2}(l_i)} \|P^\mathrm{T} x_i - P^\mathrm{T} x_j\|^2, \tag{10}$$

where S_c indicates the intra-class compactness, which is represented as the sum of distances between each point and its neighbors of the same class and $N_{k_1}^+(i)$ is the index set of the k_1 nearest neighbors of the sample x_i in the same class, S_p indicates the separability of different classes, which is characterized as the sum of distances between the marginal points and their neighboring points of different classes and $P_{k_2}(l)$ is a set of data pairs that are the k_2 nearest pairs among the set $\{(i,j), l_i = l, l_j \neq l\}$, and W is the weight matrix of the ℓ^1-graph. Similar to (6), the optimum can be obtained via the generalized eigenvalue decomposition method, and the derived projection matrix P is then used for dimensionality reduction and consequent classification.

4 Experiments

In this section, we extensively evaluate the effectiveness of our ℓ^1-graph on three learning tasks, including data clustering, subspace learning, and semi-supervised learning. For comparison purpose, the classical k-nn graph and ϵ-ball graph with different edge weight assignment approaches are implemented as the baselines. For all algorithms based on k-nn and ϵ-ball graphs, the reported results are based on the best tuned parameters k and ϵ.

4.1 Data Sets

For all the experiments, we evaluate on three databases. The USPS handwritten digit database [12] includes 10 classes (0–9 digit characters) and 11,000 samples in total. We randomly select 200 samples for each digit for the experiments, and all of these images are normalized to the size of 32×32 pixels. The forest covertype database (http://kdd.ics.uci.edu/databases/covertype/covertype.data.html/) was collected for predicting forest cover type from cartographic variables. It includes seven classes and 5,81,012 samples in total. We randomly select 100 samples for each type in the following experiments for computational efficiency. The Extended YALE-B database [14] contains 38 individuals and around 64 near frontal images under different illuminations per individual, where the face images are manually cropped and normalized to the size of 32×32 pixels. All the images were taken against a dark homogeneous background with the subjects in an upright and frontal position.

4.2 Spectral Clustering

For spectral clustering, we compare our ℓ^1-graph with Gaussian kernel graph [19], LE-graphs (used in Laplacian Eigenmaps [3] algorithm), LLE-graphs (used in LLE [17]), and also with the K-means clustering based on the low-dimensional representations from principal component analysis (PCA) [13]. Two evaluation metrics, accuracy (AC) and normalized mutual information (NMI) [24], are used for performance evaluations. Suppose L is the ground truth sample label vector and $\hat{L}$ is the label vector predicted from the clustering algorithms, AC is defined as

$$\mathrm{AC} = \frac{1}{N} \sum_{i=1}^{N} \delta(L(i), M_{(L,\hat{L})}(i)) \tag{11}$$

where N denotes the total number of samples, $\delta(a,b)$ is the Kronecker delta function ($\delta(a,b)$ equals to 1 if $a = b$, and 0 otherwise), $M_{(L,\hat{L})}$ is the resulting vector by permuting $\hat{L}$ in order to best match L.[2] The Kuhn–Munkres algorithm can be used to obtain the best mapping [6]. To avoid the clustering label permutation problem, we use normalized mutual information (NMI) as a second metric,

$$\mathrm{NMI}(L,\hat{L}) = \frac{\mathrm{MI}(L,\hat{L})}{\max(H(L), H(\hat{L}))}, \tag{12}$$

where $\mathrm{MI}(L,\hat{L})$ is the mutual information between L and $\hat{L}$,

[2] After clustering, the cluster id assignment for the predicted clusters are arbitrary. In order to compare with the ground truth, we need to find the assignment vector that best matches the ground truth labels.

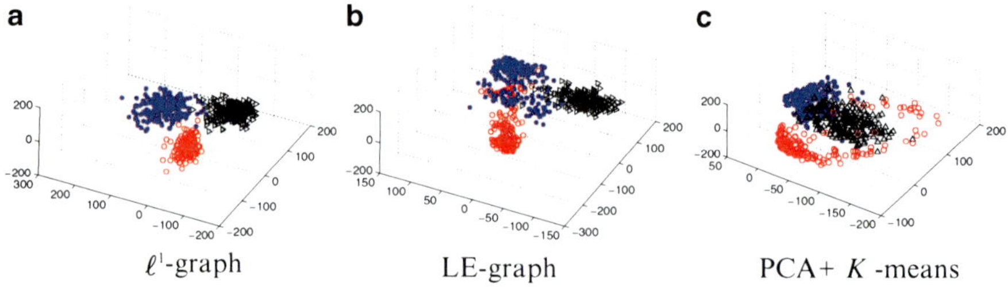

Fig. 5 Visualization of the clustering results with (**a**) ℓ^1-graph, (**b**) LE-graph, and (**c**) PCA for three clusters (handwritten digits 1, 2 and 3 in the USPS database)

$$\mathrm{MI}(L,\hat{L}) = \sum_{l\in L}\sum_{\hat{l}\in\hat{L}} p(l,\hat{l})\log\left(\frac{p(l,\hat{l})}{p(l)p(\hat{l})}\right), \tag{13}$$

and $H(L)$ and $h(\hat{L})$ denote the entropies of L and $\hat{L}$, respectively. It is obvious that NMI takes values in $[0,1]$. Unlike AC, NMI is invariant with the permutation of labels and, therefore, is more efficient to compute.

The visualization of the clustering results (digit characters 1–3 from the USPS database) for spectral clustering based on ℓ^1-graph and LE-graph and K-means are plotted in Fig. 5, where different digits are denoted by different colors. The coordinates of the points in (a) and (b) are obtained from the eigenvalue decomposition in the third step of spectral clustering 3.1. It is clear that our ℓ^1-graph can better separate different classes. Quantitative comparison results on clustering are list in Tables 1–3 for the three databases, respectively. The cluster number K indicates the number of classes used from the database as the data samples for clustering. From the listed results, three observations can be made: (1) the clustering results from ℓ^1-graph-based spectral clustering algorithm are consistently much better than those baselines in terms of both metrics; (2) Spectral clustering based on k-nn-based LLE-graph is relatively more stable compared with other baselines; and (3) ϵ-ball-based algorithms show to be generally inferior, in both accuracy and robustness, than the corresponding k-nn-based graphs, and thus for the subsequent experiments, we only report the results from k-nn-based graphs. All the results listed in the tables are from the best tuned algorithmic parameters, e.g., kernel parameter for G-graph, the number of neighbors k and neighborhood diameter ϵ for LE-graphs and LLE-graphs, and the retained PCA feature dimension for K-means.

Table 1 Clustering accuracies, measured by normalized mutual information (NMI) and accuracy (AC), for spectral clustering algorithms based on ℓ^1-graph, Gaussian kernel graph (G-graph), LE-graphs, and LLE-graphs, as well as PCA + K-means on the USPS digit database

USPS				LE-graph		LLE-graph		
Cluster #	Metric	ℓ^1-graph	G-graph	k-nn	ϵ-ball	k-nn	ϵ-ball	PCA+K-means
$K = 2$	NMI	*1.000*	0.672(110)	0.858(7)	0.627(3)	0.636(5)	0.717(4)	0.608(10)
	AC	*1.000*	0.922	0.943	0.918	0.917	0.932	0.905
$K = 4$	NMI	*0.977*	0.498(155)	0.693(16)	0.540(6)	0.606(5)	0.465(7)	0.621(20)
	AC	*0.994*	0.663	0.853	0.735	0.777	0.668	0.825
$K = 6$	NMI	*0.972*	0.370(120)	0.682(5)	0.456(6)	0.587(5)	0.427(9)	0.507(4)
	AC	*0.991*	0.471	0.739	0.594	0.670	0.556	0.626
$K = 8$	NMI	*0.945*	0.358(150)	0.568(7)	0.371(4)	0.544(12)	0.404(7)	0.462(17)
	AC	*0.981*	0.423	0.673	0.453	0.598	0.499	0.552
$K = 10$	NMI	*0.898*	0.346(80)	0.564(6)	0.424(5)	0.552(16)	0.391(4)	0.421(10)
	AC	*0.873*	0.386	0.578	0.478	0.537	0.439	0.433

Table 2 Spectral clustering results on the forest covertype database

COV				LE graph		LLE-graph		
Cluster #	Metric	ℓ^1-graph	G-graph	k-nn	ϵ-ball	k-nn	ϵ-ball	PCA+K-means
$K = 3$	NMI	*0.792*	0.651(220)	0.554(16)	0.419(6)	0.642(20)	0.475(6)	0.555(5)
	AC	*0.903*	0.767	0.697	0.611	0.813	0.650	0.707
$K = 4$	NMI	*0.706*	0.585(145)	0.533(13)	0.534(6)	0.622(20)	0.403(5)	0.522(13)
	AC	*0.813*	0.680	0.608	0.613	0.782	0.519	0.553
$K = 5$	NMI	*0.623*	0.561(240)	0.515(12)	0.451(5)	0.556(10)	0.393(7)	0.454(15)
	AC	*0.662*	0.584	0.541	0.506	0.604	0.448	0.486
$K = 6$	NMI	*0.664*	0.562(200)	0.545(6)	0.482(6)	0.602(20)	0.465(7)	0.528(8)
	AC	*0.693*	0.585	0.564	0.523	0.632	0.509	0.547
$K = 7$	NMI	*0.763*	0.621(130)	0.593(9)	0.452(6)	0.603(11)	0.319(6)	0.602(17)
	AC	*0.795*	0.642	0.629	0.498	0.634	0.394	0.631

Table 3 Spectral clustering results on the Extended YALE-B database

YALE-B				LE-graph		LLE-graph		
Cluster #	Metric	ℓ^1-graph	G-graph	k-nn	ϵ-ball	k-nn	ϵ-ball	PCA+K-means
$K = 10$	NMI	*0.738*	0.07(220)	0.420(4)	0.354(16)	0.404(3)	0.302(3)	0.255(180)
	AC	*0.758*	0.175	0.453	0.413	0.450	0.383	0.302
$K = 15$	NMI	*0.759*	0.08(380)	0.494(4)	0.475(20)	0.438(5)	0.261(5)	0.205(110)
	AC	*0.762*	0.132	0.464	0.494	0.440	0.257	0.226
$K = 20$	NMI	*0.786*	0.08(290)	0.492(2)	0.450(18)	0.454(4)	0.269(3)	0.243(110)
	AC	*0.793*	0.113	0.478	0.445	0.418	0.241	0.238
$K = 30$	NMI	*0.803*	0.09(50)	0.507(2)	0.417(24)	0.459(7)	0.283(4)	0.194(170)
	AC	*0.821*	0.088	0.459	0.383	0.410	0.236	0.169
$K = 38$	NMI	*0.776*	0.11(50)	0.497(2)	0.485(21)	0.473(8)	0.319(4)	0.165(190)
	AC	*0.785*	0.081	0.443	0.445	0.408	0.248	0.138

Note that the G-graph performs extremely bad in this case, possibly due to dramatic illumination changes on this database

4.3 Subspace Learning

In this section, classification experiments based on subspace learning are conducted on the above three databases. To make the comparison fair, for all the evaluated algorithms we first apply PCA as a preprocessing step for denoising by retaining 98% of the energy. To extensively evaluate the classification performances, on the USPS database, we randomly sampled 10, 20, 30, and 40 images from each digit for training. Similarly, on the forest covertype database, we randomly sampled 5, 10, 15, and 20 samples from each class for training, and on the Extended YALE-B database, we randomly sampled 10, 20, 30, 40, and 50 training images for each individual for training. All the remaining data samples are used for testing. We use the classification error rate to measure the performance,

$$\text{err} = 1 - \frac{\sum_{i=1}^{N_t} \delta(\hat{y}_i, y_i)}{N_t} \tag{14}$$

where $\hat{y}_i$ is the predicted data label and y_i is the ground truth label, N_t is the total number of testing samples, and $\delta(\hat{y}_i, y_i)$ is the Kronecker delta function. The best performances of each algorithm over the tuned graph parameters and feature dimensions are reported. The popular unsupervised subspace learning algorithms PCA, NPE and LPP,[3] and the supervised algorithm Fisherfaces [1] are evaluated for comparison with our unsupervised subspace learning based on our ℓ^1-graph. For LPP, we use the cosine metric in graph construction for better performance. The detailed comparison results on classification are listed in Tables 4–6 for the three databases. From these results, we observe that our ℓ^1-graph based subspace learning algorithm is much better than all the other evaluated unsupervised learning algorithms. On the forest covertype and Extended YALE-B databases, our unsupervised ℓ^1-graph even performs better than the supervised algorithm Fisherfaces. For all the classification experiments, we simply use the classical nearest neighbor classifier [1,10,11] for fair comparisons with the literature algorithms. The visualization of learned subspaces based on ℓ^1-graph and those based on PCA, LPP, and NPE are shown in Fig. 6, from which we see that the PCA bases are most similar to real faces as PCA is motivated for data reconstruction.

4.4 Semi-supervised Learning

We again use the above three databases for evaluating the effectiveness of the semi-supervised algorithm based on ℓ^1-graph by comparing with algorithms based on Gaussian-kernel graph, LE-graph, and LLE-graph. For all the semi-supervised

---

[3]Here we use the unsupervised version of NPE and LPP for fair comparison.

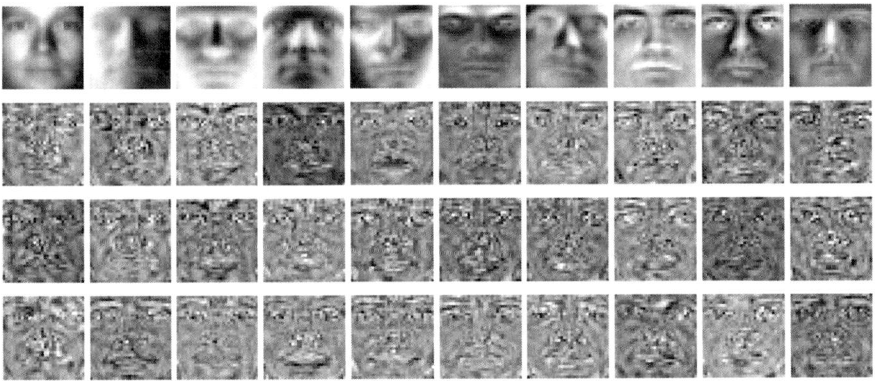

Fig. 6 Visualization of the learned subspaces. They are the first ten basis vectors of (**a**) PCA, (**b**) NPE, (**c**) LPP, and (**d**) ℓ^1-graph calculated from the face images in YALE-B database

Table 4 USPS digit recognition error rates (%) for different subspace learning algorithms

USPS	Unsupervised				Supervised
Train #	PCA	NPE	LPP	ℓ^1-graph	Fisherfaces
10	37.21(17)	33.21(33)	30.54(19)	21.91(13)	*15.82*(9)
20	30.59(26)	27.97(22)	26.12(19)	18.11(13)	*13.60*(9)
30	26.67(29)	23.46(42)	23.19(26)	16.81(15)	*13.59*(7)
40	23.25(25)	20.86(18)	19.92(32)	14.35(19)	*12.29*(7)

The numbers in the parentheses are the feature dimensions retained with the best accuracies

Table 5 Forest cover recognition error rates (%) for different subspace learning algorithms

COV	Unsupervised				Supervised
Train #	PCA	NPE	LPP	ℓ^1-graph	Fisherfaces
5	33.23(17)	28.80(6)	35.09(12)	*23.36*(6)	23.81(6)
10	27.29(18)	25.56(11)	27.30(16)	*19.76*(15)	21.17(4)
15	23.75(14)	22.69(16)	23.26(34)	*17.85*(7)	19.57(6)
20	21.03(29)	20.10(10)	20.75(34)	*16.44*(6)	18.09(6)

learning algorithms, the supervised part is based on the Marginal Fisher Analysis (MFA) [22] algorithm. For fair comparisons, the parameters k_1 and k_2 in MFA, and γ are tuned for best performance. The detailed comparisons for different semi-supervised leaning algorithms, the original supervised MFA algorithm and the baseline PCA, are shown in Tables 7–9. The semi-supervised learning based on our ℓ^1-graph generally achieves the highest classification accuracy compared with semi-supervised learning based on other graphs and outperforms the supervised learning counterparts without considering the unlabeled data.

Table 6 Face recognition error rates (%) for different subspace learning algorithms on the Extended YALE-B database

YALE-B	Unsupervised				Supervised
Train #	PCA	NPE	LPP	ℓ^1-graph	Fisherfaces
10	44.41(268)	23.41(419)	24.61(234)	*14.26*(112)	*13.92*(37)
20	27.17(263)	14.62(317)	14.76(281)	*5.30*(118)	9.46(37)
30	20.11(254)	9.40(485)	8.65(246)	*3.36*(254)	12.45(34)
40	16.98(200)	5.84(506)	5.30(263)	*1.93*(143)	3.79(37)
50	12.68(366)	3.78(488)	3.02(296)	*0.75*(275)	1.64(37)

Table 7 USPS digit recognition error rates (%) for different semi-supervised, supervised, and unsupervised learning algorithms

USPS	Semi-supervised			Supervised	Unsupervised
Train #	ℓ^1-graph	LLE-graph	LE-graph	MFA [22]	PCA
10	*25.11*(33)	34.63(9)	30.74(33)	34.63(9)	37.21(17)
20	*26.94*(41)	41.38(39)	30.39(41)	41.38(39)	30.59(26)
30	*23.25*(49)	36.55(49)	27.50(49)	44.34(47)	26.67(29)
40	*19.17*(83)	30.28(83)	23.55(83)	35.95(83)	23.35(25)

The numbers in the parentheses are the feature dimensions that give the best accuracies

Table 8 Forest cover recognition error rates (%) for different semi-supervised, supervised, and unsupervised learning algorithms

COV	Semi-supervised			Supervised	Unsupervised
Train #	ℓ^1-graph	LLE-graph	LE-graph	MFA [22]	PCA
5	*22.50*(9)	29.89(5)	25.81(7)	29.89(5)	33.23(17)
10	*17.45*(10)	24.93(10)	22.74(8)	24.93(10)	27.29(18)
20	*15.00*(8)	19.17(10)	17.38(9)	19.17(10)	23.75(14)
30	*12.26*(8)	15.32(8)	13.81(10)	16.40(8)	21.03(29)

Table 9 Face recognition error rates (%) for different semi-supervised, supervised, and unsupervised learning algorithms on the Extended YALE-B database

YALE-B	Semi-supervised			Supervised	Unsupervised
Train #	ℓ^1-graph	LLE-graph	LE-graph	MFA [22]	PCA
5	*21.63*(51)	33.47(51)	33.47(51)	33.47(51)	61.34(176)
10	*9.56*(61)	18.39(33)	18.39(33)	18.39(33)	44.41(268)
20	*5.05*(57)	14.30(29)	11.26(53)	14.30(29)	27.17(263)
30	*2.92*(73)	9.15(70)	7.37(71)	11.06(70)	20.11(254)

5 Summary

In this chapter, we propose a new graph construction procedure based on the sparse representation of each individual datum with respect to the remaining data samples by ℓ^1-norm minimization, and thus the new graph is called ℓ^1-graph. The ℓ^1-graph is robust to noise, does not have the local neighborhood assumption, and is especially good at modeling high-dimensional feature spaces, where, empirically, many data distributions can be well approximated by a union of much lower dimensional linear subspaces. By seeking a sparse representation of each datum in terms of the remaining data samples, we can select the low-dimensional linear subspace it lies in. Therefore, the ℓ^1-graph conveys greater discriminating power compared with the conventional graphs based on k-nearest neighbors and ϵ-ball. We apply this ℓ^1-graph to three graph-oriented machine learning tasks, including spectral clustering, subspace learning, and semi-supervised learning. In all cases, our new ℓ^1-graph significantly outperforms the corresponding baselines on three different databases. As a generic graph, the new ℓ^1-graph can be applied in many other tasks as well, e.g., transfer learning and label propagation, which will be left for future explorations.

Acknowledgements This work is supported by U.S. ARL and ARO under grant number W911NF-09-1-0383.

References

1. Belhumeur P, Hespanda J, Kiregeman D (1997) Eigenfaces vs. fisherfaces: Recognition using class specific linear projection. IEEE Trans Pattern Anal Mach Intell 19(7):711–720
2. Belkin M, Matveeva I, Niyogi P (2004) Regularization and semi-supervised learning on large 366 graphs. In: International conference on learning theory, springer, 3120:624–638
3. Belkin M, Niyogi P (2002) Laplacian eigenmaps for dimensionality reduction and data representation. Neural Comput 15(6):1373–1396
4. Cai D, He X, Han J (2007) Semi-supervised discriminant analysis. In: IEEE international conference on computer vision, pp 1–7
5. Candès EJ (2006) Compressive sampling. In: Proceedings of the international congress of mathematicians, 3:1433–1452.
6. Chen S, Donoho D, Saunders M (2001) Atomic decomposition by basis pursuit. Soc Indust Appl Math Rev 43(1):129–159
7. Donoho D (2004) For most large underdetermined systems of linear equations the minimal ℓ^1 norm solution is also the sparsest solution. Commun Pure Appl Math 59(7):797–829
8. Elad M, Aharon M (2006) Image denoising via sparse and redundant representations over learned dictionaries. IEEE Trans Image Process 15:3736–3745
9. Hastie T, Tibshirani R, Friedman J (2008) The elements of statistical learning, 2nd edn. Statistics, Springer, Berlin
10. He X, Cai D, Yan S, Zhang H (2005) Neighborhood preserving embedding. In: IEEE international conference on computer vision, vol 2, pp 1208–1213
11. He X, Niyogi P (2003) Locality preserving projections. In: Advances in neural information processing systems, vol 16, pp 585–591

12. Hull J (1994) A database for handwritten text recognition research. IEEE Trans Pattern Anal Mach Intell 16(5):550–554
13. Joliffe I (1986) Principal component analysis. Springer, Berlin, pp 1580–1584
14. Lee K, Ho J, Kriegman D (2005) Acquiring linear subspaces for face recognition under variable lighting. IEEE Trans Pattern Anal Mach Intell 27(5):684–698
15. Meinshausen N, Bühlmann P (2006) High-dimensional graphs and variable selection with the lasso. Ann Stat 34(3):1436–1462
16. Olshausen B, Field D (1998) Sparse coding with an overcomplete basis set: A strategy employed by v1? Vision Res 37(23):3311–3325
17. Roweis S, Saul L (2000) Nonlinear dimensionality reduction by locally linear embedding. Science 290(5500):2323–2326
18. Rubinstein R, Bruckstein AM, Elad M (2010) Dictionaries for sparse representation modeling. In: Proceedings of IEEE, 98:1045–1057
19. Shi J, Malik J (2000) Normalized cuts and image segmentation. IEEE Trans Pattern Anal Mach Intell 22(8):888–905
20. Tenenbaum J, Silva V, Langford J (2000) A global geometric framework for nonlinear dimensionality reduction. Science 290(5500):2319–2323
21. Wright J, Ganesh A, Yang A, Ma Y (2009) Robust face recognition via sparse representation. IEEE Trans Pattern Anal Mach Intell, 31(2):210–227
22. Yan S, Xu D, Zhang B, Yang Q, Zhang H, Lin S (2007) Graph embedding and extensions: A general framework for dimensionality reduction. IEEE Trans Pattern Anal Mach Intell 29(1):40–51
23. Yang J, Yan S, Huang TS (2009) Ubiquitously supervised subspace learning. IEEE Trans Image Process 18(2):241–249
24. Zheng X, Cai D, He X, Ma W, Lin X (2004) Locality preserving clustering for image database. In: ACM international conference on multimedia, pp 885–891
25. Zhu X, Ghahramani Z, Lafferty J (2003) Semi-supervised learning using gaussian fields and harmonic functions. In: International conference on machine learning, pp 912–919
26. Zou H, Hastie T (2005) Regularization and variable selection via the elastic net. J Roy Stat Soc Ser B 67:301–320

Graph-Embedding Discriminant Analysis on Riemannian Manifolds for Visual Recognition

Sareh Shirazi, Azadeh Alavi, Mehrtash T. Harandi, and Brian C. Lovell

1 Introduction

Recently, several studies have utilised non-Euclidean geometry to address several computer vision problems including object tracking [17], characterising the diffusion of water molecules as in diffusion tensor imaging [24], face recognition [23,31], human re-identification [4], texture classification [16], pedestrian detection [39] and action recognition [22,43].

Among various Riemannian manifolds, structures induced from subspaces and symmetric positive definite matrices have been shown to be quite useful in computer vision. Subspaces form a non-Euclidean and curved Riemannian manifold known as a Grassmann manifold and are able to accommodate the effects of various image variations. For example, a widely used approximation for photometric invariance, under conditions of no shadowing and Lambertian reflectance, is a linear subspace [1]. Moreover, subspaces can capture the dynamic properties of videos [36].

In computer vision and machine learning disciplines, trace of covariance and kernel matrices can be seen in many ways. One notable example is the covariance descriptor introduced by Tuzel et al. [38]. Covariance descriptor is a structured representation and comes with several advantages over traditional descriptors. A single covariance matrix extracted from a region (2D regions in images or 3D in videos) is usually enough to match the region in different views and poses. Furthermore, the covariance matrix proposes a natural way of fusing multiple features which might be correlated. The diagonal entries of the covariance matrix

---

The first and second authors contributed equally to the chapter.

S. Shirazi • A. Alavi • M.T. Harandi (✉) • B.C. Lovell
NICTA, PO Box 6020, St Lucia, QLD 4067, Australia

The University of Queensland, School of ITEE, QLD 4072, Australia
e-mail: Sareh.Abolahrari@nicta.com.au; Azadeh.Alavi@nicta.com.au; Mehrtash.Harandi@nicta.com.au; lovell@itee.uq.edu.au

Y. Fu and Y. Ma (eds.), *Graph Embedding for Pattern Analysis*,
DOI 10.1007/978-1-4614-4457-2_7,

represent the variance of each feature and the non-diagonal entries represent the correlations. The noise corrupting individual samples are largely filtered out with an average filter during covariance computation. Nevertheless, the space of covariance/correlation/kernel matrices (more generally symmetric positive definite matrices) is not Euclidean; it is a Riemannian manifold of negative curvature.

Several studies show that better performance can be achieved when the geometry of the Riemannian spaces is considered to its uttermost level [14, 15, 21, 34, 36, 39]. Exploiting the geometry of space is especially important in the computer vision discipline since the notion of Euclidean space is not well supported for high-dimensional vision data (c.f., think how inaccurate distances could be on a sphere when the geometry is not considered). Inference on manifold spaces can be achieved by embedding the manifolds in higher dimensional Euclidean spaces, which can be considered as flattening the manifolds. In the literature, the most popular choice for embedding manifolds is through considering tangent spaces [36, 39]. Two bold examples are the pedestrian detection system by Tuzel et al. [39] and non-linear mean shift [10] by Subbarao et al. [34]. Nevertheless, flattening the manifold through tangent spaces is not without drawbacks. For example, only distances between points to the tangent pole are equal to true geodesic distances. This is restrictive and may lead to inaccurate modelling.

Instead of using tangent spaces to do inference on manifolds, we propose to embed Riemannian manifolds into Reproducing Kernel Hilbert Spaces (RKHS). This in turn opens the door for employing many kernel-based machine learning algorithms [29]. As such, we tackle the problem of Discriminant Analysis (DA) on Riemannian manifolds through RKHS space and propose a graph-based local DA that utilises both within-class and between-class similarity graphs to characterise intra-class compactness and inter-class separability, respectively. See Fig. 1 for a conceptual example. Our graph-based DA is inspired by findings in the Euclidean space that explain why the conventional formalism of DA is not optimal when data comprises outliers and multi-modal classes and contains outliers. Our experiments for several recognition problems show that considerable gains in discrimination accuracy can be obtained by exploiting the geometrical structure and local information on Riemannian manifolds.

2 Riemannian Geometry

In this study we are interested in two types of Riemannian manifolds, namely the Grassmann manifolds and the manifolds of Symmetric, Positive Definite matrices (SPD). Manifolds are smooth, curved surfaces embedded in higher dimensional Euclidean spaces and formally defined as follows:

Definition 1. A topological space $\mathcal{M}$ is called a manifold if:

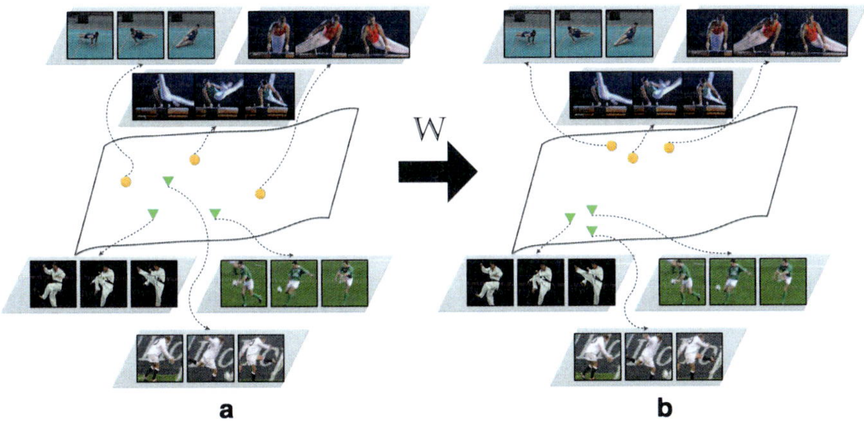

Fig. 1 A conceptual illustration of the proposed approach. **(a)** Actions can be modelled as points on the manifold $\mathcal{M}$ by linear subspaces. In this figure, two types of actions (*"kicking"* and *"swinging"*) are shown. Having a proper geodesic distance between the points on the manifold, it is possible to convert the action recognition problem into a point-to-point classification problem. **(b)** By having a kernel in hand, points on the manifold can be mapped into an optimised RKHS where not only certain local properties have been retained but also the discriminatory power between classes has been increased

- $\mathcal{M}$ is Hausdorff,[2] i.e. every pair X, Y can be separated by two disjoint open sets.
- $\mathcal{M}$ is locally Euclidean, that is, for every $X \in \mathcal{M}$ there exists an open set $U \subset \mathcal{M}$ with $X \in \mathcal{U}$ and an open set $V \subset \mathbb{R}^n$ with a homeomorphism $\varphi : U \to V$.

Riemannian manifolds are analytical manifolds endowed with a distance measure which allows the measurement of similarity or dissimilarity (close or distant) of points. The geodesic distance between two points $X, Y \in \mathcal{M}$, denoted by $d_g(X, Y)$, is defined as the minimum length over all possible smooth curves between X and Y. A geodesic curve is a curve that locally minimises the distance between points.

Symmetric positive definite matrices of size $D \times D$, e.g. non-singular covariance matrices, form a connected Riemannian manifold (Sym_D^+). The geodesic distance between two points X and Y on Sym_D^+ can be computed as

$$d_G(X, Y) = \operatorname{trace}\left\{\log^2\left(X^{-\frac{1}{2}} Y X^{-\frac{1}{2}}\right)\right\} \tag{1}$$

In (1), $\log(\cdot)$ is matrix logarithm operator and can be computed through singular value decomposition (SVD). More specifically, let $X = U \Sigma U^T$ be the SVD of the symmetric matrix X, then

$$\log(X) = U \log(\Sigma) U^T \tag{2}$$

---

[2] In a Hausdorff space, distinct points have disjoint neighbourhoods. This property is important to establish the notion of a differential manifold, as it guarantees that convergent sequences have a single limit point.

where $\log(\Sigma)$ is a diagonal matrices where the diagonal elements are equivalent to the logarithms of the diagonal elements of matrix Σ.

To formally define a Grassmann manifold and its geometry, we need to define the quotient space of a manifold. A quotient space of a manifold, intuitively speaking, is the result of "gluing together" certain points of the manifold. Formally, given $\sim_\psi$ as an equivalence relation on $\mathcal{M}$, the quotient space $\Upsilon = \mathcal{M}/\sim_\psi$ is defined to be the set of equivalence classes of elements of $\mathcal{M}$, i.e. $\Upsilon = \{[X] : X \in \mathcal{M}\} = \{[Y \in \mathcal{M} : Y \sim_\psi X] : X \in \mathcal{M}\}$.

Definition 2. A Grassmann manifold is a quotient space of the special orthogonal group[3] $\mathrm{SO}(n)$ and is defined as a set of p-dimensional linear subspaces of $\mathbb{R}^n$.

In practice an element X of $\mathcal{G}n,p$ is represented by an orthonormal basis as a $n \times p$ matrix, i.e. $X^\mathrm{T}X = I_p$. The geodesic distance between two points on the Grassmann manifold can be computed as:

$$d_\mathrm{G}(X,Y) = \|\Theta\|_2 \tag{3}$$

where $\Theta = [\theta_1, \theta_2, \ldots, \theta_p]$ is the principal angle vector, i.e.:

$$\cos(\theta_i) = \max_{\mathbf{x}_i \in X,\ \mathbf{y}_j \in Y} \mathbf{x}_i^\mathrm{T}\mathbf{y}_j \tag{4}$$

subject to $\mathbf{x}_i^\mathrm{T}\mathbf{x}_i = \mathbf{y}_i^\mathrm{T}\mathbf{y}_i = 1$, $\mathbf{x}_i^\mathrm{T}\mathbf{x}_j = \mathbf{y}_i^\mathrm{T}\mathbf{y}_j = 0$, $i \neq j$. The principal angles have the property of $\theta_i \in [0, \pi/2]$ and can be computed through SVD of $X^\mathrm{T}Y$ [11].

3 Kernel Analysis on Riemannian Manifolds

In this section, we first overview the essentials of kernel analysis on Riemannian manifolds, followed by elucidating graph embedding DA in Sect. 3.2 and how to accomplish classification in Sect. 3.3.

3.1 Background

Given a set of input/output data $\{(X_1,l_1),(X_2,l_2),\ldots,(X_N,l_N)\}$, where $X_i \in \mathcal{M}$ is a Riemannian point and $l_i \in \{1,2,\ldots,C\}$ is the corresponding class label, we are interested in optimisation problems in the form of Tikhonov regularisation [35]:

$$\max\{\mathbf{J}(\langle \mathbb{W}, \Phi(X_1)\rangle, \ldots, \langle \mathbb{W}, \Phi(X_N)\rangle, l_1, \ldots, l_N) + \lambda\Omega(\mathbb{W}) : \mathbb{W} \in \mathcal{H}\} \tag{5}$$

[3] Special orthogonal group $\mathrm{SO}(n)$ is the space of all $n \times n$ orthogonal matrices with the determinant $+1$. It is not a vector space but a differentiable manifold, i.e. it can be locally approximated by subsets of a Euclidean space.

Here, $\mathcal{H}$ is a prescribed Hilbert space of dimension h (h could be infinity) equipped with an inner product $\langle \cdot, \cdot \rangle$, $\Omega : \mathcal{H} \to \mathbb{R}$ is a regulariser, $\mathbf{J} : \left(\mathbb{R}^h\right)^N \times \mathcal{Y}^N \to \mathbb{R}$ is a cost function. For certain choices of the regulariser, solving (5) reduces to identifying N parameters and not the dimension of $\mathcal{H}$. This is more formally explained by the representer theorem [29] which states that the solution $\widehat{\mathbb{W}}$ of (5) is a linear combination of the inputs when the regulariser is the square of the Hilbert space norm. For vector Hilbert spaces, this result is simple to prove and dates back to 1970s [18]. Argyriou et al. [2] showed that the representer theorem holds for matrix Hilbert spaces as well.

Implicitly embedding Riemannian manifolds into RKHS is achieved through a Riemannian kernel. A function $k : \mathcal{M} \times \mathcal{M} \to \mathbb{R}^+$ is a Riemannian kernel provided that it is positive definite and well defined for all $X \in \mathcal{M}$.

For the Grassmann manifold $X_i \in \mathcal{G}_{D,m}$, the latter criterion means that the kernel should be invariant to various representations of the subspaces, i.e. $k(X,Y) = k(XQ_1, YQ_2)$, $\forall\ Q_1, Q_2 \in \mathbb{O}(m)$, where $\mathbb{O}(m)$ indicates orthonormal matrices of order m [14]. The repertoire of Grassmann kernels includes Binet-Cauchy [41] and projection kernels [14]. Furthermore, the first canonical correlation of two subspaces forms a pseudo kernel[4] on Grassmann manifolds [15]. The three kernels are, respectively, shown below:

$$k_{\mathrm{BC}}(XX,Y) = \det\left(X^{\mathrm{T}} Y Y^{\mathrm{T}} X\right) \tag{6}$$

$$k_{\mathtt{proj}}(X,Y) = \mathrm{Tr}\left(X^{\mathrm{T}} Y Y^{\mathrm{T}} X\right) \tag{7}$$

$$k_{\mathrm{CC}s}(X,Y) = \max_{x \in X,\ y \in Y} x^{\mathrm{T}} y \tag{8}$$

For the Sym_D^+, in [16] a pseudo kernel based on geodesic distances was devised as followed:

$$k_{\mathrm{R}}(X,Y) = \exp\{-\sigma^{-1} d_{\mathrm{G}}(X,Y)\} \tag{9}$$

where $d_{\mathrm{G}}(X,Y)$ is obtained using (1). Very recently, Sra et al. introduced the Stein kernel using *Bregman* matrix divergence as follows [33]:

$$k(X,Y) = \mathrm{e}^{-\sigma S(X,Y)} = 2^{d\sigma} \frac{\sqrt{\det(X)^{\sigma} \det(Y)^{\sigma}}}{\det(X+Y)^{\sigma}} \tag{10}$$

In (10), $S(X,Y)$ is the symmetric Stein divergence and defined as:

$$S(X,Y) \triangleq \log\left(\det\left(\frac{X+Y}{2}\right)\right) - \frac{1}{2}\log\left(\det(XY)\right), \text{ for } X, Y \succ 0 \tag{11}$$

[4] A pseudo kernel is a function where the positive definiteness is not guaranteed to be satisfied for whole range of the function's parameters. Nevertheless, it is possible to convert a pseudo kernel into sa true kernel, as discussed, for example, in [9].

3.2 Graph Embedding Discriminant Analysis on Riemannian Manifolds

A graph (V,G) in our context refers to a collection of vertices or nodes, V, and a collection of edges that connect pairs of vertices. We note that G is a symmetric matrix with elements describing the similarity between pairs of vertices. Moreover, the diagonal matrix D and the Laplacian matrix L of a graph are defined as $L = D - G$, with the diagonal elements of D obtained as $D(i,i) = \sum_j G(i,j)$.

Given N labelled points $\mathbb{X} = \{(X_i, l_i)\}_{i=1}^N$ from the underlying Riemannian manifold $\mathcal{M}$, where $X_i \in \mathbb{R}^{D\times m}$ and $l_i \in \{1,2,\dots,C\}$, with C denoting the number of classes, the local geometrical structure of $\mathcal{M}$ can be modelled by building a within-class similarity graph G_w and a between-class similarity graph G_b. The simplest forms of G_w and G_b are based on the nearest neighbour graphs defined below:

$$G_w(i,j) = \begin{cases} 1, \text{ if } X_i \in N_w(X_j) \text{ or } X_j \in N_w(X_i) \\ 0, \qquad \text{otherwise} \end{cases} \tag{12}$$

$$G_b(i,j) = \begin{cases} 1, \text{ if } X_i \in N_b(X_j) \text{ or } X_j \in N_b(X_i) \\ 0, \qquad \text{otherwise} \end{cases} \tag{13}$$

In (12), $N_w(X_i)$ is the set of v_w neighbours $\{X_i^1, X_i^2, \dots, X_i^v\}$, sharing the same label as l_i. Similarly in (13), $N_b(X_i)$ contains v_b neighbours having different labels. We note that more complex similarity graphs, like heat kernel graphs, can also be used to encode distances between points on Riemannian manifolds [27].

Our aim is to simultaneously maximise a measure of discriminatory power and preserve the geometry of points. This can be formalised by finding $\mathbb{W} : \Phi(X_i) \to Y_i$ such that the connected points of G_w are placed as close as possible, while the connected points of G_b are moved as far as possible. As such, a mapping must be sought by optimising the following two objective functions:

$$f_1 = \min \frac{1}{2} \sum_{i,j} \|Y_i - Y_j\|^2 G_w(i,j) \tag{14}$$

$$f_2 = \max \frac{1}{2} \sum_{i,j} \|Y_i - Y_j\|^2 G_b(i,j) \tag{15}$$

Equation (14) punishes neighbours in the same class if they are mapped far away, while (15) punishes points of different classes if they are mapped close together.

According to the representer theorem [29], the solution $\mathbb{W} = [\Gamma_1 | \Gamma_2 | \cdots | \Gamma_r]$, can be expressed as a linear combination of data points, i.e. $\Gamma_i = \sum_{j=1}^N w_{i,j} \phi(X_j)$. More specifically:

$$Y_i = \left(\langle \Gamma_1, \phi(X_i) \rangle, \langle \Gamma_2, \phi(X_i) \rangle, \dots, \langle \Gamma_r, \phi(X_i) \rangle\right)^{\mathrm{T}} \tag{16}$$

Since $\langle \Gamma_l, \phi(X_i) \rangle = \sum_{j=1}^{N} w_{l,j} \operatorname{Tr}\left(\phi(X_j)^{\mathrm{T}} \phi(X_i)\right) = \sum_{j=1}^{N} w_{l,j} k(X_j, X_i)$, $Y_i = W^{\mathrm{T}} K_i$, with $K_i = (k(X_i,X_1), k(X_i,X_2), \ldots, k(X_i,X_N))^{\mathrm{T}}$ and

$$W = \begin{pmatrix} w_{1,1} & w_{1,2} & \cdots & w_{1,r} \\ w_{2,1} & w_{2,2} & \cdots & w_{2,r} \\ \vdots & \vdots & \vdots & \vdots \\ w_{N,1} & w_{N,2} & \cdots & w_{N,r} \end{pmatrix}$$

Plugging this back into (14) results in:

$$\begin{aligned} &\frac{1}{2} \sum_{i,j} \|Y_i - Y_j\|^2 G_w(i,j) \\ &= \frac{1}{2} \sum_{i,j} \|W^{\mathrm{T}} K_i - W^{\mathrm{T}} K_j\|^2 G_w(i,j) \\ &- \sum_{i} \operatorname{Tr}\left(W^{\mathrm{T}} K_i K_i^{\mathrm{T}} W\right) G_w(i,i) - \sum_{i,j} \operatorname{Tr}\left(W^{\mathrm{T}} K_j K_i^{\mathrm{T}} W\right) G_w(i,j) \\ &= \operatorname{Tr}\left(W^{\mathrm{T}} \mathbb{K} D_w \mathbb{K}^{\mathrm{T}} W\right) - \operatorname{Tr}\left(W^{\mathrm{T}} \mathbb{K} G_w \mathbb{K}^{\mathrm{T}} W\right) \end{aligned} \tag{17}$$

where $\mathbb{K} = [K_1 | K_2 | \cdots | K_N]$. Considering that $L_b = D_b - W_b$, in a similar manner it can be shown that (15) can be simplified to:

$$\begin{aligned} &\frac{1}{2} \sum_{i,j} \|Y_i - Y_j\|^2 G_b(i,j) \\ &= \operatorname{Tr}\left(W^{\mathrm{T}} \mathbb{K} D_b \mathbb{K}^{\mathrm{T}} W\right) - \operatorname{Tr}\left(W^{\mathrm{T}} \mathbb{K} G_b \mathbb{K}^{\mathrm{T}} W\right) \\ &= \operatorname{Tr}\left(W^{\mathrm{T}} \mathbb{K} L_b \mathbb{K}^{\mathrm{T}} W\right) \end{aligned} \tag{18}$$

To solve (14) and (15) simultaneously, we need to add the following normalising constraint to the problem:

$$\operatorname{Tr}\left(W^{\mathrm{T}} \mathbb{K} D_w \mathbb{K}^{\mathrm{T}} W\right) = 1 \tag{19}$$

This constraint enables us to convert the minimisation problem (14) into a maximisation one. Consequently, both equations can be combined into one maximisation problem. Moreover, as we will see later, the imposed constraint acts as a norm regulariser in the original Tikhonov problem (5), thus satisfying the necessary condition of the representer theorem.

Plugging (19) into (14) results in:

$$\begin{aligned} &\min \left\{ \operatorname{Tr}\left(W^{\mathrm{T}} \mathbb{K} D_w \mathbb{K}^{\mathrm{T}} W\right) - \operatorname{Tr}\left(W^{\mathrm{T}} \mathbb{K} G_w \mathbb{K}^{\mathrm{T}} W\right) \right\} \\ &= \min \left\{ 1 - \operatorname{Tr}\left(W^{\mathrm{T}} \mathbb{K} G_w \mathbb{K}^{\mathrm{T}} W\right) \right\} \\ &= \max \left\{ \operatorname{Tr}\left(W^{\mathrm{T}} \mathbb{K} G_w \mathbb{K}^{\mathrm{T}} W\right) \right\} \end{aligned} \tag{20}$$

subject to the constraint shown in (19). As a result, the max versions of (14) and (15) can be merged by the Lagrangian method as follows:

$$\max\left\{\operatorname{Tr}\left(W^{\mathrm{T}}\mathbb{K}(L_b+\beta G_w)\mathbb{K}^{\mathrm{T}}W\right)\right\}$$
$$\text{subject to}\ \ \operatorname{Tr}\left(W^{\mathrm{T}}\mathbb{K}D_w\mathbb{K}^{\mathrm{T}}W\right)=1 \tag{21}$$

where β is a Lagrangian multiplier. The solution to the optimisation in (21) can be sought as the r largest eigenvectors of the following generalised eigenvalue problem:

$$\mathbb{K}\left\{L_b+\beta G_w\right\}\mathbb{K}^{\mathrm{T}}W=\lambda\mathbb{K}D_w\mathbb{K}^{\mathrm{T}}W \tag{22}$$

We note that in (22), the imposed constraint (19) serves as a norm regulariser and satisfies the representer theorem condition. Algorithm 1 assembles all the above details into pseudo-code for Riemannian Graph Embedding Discriminant Analysis (RGDA) training algorithm.

3.3 *Classification*

Upon acquiring the mapping W, the matching problem over Riemannian manifolds is reduced to classification in vector spaces. More precisely, for any query image set X_q, a vector representation using the kernel function and the mapping W is acquired, i.e. $\mathbf{V}_q=W^{\mathrm{T}}\mathbf{K}_q$, where $\mathbf{K}_q=\left(\langle\phi(X_1),\phi(X_q)\rangle,\langle\phi(X_2),\phi(X_q)\rangle,\ldots,\right.$ $\left.\langle\phi(X_N),\phi(X_q)\rangle\right)^{\mathrm{T}}$. Similarly, gallery points X_i are represented by r dimensional vectors $\mathbf{V}_i=W^{\mathrm{T}}\mathbf{K}_i$ and classification methods such as Nearest-Neighbour or Support Vector Machines [7] can be employed to label X_q.

4 Experiments

In this section we investigate the performance of the proposed RGDA method on several classification tasks, including face and object recognition, action recognition, texture classification and person re-identification. In the sequel, we first study RGDA using Grassmann geometry followed by evaluating RGDA over SPD manifolds.

4.1 *Experiments on Grassmann Manifolds*

Yamaguchi et al. [42] addressed the problem of face recognition from sets of face images and proposed to model set samples through a linear subspace. Image-sets can be seen as extension of videos since the sequential information is not considered.

Algorithm 1: Pseudocode for training Riemannian graph-embedding discriminant analysis (RGDA).

Input:

- Training set $\mathbb{X} = \{(X_i, l_i)\}_{i=1}^{N}$ from the underlying Riemannian manifold where $l_i \in \{1, 2, \cdots, C\}$, and C denoting the number of classes. For the Grassmann manifold $\mathcal{G}D, m$, $X_i \in \mathbb{R}^{D \times m}$ is a subspace. For the SPD manifold Sym_D^+, $X_i \in \mathbb{R}^{D \times D}$ is a SPD matrix.
- A kernel function k_{ij}, for measuring the similarity between two points on the Riemannian manifold

Output: The projection matrix $W = [\Upsilon_1 | \Upsilon_2 | \cdots | \Upsilon_r]$,

Compute the Gram matrix $[\mathbb{K}]_{ij}$ for all X_i, X_j
for $i = 1 \rightarrow N - 1$ **do**
  for $j = i + 1 \rightarrow N$ **do**
    Compute the geodesic distances $d_g(i, j)$ between X_i and X_j.
    $d_g(j, i) = d_g(i, j)$
  end for
end for
$G_w \leftarrow 0_{N \times N}$
$G_b \leftarrow 0_{N \times N}$ {% Use the obtained $d_g(i, j)$ to determine neighbourhoods in the following loop.}
for $i = 1 \rightarrow N$ **do**
  if (X_j is in the first k_w nearest neighbours of X_i) **and** ($l_j == l_i$) **then**
    $G_w(i, j) \leftarrow 1$
    $G_w(j, i) \leftarrow 1$
  end if
  if (X_j is in the first k_b nearest neighbours of X_i) **and** ($l_j \neq l_i$) **then**
    $G_b(i, j) \leftarrow 1$
    $G_b(j, i) \leftarrow 1$
  end if
end for
$D_w \leftarrow 0_{N \times N}$
$D_b \leftarrow 0_{N \times N}$
$D_w(i, i) \leftarrow \sum_j G_w(i, j)$
$D_b(i, i) \leftarrow \sum_j G_b(i, j)$
$L_b \leftarrow D_b - G_b$
$\{\Upsilon_i, \tilde{\lambda}_i\}_{i=1}^{r} \leftarrow$ generalised eigenvectors and eigenvalues of $\mathbb{K}\{L_b + \beta G_w\}\mathbb{K}^T W = \lambda \mathbb{K} D_w \mathbb{K}^T W$ $\{(\tilde{\lambda}_1 \geq \tilde{\lambda}_2 \geq \cdots \tilde{\lambda}_r)\}$ {In Matlab and Octave, The generalised eigenvalue problem $Av = \lambda Bv$ can be solved by the command eig(A, B)}

Modelling image-sets by linear subspaces has been shown to deliver improved performance in the presence of practical issues such as misalignment as well as variations in pose and illumination [14, 15, 41]. An image-set $\mathbb{F} = \{\mathbf{f}(t)\}_{t=1}^{\tau}; \mathbf{f}(t) \in \mathbb{R}^n$, where $\mathbf{f}_i$ is the vectorised representation of image i, can be represented on the Grassmann manifold through any orthogonalisation procedure like SVD. More specifically, let $\mathbb{F} = UDV^{\mathrm{T}}$ be the SVD of $\mathbb{F}$. The first p columns of U represents an optimised subspace of order p (in the mean square sense) for $\mathbb{F}$ and can be seen as point on manifold $\mathcal{G}n, p$.

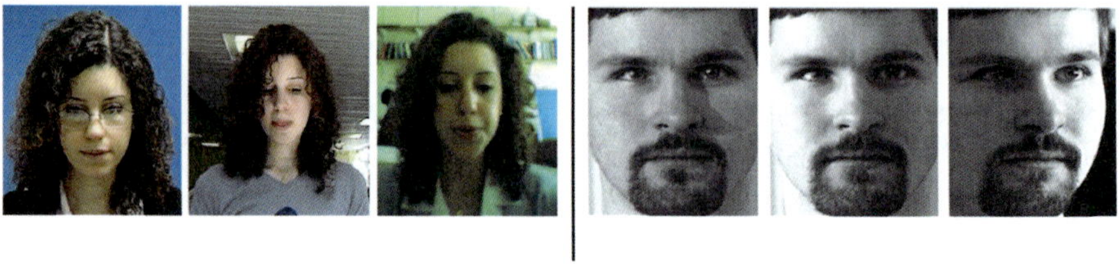

Fig. 2 Examples of appearance variations in (**a**) BANCA, and (**b**) CMU-PIE. The variations include image quality, pose, illumination and expression

Here we consider three image-set classification problems on Grassmann manifold: face, object and action recognition.

4.1.1 Face and Object Recognition

For the face recognition task we used the CMU-PIE [30] and BANCA [3] datasets. CMU-PIE contains images of 68 people captured under 13 poses, 43 illuminations conditions, and with 4 expressions. In our experiments, near frontal poses (c05, c07, c09, c27, c29) were used. See Fig. 2 for examples of the variations. We generated 204 image sets as training data and 204 image sets as test data. Images were cropped to the internal part of the face (i.e. closely cropped, no background) and downsampled to 64×64 pixels.

BANCA contains image sets for 52 people (26 males and 26 female). For each person video recordings were made under various conditions (illumination, pose and camera variations), while the person was talking. In each condition two recordings were made per person and 5 images were extracted from each video. We generated 150 image sets as training data and 150 sets as test data. All faces were closely cropped and resized to 64×64.

For the object recognition task, we used the ETH-80 dataset [20] which contains images of eight object categories: apples, cows, cups, dogs, horses, pears, tomatoes and cars. Each category includes ten object subcategories (e.g. various dogs) in 41 orientations, resulting in 410 images per category. Examples are shown in Fig. 3. We resized the images to 64×64. Unlike the face images mentioned above, the background was kept. We generated 24 image sets as gallery data and 56 sets as probe data. Following [8, 14, 41], normalised pixel intensities were used as image features.

The proposed RGDA algorithm was compared against: (1) the kernel version of Affine Hull Method (KAHM) [8] and (2) Grassmann Discriminant Analysis (GDA) [14].

In KAHM images are considered as points in a linear or affine feature space, while image sets are characterised by a convex geometric region (affine or convex hull) spanned by their feature points. GDA can be considered as an extension of

Fig. 3 **(a)** Examples from the eight object categories in the ETH-80 dataset; **(b)** examples of various classes within an object category

Table 1 Average correct recognition rate for image set matching using KAHM [8], GDA [14], and the RGDA approach

Method	CMU-PIE	BANCA	ETH-80
KAHM [8]	46.60 (6.6)	19.87 (1.2)	69.11 (5.1)
GDA [14]	24.07 (9.8)	46.60 (2.9)	83.21 (5.7)
RGDA	68.47 (6.8)	63.00 (2.5)	*91.96 (3.1)*

The standard deviation is shown in brackets

kernel discriminant analysis over Grassmann manifolds [14]. In GDA, a transform over the Grassmann manifold is learned to simultaneously maximise a measure of inter-class distances and minimise intra-class distances. RGDA can be considered as an extension of GDA, where a local discriminant transform over Grassmann manifolds is learned. This is achieved by incorporating local similarities/dissimilarities through within-class and between-class similarity graphs.

The results are presented in Table 1. For each dataset (CMU-PIE, BANCA and ETH-80), we created ten random splits of training and testing sets and reported the mean and standard deviation. Bar one case, the results indicate that the proposed GSR approach obtains the highest performance. The improvement over KAHM and GDA is especially remarkable on the BANCA and CMU-PIE datasets.

4.1.2 Action Recognition

The ballet dataset contains 44 real video sequences of 8 actions collected from an instructional ballet DVD [40].[5] The dataset consists of 8 complex motion patterns performed by 3 subjects. The actions include: *'hand opening'*, *'leg swinging'*, *'jumping'*, *'turning'*, *'hopping'* and *'standing still'*. Figure 4 shows samples. This dataset is very challenging due to the significant intra-class variations in terms of speed, spatial and temporal scale, clothing and movement variations.

Available samples of each action were randomly split into training and testing set (the number of actions in both training and testing sets were fairly even). The process of random splitting was repeated ten times and the average classification

[5]The study in [40] addresses the problem of recognising actions in still images, which is different from the work presented here.

Fig. 4 Examples of the Ballet dataset

Table 2 Recognition accuracy (in %) along its standard deviation for the Ballet dataset using Grassmann geodesic distance [36], Kernel Affine Hull method (KAHM) [8], Grassmann Discriminant Analysis (GDA) [14] and the proposed RGDA approach

Method	Geodesic distance	KAHM [8]	GDA [14]	RGDA
Recognition accuracy	$77.34\% \pm 1.8$	$79.71\% \pm 2.3$	$78.05\% \pm 2.9$	$83.08\% \pm 1.8$

accuracy is reported in Table 2. For comparison, the RGDA algorithm is contrasted with geodesic distance on Grassmann manifolds (3), KAHM [8] and GDA [14]. In this experiment, for Grassmann-based analysis, actions were modelled by image-sets of order 10. Studying Table 2 reveals that the RGDA algorithm obtains the highest accuracy and outperforms state-of-the-art methods KAHM and GDA significantly.

4.2 Experiments on SPD Manifolds

Mathematically, a covariance descriptor can be defined as follows: Let $\{\mathbf{f}_i\}_{i=1}^{N}; \mathbf{f}_i \in \mathbb{R}^n$ be the feature vectors from the region of interest of an image or video, then the covariance descriptor of this region $C \in \mathrm{Sym}_D^+$ is defined as:

$$C = \frac{1}{N}\sum_{i=1}^{N}(\mathbf{f}_i - \mathbf{m})(\mathbf{f}_i - \mathbf{m})^{\mathrm{T}} \tag{23}$$

where $\mathbf{m}$ is the mean feature vector. In the following text, we study how covariance descriptors and the induced geometry can be exploited for face recognition, texture classification and people re-identification.

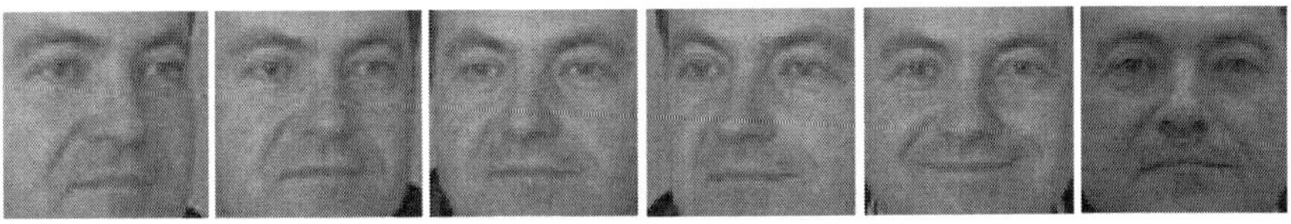

Fig. 5 Examples of closely cropped faces from the FERET 'b' subset

4.2.1 Face Recognition

For the face recognition task, we considered the subset 'b' of the FERET dataset [25]. This subset includes 1,400 images from 198 subjects. Each image is closely cropped to merely include the face and then downsampled to 64×64. Figure 5 shows examples of the FERET dataset.

To evaluate the performance, we created three tests with various pose angles. In all the tests, training data consisted of the images labeled as 'bj', 'bk' and 'bf' (i.e. frontal image with illumination, expression and small pose variations). Images marked as 'bd', 'be' and 'bg'(i.e. non-frontal images) were used as three separate test sets. In our method, each face image is represented by a 43×43 covariance matrix as a point on the Riemannian manifold. To this end, for every pixel $I(x,y)$, we then computed $G_{u,v}(x,y)$ as the response of a 2D Gabor wavelet [19], centered at x,y with orientation u and scale v. To be specific, we considered the number of scales and orientations to be 5 and 8, respectively.

$$G_{u,v}(x,y) = \frac{k_v^2}{4\pi^2} \sum\nolimits_{t,s} e^{-\frac{k_v^2}{8\pi^2}\left((x-s)^2+(y-t)^2\right)} \left(e^{ik_v((x-t)\cos(\theta_u)+(y-s)\sin(\theta_u))} - e^{-2\pi^2}\right)$$

with $k_v = \frac{1}{\sqrt{2^{v-1}}}$ and $\theta_u = \frac{\pi u}{8}$. Then the feature vector is defined as following:

$$F_{x,y} = [\, I(x,y),\ x,\ y,\ |G_{0,0}(x,y)|,\ \ldots,\ |G_{0,7}(x,y)|,\ |G_{1,0}(x,y)|,\ \ldots,\ |G_{4,7}(x,y)|\,]$$

Table 3 shows a comparison of RGDA against three Euclidean space approaches, PCA [7], KPCA [7], and LDA [7], applied on Gabor features. The results show that the proposed approach outperforms PCA with considerably better results. Furthermore, the results illustrate that the overall performance of RGDA is better by a notable margin. In addition, although images labelled with 'bg' and 'bd' represent the same pose variation (in different directions), results indicate a better performance for all the algorithms on 'bg'. Training data in 'bf' includes face images with -15 degree pose angle which is closer to the pose angle of test data in 'bg' comparing with the ones in 'bd'. This explains the superior performance of 'bg'.

Table 3 Recognition accuracy (in %) for the face recognition task using PCA [37], LDA [6], and the proposed RGDA approach

	Gabor + PCA	Gabor + KPCA	Gabor + LDA	RGDA (proposed)
bd	24.50	42.00	61.50	*78.00*
be	52.00	73.50	92.00	*98.50*
bg	74.00	94.00	*99.00*	98.50
average	50.16	69.80	84.16	*91.67*

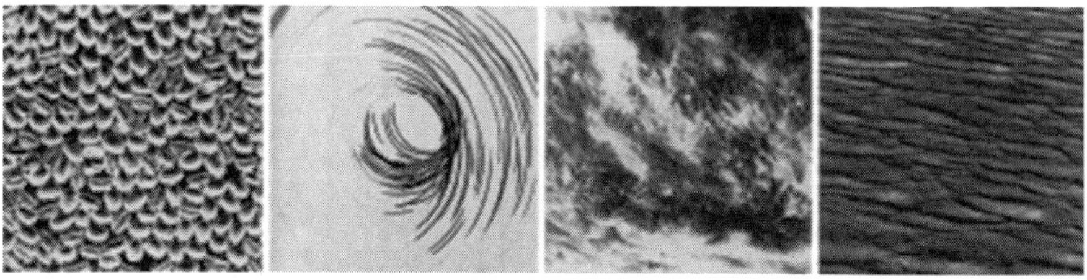

Fig. 6 Samples of Brodatz texture dataset [26]

4.2.2 Texture Classification

To examine RGDA's performance on classification using the Brodatz texture dataset [26] (Examples are shown in Fig. 7, we have Followed the test protocol advised in [32]. Nine test scenarios with various number of classes were generated. The test scenarios included 5-texture ('5c', '5m', '5v', '5v2', '5v3'), 10-texture ('10', '10v') and 16-texture ('16c', '16v') mosaics. To create a Riemannian manifold, first step was downsampeling each image to 256×256, followed by splitting them into 64 regions of size 32×32.

The feature vector for each pixel $I(x, y)$ is defined as:

$$F(x,y) = \left[I(x,y), \left| \frac{\partial I}{\partial x} \right|, \left| \frac{\partial I}{\partial y} \right|, \left| \frac{\partial^2 I}{\partial x^2} \right|, \left| \frac{\partial^2 I}{\partial y^2} \right| \right]$$

Each region is described by a 5×5 covariance descriptor computed based on these features. For each test scenario, 25 covariance matrices per class were randomly selected to construct training data and the rest was used for testing. The random selection of training/testing data was repeated 20 times. Finally, for any covariance descriptor we find the nearest neighbour from the training set and, respectively, assign the corresponding image class to it.

Figure 7 compares the proposed RGDA method against and TSC [31]. The results indicate that the proposed RGDA achieves better performance on all the tests except for the '5c' test.

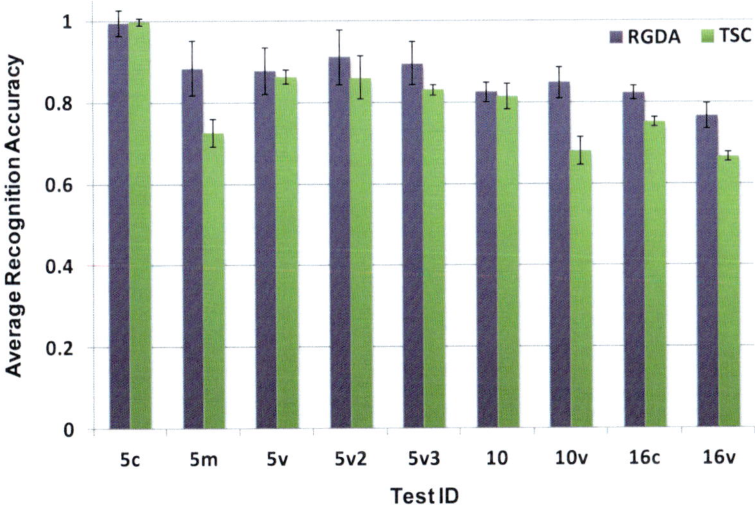

Fig. 7 Performance on the Brodatz texture dataset [26] for Tensor Sparse Coding (TSC) [32] and the proposed REDA approach. The *black bars* indicate standard deviation

Fig. 8 Examples of pedestrians in the ETHZ dataset

4.2.3 Person Re-identification

In this section we test the performance of RGDA method for person reidentification task on the modified ETHZ dataset [28]. The original version of this dataset was captured from a moving camera [12], and it has been used for human detection. The main challenging aspects of ETHZ dataset are variations in pedestrian?s appearances and occlusions. Some sample images of the ETHZ dataset are shown in Fig. 9.

This dataset contains three video sequences. Table 4 summarises the information about this dataset.

We downsampled all the images to 64×32. For each subject, training set consisted of ten randomly selected images and the rest used for the test set. To generalise the practical assessment of the algorithm, random selection of the training and testing data was repeated 20 times.

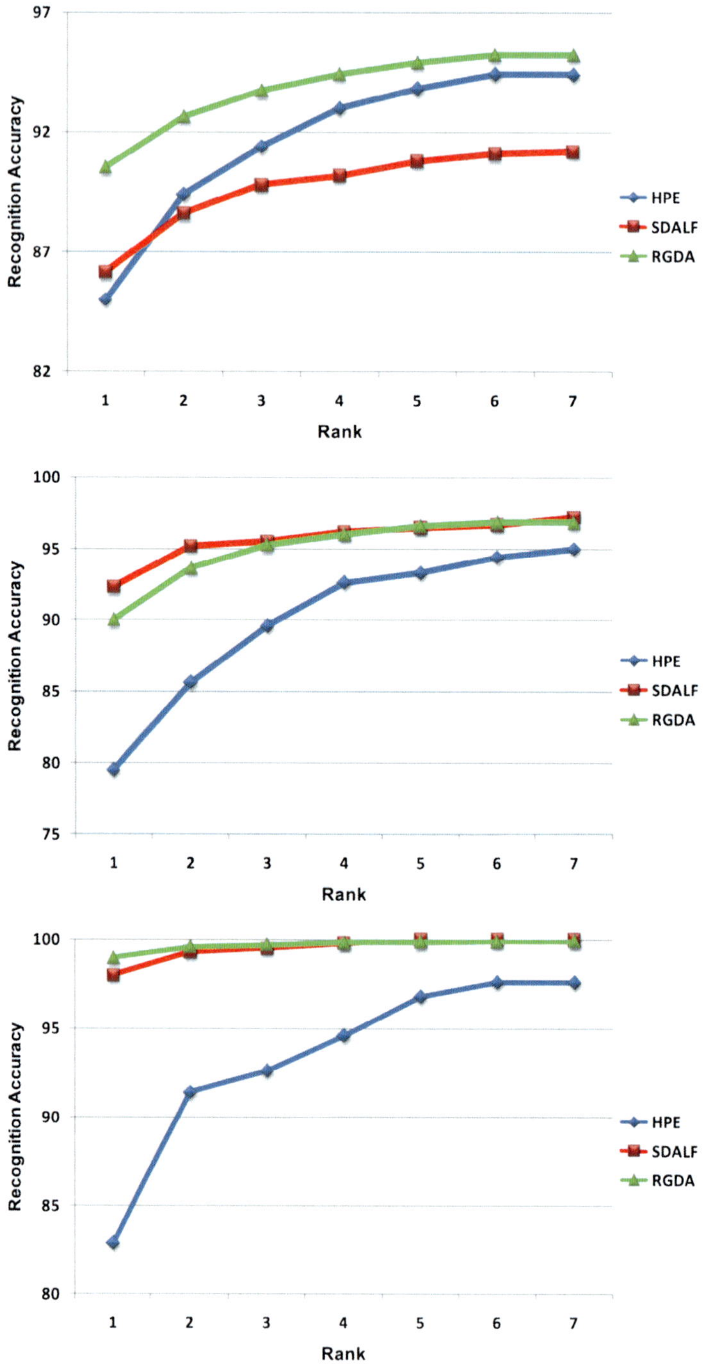

Fig. 9 Performance on Sequences 1, 2, and 3 of the ETHZ dataset (*top, middle and bottom* panels, respectively), in terms of Cumulative Matching Characteristic curves. The proposed RGDA method is compared with Histogram Plus Epitome (HPE) [5], Symmetry-Driven Accumulation of Local Features (SDALF) [13]

Table 4 The ETHZ dataset

	SEQ 1	SEQ 2	SEQ 3
Num of people	83	35	28
Total Num of images	4,857	1,936	1,762

To create points on the Riemannian manifold, a feature vector was formed for each pixel using the position of the pixel (x and y), the corresponding colour information ($R_{x,y}$, $G_{x,y}$ and $B_{x,y}$) and the gradient and Laplacian for colour C, defined as $C'_{x,y} = \left[|\partial C/\partial x|, |\partial C/\partial y|\right]$ and $C''_{x,y} = \left[|\partial^2 C/\partial x^2|, |\partial^2 C/\partial y^2|\right]$, respectively. Then the representative of each image is the covariance matrix using the following feature:

$$F_{x,y} = \left[\, x,\ y,\ R_{x,y},\ G_{x,y},\ B_{x,y},\ R'_{x,y},\ G'_{x,y},\ B'_{x,y},\ R''_{x,y},\ G''_{x,y},\ B''_{x,y} \,\right]$$

We compared the proposed RGDA with Histogram Plus Epitome (HPE) [5] and Symmetry-Driven Accumulation of Local Features (SDALF) [13]. The evaluation is done in terms of cumulative matching characteristic (CMC) curves. The CMC curve plots the percentage of the test queries whose correct match is within the top n closest matches. Based on the results shown in Fig. 9, the proposed approach achieves the highest accuracy on sequences 1 and 2. For sequence 3, RGDA obtains a very similar performance to SDALF while HPE scores the lowest.

5 Summary

In this work, we showed how discriminant analysis can be reformulated on two non-Euclidean spaces, namely the Grassmann and SPD manifolds. Inference on manifold spaces can be achieved by embedding the manifolds in higher dimensional Euclidean spaces, which can be considered as flattening the manifolds. In this work we propose to embed Riemannian manifolds into Reproducing Kernel Hilbert Spaces (RKHS). This in turn opens the door for employing many kernel-based machine learning algorithms [29]. As such, we tackle the problem of Discriminant Analysis (DA) on Riemannian manifolds through RKHS space and propose a graph-based local DA that utilises both within-class and between-class similarity graphs to characterise intra-class compactness and inter-class separability, respectively.

Thorough experiments on face and object recognition, action recognition, texture classification and person re-identification showed that notable improvements in discrimination accuracy can be obtained through graph-embedding analysis.

Acknowledgements This project is supported by a grant from the Australian Government Department of the Prime Minister and Cabinet. NICTA is funded by the Australian Government's Backing Australia's Ability initiative, in part through the Australian Research Council. The first and second authors contributed equally.

References

1. Adini Y, Moses Y, Ullman S (1997) Face recognition: The problem of compensating for changes in illumination direction. IEEE Trans Pattern Anal Mach Intell 19(7):721–732
2. Argyriou A, Micchelli CA, Pontil M (2009) When is there a representer theorem? vector versus matrix regularizers. J Mach Learn Res 10:2507–2529
3. Bailly-Baillière E, Bengio S, Bimbot F, Hamouz M, Kittler J, Mariéthoz J, Matas J, Messer K, Popovici V, Porée F, Ruiz B, Thiran JP (2003) The BANCA database and evaluation protocol. In: AVBPA. Lecture notes in computer science (LNCS), springer pp 1057–1057
4. Bak S, Corve E, Brmond F, Thonnat M (2011) Boosted human re-identification using riemannian manifolds. Image and Vision Comput, Elsevier
5. Bazzani L, Cristani M, Perina A, Farenzena M, Murino V (2010) Multiple-shot person re-identification by hpe signature. In: Proceedings of the 2010 20th international conference on pattern recognition. IEEE Computer Society, Silver Spring, pp 1413–1416
6. Belhumeur P, Hespanha J, Kriegman D (1997) Eigenfaces vs. fisherfaces: Recognition using class specific linear projection. IEEE Trans Pattern Anal Mach Intell 19(7):711–720
7. Bishop CM (2006) Pattern recognition and machine learning. Springer, Berlin
8. Cevikalp H, Triggs B (2010) Face recognition based on image sets. In: IEEE conference on computer vision and pattern recognition (CVPR). IEEE pp 2567–2573
9. Chen Y, Garcia EK, Gupta MR, Rahimi A, Cazzanti L (2009) Similarity-based classification: Concepts and algorithms. J Mach Learn Res 10:747–776
10. Comaniciu D, Meer P (2002) Mean shift: A robust approach toward feature space analysis. IEEE Trans Pattern Anal Mach Intell 24(5):603–619
11. Edelman A, Arias TA, Smith ST (1999) The geometry of algorithms with orthogonality constraints. SIAM J Matrix Anal Appl 20(2):303–353
12. Ess A, Leibe B, Van Gool L (2007) Depth and appearance for mobile scene analysis. In: IEEE 11th international conference on computer vision, 2007. ICCV 2007. IEEE, New York, pp 1–8
13. Farenzena M, Bazzani L, Perina A, Murino V, Cristani M (2010) Person re-identification by symmetry-driven accumulation of local features. In: 2010 IEEE conference on computer vision and pattern recognition (CVPR). IEEE, New York, pp 2360–2367
14. Hamm J, Lee DD (2008) Grassmann discriminant analysis: a unifying view on subspace-based learning. In: Proceedings of the 25th international conference on Machine learning ACM pp 376–383
15. Harandi MT, Sanderson C, Shirazi S, Lovell BC (2011) Graph embedding discriminant analysis on Grassmannian manifolds for improved image set matching. In: IEEE conference on computer vision and pattern recognition (CVPR). IEEE pp 2705–2712
16. Harandi MT, Sanderson C, Wiliem A, Lovell BC (2012) Kernel analysis over Riemannian manifolds for visual recognition of actions, pedestrians and textures. In: IEEE workshop on the applications of computer vision (WACV), IEEE pp 433–439
17. Hu W, Li X, Luo W, Zhang X, Maybank S, Zhang Z (2012) Single and multiple object tracking using log-euclidean riemannian subspace and block-division appearance model. IEEE Trans Pattern Anal Mach Intell. IEEE
18. Kimeldorf GS, Wahba G (1970) A correspondence between bayesian estimation on stochastic processes and smoothing by splines. Ann Math Stat 41:495–502
19. Lee T (1996) Image representation using 2d gabor wavelets. IEEE Trans Pattern Anal Mach Intell 18(10):959–971
20. Leibe B, Schiele B (2003) Analyzing appearance and contour based methods for object categorization. In: IEEE Conf. Computer Vision and Pattern Recognition (CVPR), vol 2, pp 409–415
21. Lui YM (2012) Advances in matrix manifolds for computer vision. Image and Vision Computing, **30**(6), Elsevier, 380-388
22. O'Hara S, Lui YM, Draper BA (2011) Using a product manifold distance for unsupervised action recognition. Image and Vision Computing Elsevier

23. Pang Y, Yuan Y, Li X (2008) Gabor-based region covariance matrices for face recognition. IEEE Trans Circ Syst Video Technol 18(7):989–993
24. Pennec X (2006) Intrinsic statistics on Riemannian manifolds: Basic tools for geometric measurements. J Math Imag Vision 25(1):127–154
25. Phillips P, Moon H, Rizvi S, Rauss P (2000) The feret evaluation methodology for face-recognition algorithms. IEEE Trans Pattern Anal Mach Intell 22(10):1090–1104
26. Randen T, Husoy J (1999) Filtering for texture classification: A comparative study. IEEE Trans Pattern Anal Mach Intell 21(4):291–310
27. Rosenberg S (1997) The Laplacian on a Riemannian manifold: An introduction to analysis on manifolds. Cambridge University Press, Cambridge
28. Schwartz W, Davis L (2009) Learning discriminative appearance-based models using partial least squares. In: XXII Brazilian symposium on computer graphics and image processing (SIBGRAPI), 2009. IEEE, New York, pp 322–329
29. Shawe-Taylor J, Cristianini N (2004) Kernel methods for pattern analysis. Cambridge University Press, Cambridge
30. Sim T, Baker S, Bsat M (2003) The CMU pose, illumination, and expression database. IEEE Trans Pattern Analysis and Machine Intelligence 25(12), 1615–1618
31. Sivalingam R, Boley D, Morellas V, Papanikolopoulos N (2010a) Tensor sparse coding for region covariances. In: Computer VisionECCV 2010 pp. 722–735. Springer
32. Sivalingam R, Boley D, Morellas V, Papanikolopoulos N (2010b) Tensor sparse coding for region covariances. In: Computer vision–ECCV 2010, pp 722–735 springer
33. Sra S (2012) Positive definite matrices and the symmetric Stein divergence. Preprint: [arXiv:1110.1773]
34. Subbarao R, Meer P (2009) Nonlinear mean shift over Riemannian manifolds. Int J Comput Vision 84(1):1–20
35. Tikhonov AN, Arsenin VY (1977) Solutions of Ill-posed problems. V.H. Winston & Sons, Washington, D.C.; Wiley, New York
36. Turaga P, Veeraraghavan A, Srivastava A, Chellappa R (2011) Statistical computations on grassmann and stiefel manifolds for image and video-based recognition. IEEE Trans Pattern Anal Mach Intell 33(11):2273–2286
37. Turk M, Pentland A (1991) Eigenfaces for recognition. J Cognit Neurosci 3(1):71–86
38. Tuzel O, Porikli F, Meer P (2006) Region covariance: A fast descriptor for detection and classification. In: Leonardis A, Bischof H, Pinz A (eds) European conference on computer vision (ECCV). Lecture notes in computer science, vol 3952. Springer, Berlin, pp 589–600
39. Tuzel O, Porikli F, Meer P (2008) Pedestrian detection via classification on Riemannian manifolds. IEEE Trans Pattern Anal Mach Intell (PAMI) 30:1713–1727
40. Wang Y, Mori G (2009) Human action recognition by semilatent topic models. IEEE Trans Pattern Anal Mach Intell 31(10):1762–1774
41. Wolf L, Shashua A (2003) Learning over sets using kernel principal angles. J Mach Learn Res 4:913–931
42. Yamaguchi O, Fukui K, Maeda KI (1998) Face recognition using temporal image sequence. In: IEEE international conference on automatic face and gesture recognition, Washington, DC, 1998, pp 318–323
43. Yuan C, Hu W, Li X, Maybank S, Luo G (2010) Human action recognition under log-euclidean riemannian metric. In: Asian conference on computer vision (ACCV) Lecture notes in computer science, vol 5994. Springer, Berlin, pp 343–353

A Flexible and Effective Linearization Method for Subspace Learning

Feiping Nie, Dong Xu, Ivor W. Tsang, and Changshui Zhang

1 Introduction

In the past decades, a large number of subspace learning or dimension reduction methods [2, 16, 20, 32, 34, 37, 44] have been proposed. Principal component analysis (PCA) [32] pursues the directions of maximum variance for optimal reconstruction. Linear discriminant analysis (LDA) [2], as a supervised algorithm, aims to maximize the inter-class scatter and at the same time minimize the intra-class scatter. Due to utilization of label information, LDA is experimentally reported to outperform PCA for face recognition, when sufficient labeled face images are provided [2].

To discover the intrinsic manifold structure of the data, nonlinear dimension reduction algorithms such as ISOMAP [31], Locally linear embedding (LLE) [25], Laplacian eigenmap (LE) [3], and local spline embedding (LSE) [35] were recently developed. Yan *et al.* [37] recently demonstrated that several dimension reduction algorithms (e.g., PCA, LDA, ISOMAP, LLE, LE) can be unified within a proposed graph-embedding framework, in which the desired statistical or geometric data properties are encoded as graph relationships. Recently, Zhang et al. [42–44] further reformulated many dimension reduction algorithms into a unified patch alignment framework with the same trick in [46]. Based on their patch alignment framework, a new subspace learning method called discriminative locality alignment (DLA) was also proposed [42, 44].

F. Nie (✉)
University of Texas, Arlington
e-mail: feipingnie@gmail.com

D. Xu • I.W. Tsang
Nanyang Technological University, Singapore
e-mail: DongXu@ntu.edu.sg; IvorTsang@ntu.edu.sg

C. Zhang
Tsinghua University, Beijing 100084, China
e-mail: zcs@mail.tsinghua.edu.cn

Y. Fu and Y. Ma (eds.), *Graph Embedding for Pattern Analysis*,
DOI 10.1007/978-1-4614-4457-2_8,

While supervised learning algorithms generally outperform unsupervised learning algorithms, the collection of labeled training data in supervised learning requires expensive human labor [9, 48]. Meanwhile, it is much easier to obtain unlabeled data. To utilize a large amount of unlabeled data as well as a relatively limited amount of labeled data for better classification, semi-supervised learning methods such as Transductive SVM [33], Co-Training [5], and graph-based techniques [1,4,7,28,29,38,41,47,49] were developed and demonstrated promising results for different tasks. However, most semi-supervised learning methods such as described in [5, 12, 33, 47, 49] were developed for the problem of classification. The Manifold Regularization (MR) method [4, 28, 29] can be also used for various learning problems. In practice, MR extended Regression and SVM, respectively, to the semi-supervised learning methods Laplacian regularized least squares (LapRLS) and Laplacian Support Vector Machines (LapSVM) by adding a geometrically based Laplacian regularization term. Recently, Cai et al. [7] and Song et al. [30] independently extended LDA to semi-supervised discriminant analysis (SDA).

For the transductive learning method such as ISOMAP, LLE, LE, LSE, GFHF and LGC, we only obtain the low-dimensional coordinates or the predicted labels $F \in \mathbb{R}^{m \times c}$ for the m training data. One of the major problems in transductive learning is the out-of-sample problem, i.e., they do not yield a method for mapping new data points that are not included in the training set. An effective method to solve this problem is the linearization technique, which leads to subspace learning (or dimensionality reduction) methods. For example, He et al. [13] developed the Locality Preserving Projections (LPP) method, in which the linear projection function is used for mapping new data. A strict and improved LPP can be found in [22].

In this chapter, we introduce three linearization methods for subspace learning, including rigid constrained method, two-step method with regression and a flexible method. The rigid constrained method is overstrict since it imposes a constraint that $F = X^{\mathrm{T}}W + \mathbf{1}b^{\mathrm{T}}$, i.e., the nonlinear F must be the same as the linear model $X^{\mathrm{T}}W + \mathbf{1}b^{\mathrm{T}}$. The two-step linearization method relies heavily on the result F of the first step. The flexible linearization method overcome these drawbacks. Specifically, we set the prediction labels as $F = h(X) + F_0$, where $h(X) = X^{\mathrm{T}}W + \mathbf{1}b^{\mathrm{T}}$ is a regression function for mapping new data points and F_0 is the regression residue modeling the mismatch between F and $h(X)$. The flexible linearization method aims to optimize the prediction labels F, the linear regression function $h(X)$ and the flexible regression residue F_0 simultaneously. It is interesting to see that the first two linearization methods are the two extreme cases of the flexible linearization method.

Based on the flexible linearization method, we introduce a new framework for semi-supervised and unsupervised subspace learning, referred to as *F*lexible *M*anifold *E*mbedding (*FME*), and *FME/U*, respectively [23]. FME can effectively utilize label information from labeled data as well as the manifold structure from both labeled and unlabeled data, and can better deal with the samples which reside on a nonlinear manifold. We show the FME(FME/U) is a general framework, many prior works, including the general graph embedding framework [37] are special cases of this framework.

2 Brief Review of the Prior Work

We briefly review the prior semi-supervised learning work: Local and Global Consistency (LGC) [47], Gaussian Fields and Harmonic Functions (GFHF) [49], Manifold Regularization (MR) [4,28,29], and Semi-Supervised Discriminant Analysis (SDA) [7, 30]. We denote the sample set as $X = [x_1, x_2, \ldots, x_n, x_{n+1}, \ldots, x_m] \in \mathbb{R}^{f \times m}$, where $x_i|_{i=1}^{n}$ and $x_i|_{i=n+1}^{m}$ are labeled and unlabeled data, respectively. For labeled data $x_i|_{i=1}^{n}$, the labels are denoted as $y_i \in \{1, 2, \ldots, c\}$, where c is the total number of classes. We also define a binary label matrix $Y \in \mathbb{B}^{m \times c}$ with $Y_{ij} = 1$ if x_i has label $y_i = j$; $Y_{ij} = 0$, otherwise. Let us denote $G = \{X, S\}$ as an undirected weighted graph with vertex set X and similarity matrix $S \in \mathbb{R}^{m \times m}$, in which each element S_{ij} of the real symmetric matrix S represents the similarity of a pair of vertices. The graph Laplacian matrix $L \in \mathbb{R}^{m \times m}$ is denoted as $L = D - S$, where D is a diagonal matrix with the diagonal elements as $D_{ii} = \sum_j S_{ij}, \forall\, i$. The normalized graph Laplacian matrix is represented as $\tilde{L} = D^{-\frac{1}{2}} L D^{-\frac{1}{2}} = I - D^{-\frac{1}{2}} S D^{-\frac{1}{2}}$, where I is an identity matrix. We also denote $\mathbf{0}, \mathbf{1} \in \mathbb{R}^{m \times 1}$ as a vector with all elements as 0 and a vector with all elements as 1, respectively.

2.1 *Local and Global Consistency and Gaussian Fields and Harmonic Functions*

LGC [47] and GFHF [49] estimate a prediction label matrix $F \in \mathbb{R}^{m \times c}$ on the graph with respect to the label fitness (i.e., F should be close to the given labels for the labeled nodes) and the manifold smoothness (i.e., F should be smooth on the whole graph of both labeled and unlabeled nodes). Let us denote $F_{i.}$ and $Y_{i.}$ as the i-th row of F and Y. As shown in [47–49], LGC and GFHF minimize the objective function $g_{\mathrm{L}}(F)$ and $g_{\mathrm{G}}(F)$, respectively:

$$g_{\mathrm{L}}(F) = \frac{1}{2} \sum_{i,j=1}^{m} \left\| \frac{F_{i.}}{\sqrt{D_{ii}}} - \frac{F_{j.}}{\sqrt{D_{jj}}} \right\|^2 S_{ij} + \lambda \sum_{i=1}^{m} \|F_{i.} - Y_{i.}\|^2,$$

$$g_{\mathrm{G}}(F) = \frac{1}{2} \sum_{i,j=1}^{m} \|F_{i.} - F_{j.}\|^2 S_{ij} + \lambda_\infty \sum_{i=1}^{n} \|F_{i.} - Y_{i.}\|^2, \tag{1}$$

where $\|A\|^2 = \mathrm{trace}(A^{\mathrm{T}} A)$, the coefficient λ balances the label fitness and the manifold smoothness, and λ_∞ is a very large number such that $\sum_{i=1}^{n} \|F_{i.} - Y_{i.}\|^2 = 0$, or $F_{i.} = Y_{i.}\ \forall i = 1, 2, \ldots, n$ [48]. Notice that the objective functions $g_{\mathrm{L}}(F)$ and $g_{\mathrm{G}}(F)$ in (1) share the same formulation:

$$\mathrm{Tr}(F^{\mathrm{T}} M F) + \mathrm{Tr}(F - Y)^{\mathrm{T}} U (F - Y), \tag{2}$$

where $M \in \mathbb{R}^{m \times m}$ is a graph Laplacian matrix and $U \in \mathbb{R}^{m \times m}$ is a diagonal matrix.

In LGC [47], M is the normalized graph Laplacian matrix $\tilde{L}$ and U is a diagonal matrix with all elements as λ. In GFHF [49], $M = L$ and U is also a diagonal matrix with the first n and the rest $m-n$ diagonal elements as λ_∞ and 0, respectively.

2.2 Manifold Regularization

The manifold regularization [4, 28, 29] extends many existing algorithms, such as ridge regression and SVM to their semi-supervised learning methods by adding a geometrically based regularization term. We take LapRLS/L as an example to briefly review MR methods. Let us define a linear regression function $h(x_i) = W^{\mathrm{T}}x_i + b$, where $W \in \mathbb{R}^{f\times c}$ is the projection matrix and $b \in \mathbb{R}^{c\times 1}$ is the bias term. LapRLS/L [29] minimizes the ridge regression errors and simultaneously preserves the manifold smoothness, namely:

$$\begin{aligned} g_M(W,b) &= \lambda_A \|W\|^2 + \lambda_I \mathrm{Tr}(W^{\mathrm{T}}XLX^{\mathrm{T}}W) \\ &\quad + \frac{1}{n}\sum_{i=1}^{n} \|W^{\mathrm{T}}x_i + b - Y_{i.}^{\mathrm{T}}\|^2, \end{aligned} \tag{3}$$

where the two coefficients λ_A and λ_I balance the norm of W, the manifold smoothness and the regression error.

2.3 Semi-supervised Discriminant Analysis

The core assumption in semi-supervised discriminant analysis (SDA) [7, 30] is still the manifold smoothness assumption, namely, nearby points will have similar representations in the lower-dimensional space. We define $X_l = [x_1, x_2, \ldots, x_n]$ as the data matrix of labeled data, and denote the number of the labeled samples in the i-th class as n_i. Let us denote two graph similarity matrices $\tilde{S}^w, \tilde{S}^b \in \mathbb{R}^{n\times n}$, where $\tilde{S}^w_{ij} = \delta_{y_i,y_j}/n_{y_i}$, $\tilde{S}^b_{ij} = \frac{1}{n} - \tilde{S}^w_{ij}$. The corresponding Laplacian matrices of $\tilde{S}^w$, $\tilde{S}^b$ are represented as $\tilde{L}_w$ and $\tilde{L}_b$, respectively. According to [37], the intra-class scatter S_w and the inter-class scatter S_b of LDA can be rewritten as $S_w = \sum_{i=1}^{n}(x_i - \bar{x}_{y_i})(x_i - \bar{x}_{y_i})^{\mathrm{T}} = X_l\tilde{L}_wX_l^{\mathrm{T}}$, and $S_b = \sum_{l=1}^{c} n_c(\bar{x}_l - \bar{x})(\bar{x}_l - \bar{x})^{\mathrm{T}} = X_l\tilde{L}_bX_l^{\mathrm{T}}$, where $\bar{x}_l$ is the mean of the labeled samples in the l-th class and $\bar{x}$ is the mean of all the labeled samples. The objective function in SDA is then formulated as:

$$g_S(W) = \frac{|W^{\mathrm{T}}X_l\tilde{L}_bX_l^{\mathrm{T}}W|}{|W^{\mathrm{T}}(X_l(\tilde{L}_w + \tilde{L}_b)X_l^{\mathrm{T}} + \alpha XLX^{\mathrm{T}} + \beta I)W|}, \tag{4}$$

where $L \in \mathbb{R}^{m\times m}$ is the graph Laplacian matrix for both labeled and unlabeled data, and α and β are two parameters to balance three terms. A new improved version of the SDA with the trace ratio criterion [15] can be found in [14].

3 Linearization Methods for Subspace Learning

In the transductive learning setting such as the clustering, manifold learning (nonlinear dimensionality reduction) and label propagation (GFHF and LGC), we only obtain the $F \in \mathbb{R}^{m\times c}$ for the m training data, where $F_{i.} \in \mathbb{R}^{1\times c}$ is the predicted label (clustering and label propagation) or the low-dimensional coordinate (manifold learning) of the i-th training data point. Usually, we solve an optimization problem to obtain the F:

$$\min_{F\in\mathbb{C}} f(F) \tag{5}$$

where $\mathbb{C}$ is a certain constraint on F and $f(F)$ is a certain objective function on F. One of the major problems in transductive learning is the out-of-sample problem, i.e., it cannot handle new data points that are not included in the training set, which limits its applications in practice. An effective method to solve this problem is the linearization technique, which leads to subspace learning (or dimensionality reduction) methods.

3.1 *Linearization Method with Rigid Constraint*

Assuming the solution F lies on a subspace spanned by the training data points $x_i|_{i=1}^m$, i.e. there is a linear mapping (projection) matrix $W \in \mathbb{R}^{f\times c}$ and a bias $b \in \mathbb{R}^{c\times 1}$ such that $F = X^{\mathrm{T}}W + \mathbf{1}b^{\mathrm{T}}$. Under this assumption, the problem (5) becomes

$$\min_{W,b,X^{\mathrm{T}}W+\mathbf{1}b^{\mathrm{T}}\in\mathbb{C}} f(X^{\mathrm{T}}W + \mathbf{1}b^{\mathrm{T}}). \tag{6}$$

In subspace learning, a Tikhonov regularization term $||W||^2$ is usually added to avoid overfitting. Thus, the problem (6) becomes

$$\min_{W,b,X^{\mathrm{T}}W+\mathbf{1}b^{\mathrm{T}}\in\mathbb{C}} f(X^{\mathrm{T}}W + \mathbf{1}b^{\mathrm{T}}) + \gamma||W||^2. \tag{7}$$

We can obtain a projection matrix W by solving (7). For any data points $x \in \mathbb{R}^{f\times 1}$, we can obtain its predicted label or low-dimensional coordinate by the linear mapping $W^{\mathrm{T}}x + b$.

Recall that manifold regularization extends a supervised subspace learning method solving $\min_{W,b} g(X_l^{\mathrm{T}}W + \mathbf{1}b^{\mathrm{T}})$ (X_l denotes the labeled training data) to its semi-supervised counterpart by $\min_{W,b} g(X_l^{\mathrm{T}}W + \mathbf{1}b^{\mathrm{T}}) + \gamma\mathrm{Tr}(W^{\mathrm{T}}XMX^{\mathrm{T}}W)$ (X denotes the labeled and unlabeled training data). It is interesting to note that one of manifold regularization, LapRLS/L, can be viewed as a linearization of the transductive learning problem (2) by the linearization method in (7). That is to say, based on the linearization method in (7), LapRLS/L [29] is the out-of-sample extension to the label propagation method in (2), *parallel to* that LPP [13] is the out-of-sample extension to the manifold learning method, Laplacian eigenmap [3].

3.2 Linearization Method with Two-Step Regression

Another linearization method is a two-step method based on linear regression [6,24]. After the $F \in \mathbb{R}^{m\times c}$ is obtained by solving the transductive learning problem (5), the linear mapping (projection) matrix $W \in \mathbb{R}^{f\times c}$ and a bias $b \in \mathbb{R}^{c\times 1}$ is learned to best approximate the F by solving the following regularized least squares linear regression problem:

$$\min_{W,b} \|X^{\mathrm{T}}W + \mathbf{1}b^{\mathrm{T}} - F\|^2 + \gamma\|W\|^2. \tag{8}$$

One of the advantages of this linearization method is that it is efficient since there is fast algorithm to solve the regularized least squares linear regression problem using conjugate gradient method.

3.3 A New and Flexible Linearization Method

Linearization method with rigid constraint is overstrict since it assumes the nonlinear F is exactly equal to the linear model $X^{\mathrm{T}}W + \mathbf{1}b^{\mathrm{T}}$, while the linearization method with two-step regression relies heavily on the first step result F. To overcome these drawbacks, we propose a new and flexible linearization method. Specifically, as shown in Fig. 1, we set the prediction labels as $F = h(X) + F_0$, where $h(X) = X^{\mathrm{T}}W + \mathbf{1}b^{\mathrm{T}}$ is a linear regression function for mapping new data points and F_0 is the regression residue modeling the mismatch between F and $h(X)$, then simultaneously optimize the $F \in \mathbb{C}, F_0, W, b$ by solving the following problem:

$$\min_{F\in\mathbb{C},F_0,W,b} f(F) + \mu(\|W\|^2 + \gamma\|F_0\|^2), \tag{9}$$

The problem can be equivalently written as:

$$\min_{F\in\mathbb{C},W,b} f(F) + \mu(\|W\|^2 + \gamma\|X^{\mathrm{T}}W + \mathbf{1}b^{\mathrm{T}} - F\|^2), \tag{10}$$

It is easily to see that when $\mu \to 0$ and $\mu\gamma \to \infty$, the flexible linearization method is reduced to the linearization method with rigid constraint, and when $\mu \to 0$ and $0 < \gamma < \infty$, the flexible linearization method is reduced to the linearization method with two-step regression. The linearization method with rigid constraint and the linearization method with two-step regression seems unrelated from their formulations, it is interesting that these two linearization methods are the two extreme cases of the flexible linearization method ($\mu\gamma \to \infty$ and $\mu\gamma \to 0$).

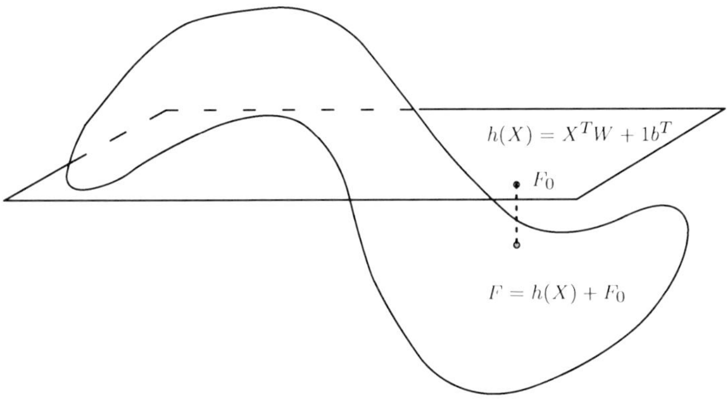

Fig. 1 Illustration of the flexible linearization method. This method aims to optimize the prediction labels F, the linear regression function $h(X)$, and the regression residue F_0 simultaneously. The regression residue F_0 measures the mismatch between F and $h(X)$

4 Flexible Manifold Embedding with the Flexible Linearization Method

In this section, based on the flexible linearization method, we introduce a new framework for semi-supervised and unsupervised subspace learning, referred to as *F*lexible *M*anifold *E*mbedding (*FME*), and *FME/U*, respectively [23].

It is worth mentioning that FME and FME/U are linear methods, which are fast and suitable for practical applications such as face, object, and text classification problems. Note that as in most linear subspace learning methods, FME and FME/U can be conducted in the Reproducing Kernel Hilbert Space (RKHS), which give rise to *Kernel FME (KFME)* and *Kernel FME/U (KFME/U)* [40], and can also be easily extended to two-dimensional or high-order tensor FME and FME/U.

4.1 Semi-supervised Flexible Manifold Embedding

As in (9), we assume that $F = h(X) + F_0 = X^{\mathrm{T}}W + \mathbf{1}b^{\mathrm{T}} + F_0$, where $F_0 \in \mathbb{R}^{m \times c}$ is the regression residue modeling the mismatch between F and $h(X)$. FME aims to optimize the prediction labels F, the regression residue F_0, and the linear regression function $h(X)$ simultaneously:

$$(F^*, F_0^*, W^*, b^*) = \arg\min_{F,F_0,W,b} \mathrm{Tr}(F-Y)^{\mathrm{T}}U(F-Y) + \mathrm{Tr}(F^{\mathrm{T}}MF) + \mu(\|W\|^2 + \gamma\|F_0\|^2), \quad (11)$$

where the two coefficients μ and γ are parameters to balance different terms, and $M \in \mathbb{R}^{m\times m}$ is the Laplacian matrix and $U \in \mathbb{R}^{m\times m}$ is the diagonal matrix. Note that similar idea was also discussed in the prior work [1, 27, 41] for binary classification problems. Here, we extend this idea for dimension reduction in multi-class setting, in which the class dependency can be captured by the extracted features.

Similarly as in LGC, GFHF, and LapRLS/L, the first two terms in (11) represent the label fitness and the manifold smoothness, respectively. Considering that it is meaningless to enforce the prediction labels $F_{i.}$ and the given labels $Y_{j.}$ of different samples (i.e., $j \neq i$) to be close, we set the matrix U as the diagonal matrix with the first n and the rest $m-n$ diagonal elements as 1 and 0, respectively, similarly as in LapRLS/L. In addition, the matrix M should be set as the graph Laplacian matrix in order to utilize the manifold structure (i.e., F should be as smooth as possible on the whole graph) in semi-supervised learning. While it is possible to construct the Laplacian matrix M according to different manifold learning criterions [25, 35–37, 39, 45], similarly as in GFHF and LapRLS/L, we choose the Gaussian function to calculate M, namely, $M = D - S$, where D is a diagonal matrix with the diagonal elements as $D_{ii} = \sum_j S_{ij}, \forall\, i$, and $S_{ij} = \exp(-\|x_i - x_j\|^2/t)$, if x_i (or x_j) is among k nearest neighbors of x_j (or x_i); $S_{ij} = 0$, otherwise.

The last two terms in (11) control the norm of projection matrix W and the regression residue F_0. In the current formulation of F, the regression function $h(X)$ and the regression residue F_0 are combined. In practice, our work can naturally map the new data points for dimension reduction by using the function $h(X)$. The regression residue F_0 can model the mismatch between the linear regression function $X^{\mathrm{T}}W + \mathbf{1}b^{\mathrm{T}}$ and the prediction labels F. Compared with LapRLS/L, we do not force the prediction labels F to lie in the space spanned by all the samples X. Therefore, our framework is more flexible and it can better cope with the samples which reside on the nonlinear manifold. Moreover, the prior work [17] on face hallucination has demonstrated that the introduction of a local residue can lead to better reconstruction of face images.

Replacing F_0 with $F - X^{\mathrm{T}}W - \mathbf{1}b^{\mathrm{T}}$, we have:

$$
\begin{aligned}
(F^*, W^*, b^*) = \arg\min_{F,W,b} & \operatorname{Tr}(F-Y)^{\mathrm{T}}U(F-Y) \\
& + \operatorname{Tr}(F^{\mathrm{T}}MF) + \mu(\|W\|^2 + \gamma\|X^{\mathrm{T}}W + \mathbf{1}b^{\mathrm{T}} - F\|^2), \qquad (12)
\end{aligned}
$$

From then on, we refer to the objective function in (12) as $g(F, W, b)$. First, we prove that the optimization problem in (12) is jointly convex with respect to F, W and b.

Theorem 1. *Denote $U, M \in \mathbb{R}^{m\times m}$, $F, Y \in \mathbb{R}^{m\times c}$, $W \in \mathbb{R}^{f\times c}$, $b \in \mathbb{R}^{c\times 1}$. If matrices U and M are positive semi-definite, $\mu \geq 0$ and $\gamma \geq 0$, then $g(F, W, b) = \operatorname{Tr}(F - Y)^{\mathrm{T}}U(F-Y) + \operatorname{Tr}(F^{\mathrm{T}}MF) + \mu(\|W\|^2 + \gamma\|X^{\mathrm{T}}W + \mathbf{1}b^{\mathrm{T}} - F\|^2)$ is jointly convex with respect to F, W and b.*

Proof. In function $g(F, W, b)$, we remove the constant term $\operatorname{Tr}(Y^{\mathrm{T}}UY)$, then $g(F, W, b)$ can be rewritten in matrix form as:

$$g(F,W,b) = \mathrm{Tr}\begin{bmatrix} F \\ W \\ b^{\mathrm{T}} \end{bmatrix}^{\mathrm{T}} P \begin{bmatrix} F \\ W \\ b^{\mathrm{T}} \end{bmatrix} - \mathrm{Tr}\begin{bmatrix} F \\ W \\ b^{\mathrm{T}} \end{bmatrix}^{\mathrm{T}} \begin{bmatrix} 2UY \\ 0 \\ 0 \end{bmatrix},$$

where

$$P = \begin{bmatrix} \mu\gamma I + M + U & -\mu\gamma X^{\mathrm{T}} & -\mu\gamma \mathbf{1} \\ -\mu\gamma X & \mu I + \mu\gamma X X^{\mathrm{T}} & \mu\gamma X \mathbf{1} \\ -\mu\gamma \mathbf{1}^{\mathrm{T}} & \mu\gamma \mathbf{1}^{\mathrm{T}} X^{\mathrm{T}} & \mu\gamma m \end{bmatrix}.$$

Thus in order to prove that $g(F,W,b)$ is jointly convex with respect to F, W and b, we only need to prove that the matrix P is positive semi-definite.

For any vector $z = [z_1^{\mathrm{T}}, z_2^{\mathrm{T}}, z_3]^{\mathrm{T}} \in \mathbb{R}^{(m+f+1)\times 1}$, where $z_1 \in \mathbb{R}^{m\times 1}$ $z_2 \in \mathbb{R}^{f\times 1}$, and z_3 is a scalar, we have

$$\begin{aligned} & z^{\mathrm{T}} P z \\ &= z_1^{\mathrm{T}}(\mu\gamma I + M + U)z_1 - 2\mu\gamma z_1^{\mathrm{T}} X^{\mathrm{T}} z_2 - 2\mu\gamma z_1^{\mathrm{T}} \mathbf{1} z_3 \\ &\quad + z_2^{\mathrm{T}}(\mu I + \mu\gamma X X^{\mathrm{T}}) z_2 + 2\mu\gamma z_2^{\mathrm{T}} X \mathbf{1} z_3 + \mu\gamma m z_3^{\mathrm{T}} z_3 \\ &= z_1^{\mathrm{T}}(M+U)z_1 + \mu z_2^{\mathrm{T}} z_2 + \mu\gamma(z_1^{\mathrm{T}} z_1 - 2 z_1^{\mathrm{T}} X^{\mathrm{T}} z_2 \\ &\quad - 2 z_1^{\mathrm{T}} \mathbf{1} z_3 + z_2^{\mathrm{T}} X X^{\mathrm{T}} z_2 + 2 z_2^{\mathrm{T}} X \mathbf{1} z_3 + m z_3^{\mathrm{T}} z_3) \\ &= z_1^{\mathrm{T}}(M+U)z_1 + \mu z_2^{\mathrm{T}} z_2 + \mu\gamma(z_1 - X^{\mathrm{T}} z_2 - \mathbf{1} z_3)^{\mathrm{T}} \\ & (z_1 - X^{\mathrm{T}} z_2 - \mathbf{1} z_3). \end{aligned}$$

So if U and M are positive semi-definite, $\mu \geq 0$ and $\gamma \geq 0$, then $z^{\mathrm{T}} P z \geq 0$ for any z, and thus P is positive semi-definite. Therefore, $g(F,W,b)$ is jointly convex with respect to F, W and b.

A much more simple proof can be found from the fact that $g(F,W,b)$ is a sum of several convex functions, thus $g(F,W,b)$ is jointly convex. □

To obtain the optimal solution, we set the derivatives of the objective function in (12) with respect to b and W equal to zero. We have:

$$\begin{aligned} b &= \frac{1}{m}(F^{\mathrm{T}}\mathbf{1} - W^{\mathrm{T}} X \mathbf{1}) \\ W &= \gamma(\gamma X H_c X^{\mathrm{T}} + I)^{-1} X H_c F = AF, \end{aligned} \tag{13}$$

where $A = \gamma(\gamma X H_c X^{\mathrm{T}} + I)^{-1} X H_c$ and $H_c = I - \frac{1}{m}\mathbf{1}\mathbf{1}^{\mathrm{T}}$ is used for centering the data by subtracting the mean of the data With W and b, we rewrite the regression function $X^{\mathrm{T}}W + \mathbf{1}b^{\mathrm{T}}$ in (12) as:

$$\begin{aligned} X^{\mathrm{T}}W + \mathbf{1}b^{\mathrm{T}} &= X^{\mathrm{T}}AF + \frac{1}{m}\mathbf{1}\mathbf{1}^{\mathrm{T}}F - \frac{1}{m}\mathbf{1}\mathbf{1}^{\mathrm{T}}X^{\mathrm{T}}AF \\ &= H_c X^{\mathrm{T}} AF + \frac{1}{m}\mathbf{1}\mathbf{1}^{\mathrm{T}}F = BF, \end{aligned} \tag{14}$$

Algorithm 1: Procedure of FME

Given a binary label matrix $Y \in \mathbb{B}^{m\times c}$ and a sample set $X = [x_1, x_2, \ldots, x_m] \in \mathbb{R}^{f\times m}$, where $x_i|_{i=1}^{n}$ and $x_i|_{i=n+1}^{m}$ are labeled and unlabeled data respectively.

1: Set M as the graph Laplacian matrix $L \in \mathbb{R}^{m\times m}$, and $U \in \mathbb{R}^{m\times m}$ as the diagonal matrix with the first n and the rest $m-n$ diagonal entries as 1 and 0 respectively.
2: Compute the optimal F with (17).
3: Compute the optimal projection matrix W with (13).

where $B = H_cX^{\mathrm{T}}A + \frac{1}{m}\mathbf{1}\mathbf{1}^{\mathrm{T}}$. Replacing W and b to $g(F,W,b)$ in (12), we arrive at:

$$\begin{aligned} F^* &= \arg\min_F \mathrm{Tr}(F-Y)^{\mathrm{T}}U(F-Y) + \mathrm{Tr}(F^{\mathrm{T}}MF) \\ &\quad + \mu(\mathrm{Tr}(F^{\mathrm{T}}A^{\mathrm{T}}AF) + \gamma\mathrm{Tr}(BF-F)^{\mathrm{T}}(BF-F)). \end{aligned}$$

By setting the derivative of this objective function with respect to F as 0, the prediction labels F are obtained by:

$$F = (U + M + \mu\gamma(B-I)^{\mathrm{T}}(B-I) + \mu A^{\mathrm{T}}A)^{-1}UY. \tag{15}$$

Using $H_cH_c = H_c = H_c^{\mathrm{T}}$ and $\mu\gamma A^{\mathrm{T}}XH_cX^{\mathrm{T}}A + \mu A^{\mathrm{T}}A = \mu\gamma A^{\mathrm{T}}XH_c = \mu\gamma H_cX^{\mathrm{T}}A$, the term $\mu\gamma(B-I)^{\mathrm{T}}(B-I) + \mu A^{\mathrm{T}}A$ in (15) can be rewritten as $\mu\gamma(A^{\mathrm{T}}X - I)H_c(X^{\mathrm{T}}A - I) + \mu A^{\mathrm{T}}A$ or $\mu\gamma A^{\mathrm{T}}XH_cX^{\mathrm{T}}A - 2\mu\gamma H_cX^{\mathrm{T}}A + \mu\gamma H_c + \mu A^{\mathrm{T}}A$. Then, we have:

$$\begin{aligned} &\mu\gamma(B-I)^{\mathrm{T}}(B-I) + \mu A^{\mathrm{T}}A \\ &= \mu\gamma H_c - \mu\gamma^2 H_cX^{\mathrm{T}}(\gamma XH_cX^{\mathrm{T}} + I)^{-1}XH_c. \end{aligned} \tag{16}$$

By defining $X_c = XH_c$, we can also calculate the prediction labels F by

$$F = (U + M + \mu\gamma H_c - \mu\gamma^2 N)^{-1}UY, \tag{17}$$

where $N = X_c^{\mathrm{T}}(\gamma X_cX_c^{\mathrm{T}} + I)^{-1}X_c = X_c^{\mathrm{T}}X_c(\gamma X_c^{\mathrm{T}}X_c + I)^{-1} = \frac{1}{\gamma}I - \frac{1}{\gamma}(\gamma X_c^{\mathrm{T}}X_c + I)^{-1}$.

4.2 Unsupervised Flexible Manifold Embedding

We introduce a simplified version for unsupervised learning by setting the diagonal elements of matrix U in (12) equal to 0. We also pursue the projection matrix W, the bias term b and the latent variable F simultaneously[1]:

[1] An alternative method is to solve $\min_{F,W,b,W^{\mathrm{T}}W=I} \mathrm{Tr}(F^{\mathrm{T}}MF) + \gamma\|X^{\mathrm{T}}W + \mathbf{1}b^{\mathrm{T}} - F\|^2)$, which is equivalent to $\min_{F,W,W^{\mathrm{T}}W=I} \mathrm{Tr}(F^{\mathrm{T}}MF) + \gamma\|X^{\mathrm{T}}W - F\|^2)$.

Algorithm 2: Procedure of FME/U

Given the unlabeled sample set as $X = [x_1, x_2, \ldots, x_m] \in \mathbb{R}^{f \times m}$.

1: Set M as the graph Laplacian matrix $L \in \mathbb{R}^{m \times m}$.
2: Compute the optimal F with (20) by generalized eigenvalue decomposition.
3: Compute the optimal projection matrix W with (13).

$$(F^*, W^*, b^*) = \arg \min_{F,W,b,\ F^{\mathrm{T}}VF=I} \mathrm{Tr}(F^{\mathrm{T}}MF) + \mu(\|W\|^2 + \gamma\|X^{\mathrm{T}}W + \mathbf{1}b^{\mathrm{T}} - F\|^2), \tag{18}$$

where V can be set as H_c, I is an identity matrix, and the coefficients μ and γ are two parameters to balance different terms.

In unsupervised learning, the variable F can be treated as the latent variable, denoting the lower dimensional representation. Similar to prior work (e.g., LE [3] and LPP [13]), we constrain that F after centering operation lies in a sphere (i.e., $F^{\mathrm{T}}VF = I$) to avoid the trivial solution $F = 0$, where we set $V = H_c$. Beside unsupervised learning, the formulation in (18) is a general formulation, which can be also used for supervised learning by using different matrices M and V. Again, FME/U naturally provides a method for mapping new data points through the regression function $h(X) = X^{\mathrm{T}}W + \mathbf{1}b^{\mathrm{T}}$. Compared with the prior linear dimension reduction algorithms (such as PCA, LDA, LPP), the rigid mapping function $F = X^{\mathrm{T}}W$ in these methods is relaxed by introducing a flexible penalty term (i.e., regression residue $\|h(X) - F\|^2$) in (18).

Similarly, by setting the derivatives of the objective function in (18) with respect to W and b to zero, W and b can be calculated by (13). Substituting W and b back in (18), then we have:

$$F^* = \arg \min_{F, F^{\mathrm{T}}H_cF=I} \mathrm{Tr}(F^{\mathrm{T}}MF) + \mu(\mathrm{Tr}(F^{\mathrm{T}}A^{\mathrm{T}}AF) + \gamma \mathrm{Tr}(BF - F)^{\mathrm{T}}(BF - F)). \tag{19}$$

According to (16), we rewrite (19) as:

$$\begin{aligned} F^* &= \arg \min_{F, F^{\mathrm{T}}H_cF=I} TrF^{\mathrm{T}}(M + \mu\gamma H_c - \mu\gamma^2 N)F \\ &= \arg \min_{F, F^{\mathrm{T}}H_cF=I} TrF^{\mathrm{T}}(M - \mu\gamma^2 N)F, \end{aligned} \tag{20}$$

where $N = X_c^{\mathrm{T}}(\gamma X_c X_c^{\mathrm{T}} + I)^{-1}X_c = X_c^{\mathrm{T}}X_c(\gamma X_c^{\mathrm{T}}X_c + I)^{-1}$. This objective function can be optimized by generalized eigenvalue decomposition [37].

4.3 Discussions with the Prior Work

In this section, we discuss the connection between FME and semi-supervised algorithms LGC [47], GFHF [49], and LapRLS/L [29]. We also discuss the connection between FME/U with graph embedding framework [37] and spectral regression [8].

4.3.1 Connection Between FME and Semi-supervised Learning Algorithms

Example 1. LGC and GFHF are two special cases of FME.

Proof. If we set $\mu = 0$, then the objective function of FME in (12) reduces to (2), which is a general formulation for both LGC and GFHF. Therefore, LGC and GFHF are special cases of FME. □

Example 2. LapRLS/L is also a special case of FME.

Proof. If we set $\mu = \frac{\lambda_A}{\lambda_I}$ and $\gamma \to \infty$ (i.e., $\mu\gamma \to \infty$) in (12), we have $F = X^{\mathrm{T}}W + \mathbf{1}b^{\mathrm{T}}$. Replacing F to (12), then we have a new formulation for FME:

$$\begin{aligned} g(W,b) = {} & \mathrm{Tr}(X^{\mathrm{T}}W + \mathbf{1}b^{\mathrm{T}})^{\mathrm{T}}M(X^{\mathrm{T}}W + \mathbf{1}b^{\mathrm{T}}) + \mu\|W\|^2 \\ & + \mathrm{Tr}(X^{\mathrm{T}}W + \mathbf{1}b^{\mathrm{T}} - Y)^{\mathrm{T}}U(X^{\mathrm{T}}W + \mathbf{1}b^{\mathrm{T}} - Y). \end{aligned} \tag{21}$$

If we further set $M = L$ and the first n and the rest $m - n$ diagonal elements of the diagonal matrix U in (21) as $\frac{1}{n\lambda_I}$ and 0, respectively, then $g(W,b)$ is equal to $\frac{1}{\lambda_I}g_M(W,b)$ in (3). That is LapRLS/L is also a special case of FME. □

4.3.2 Connection Between FME/U and Graph Embedding Framework

Recently, Yan et al. [37] proposed a general graph-embedding framework to unify a large family of dimension reduction algorithms (such as PCA, LDA, ISOMAP, LLE, and LE). As shown in [37], the statistical or geometric properties of a given algorithm are encoded as graph relationships, and each algorithm can be considered as direct graph embedding, linear graph embedding, or other extensions. The objective function in direct graph embedding is:

$$F^* = \arg \min_{F,\, F^{\mathrm{T}}VF = I} \mathrm{Tr}(F^{\mathrm{T}}MF), \tag{22}$$

where V is another graph Laplacian matrix (e.g., the centering matrix H_c) such that $V\mathbf{1} = \mathbf{0}$ and $\mathbf{1}^{\mathrm{T}}V = \mathbf{0}^{\mathrm{T}}$.

While direct graph embedding computes a low-dimensional representation F for the training samples, it does not provide a method to map new data points.

For mapping out-of-sample data points, linearization and other extensions (e.g., kernelization and tensorization) are also proposed in [37]. Assuming a rigid linear mapping function $F = X^{\mathrm{T}}W + \mathbf{1}b^{\mathrm{T}}$, the objective function in linear graph embedding is formulated as:

$$\begin{aligned} W^* &= \arg \min_{W,\ (X^{\mathrm{T}}W+\mathbf{1}b^{\mathrm{T}})^{\mathrm{T}}V(X^{\mathrm{T}}W+\mathbf{1}b^{\mathrm{T}})=I} \\ &\quad \mathrm{Tr}(X^{\mathrm{T}}W + \mathbf{1}b^{\mathrm{T}})^{\mathrm{T}}M(X^{\mathrm{T}}W + \mathbf{1}b^{\mathrm{T}}), \\ &= \arg \min_{W,\ W^{\mathrm{T}}XVX^{\mathrm{T}}W=I} \mathrm{Tr}(W^{\mathrm{T}}XMX^{\mathrm{T}}W). \end{aligned} \tag{23}$$

Example 3. Direct graph embedding and its linearization are special cases of FME/U.

Proof. If we set μ in (18) as 0, then the objective function of FME/U reduces to the formulation of direct graph embedding in (22).

When $\mu \to 0$ and $\mu\gamma \to \infty$ in (18), then we have $F = X^{\mathrm{T}}W + \mathbf{1}b^{\mathrm{T}}$. Replacing F to (18) then the objective function of FME/U reduces to the formulation of linear graph embedding in (23).

Therefore, direct graph embedding and its linearization are special cases of FME/U. □

Note that one recently published semi-supervised dimension reduction method, transductive component analysis (TCA) [18], is closely related to the FME/U. However, TCA is a special case of Graph Embedding Framework [37], in which the matrix M is a weighted sum of two matrices M_1 and M_2, i.e. $M = M_1 + \beta M_2$, where $\beta > 0$ is a trade-off parameter to control the importance between the two matrices. The first matrix $M_1 = (I + \alpha L)^{-1}(\alpha L)$ models two terms related to the manifold regularization and the embedding (similarly as in (18)), where $\alpha > 0$ is a parameter to balance two terms. The second matrix M_2 models the average margin criterion of the distance constraints for labeled data. Moreover, the prediction label matrix is constrained as $F = X^{\mathrm{T}}W$. In comparison, the FME and FME/U do not constrain $F = X^{\mathrm{T}}W$ on the prediction labels or the lower-dimensional representation. For semi-supervised setting, the (17) in FME can be solved by a linear system, which is much more efficient than solving the eigenvalue decomposition problem as in TCA and many other dimension reduction methods [2, 7, 13].

4.3.3 Connection Between FME/U and Spectral Regression

Cai et al. [8] recently proposed a two-step method, referred to as spectral regression (SR), to solve the projection matrix W for mapping new data points. Firstly, the optimal solution F of (22) is solved. And then, the optimal projection matrix W is computed by solving a regression problem:

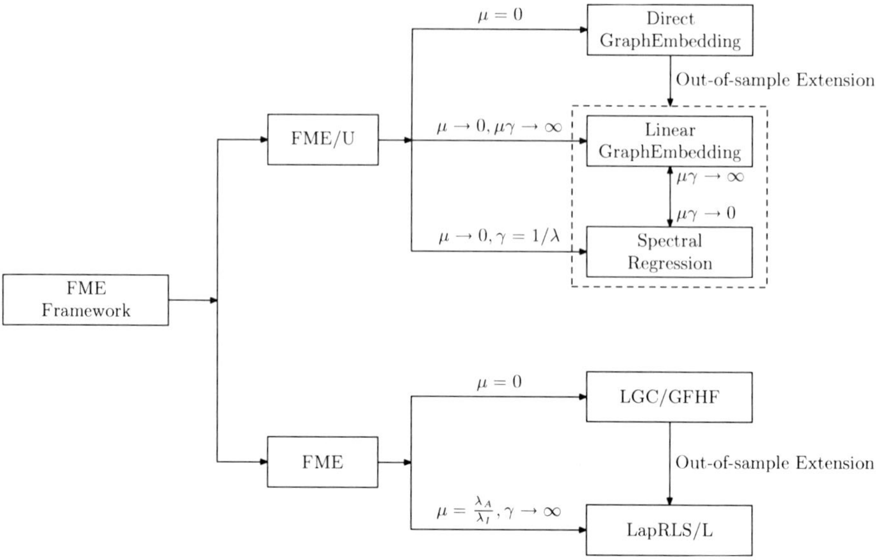

Fig. 2 The relationship of FME framework and other related methods

$$W^* = \arg\min_W \|X^T W + \mathbf{1}b^T - F\|^2 + \lambda \|W\|^2. \tag{24}$$

Example 4. Spectral Regression is also a special case of FME/U.

Proof. When $\mu \to 0$ and $\gamma = \frac{1}{\lambda}$ (i.e., $\mu\gamma \to 0$) in (18), then (18) reduces to (22), namely, we need to solve F at first. Then the objective function in (18) is converted to (24) to solve W. Note that the optimal W^* of the objective function of SR (i.e., (24)) is $W^* = (XH_cX^T + \lambda I)^{-1}XH_cF$, which is equal to W^* from FME/U (See (13)). Therefore, spectral regression is also a special case of FME/U. □

4.3.4 Discussion

The relationships of FME framework with other related methods are shown in Fig. 2. Direct graph embedding [37] has unified a large family of dimension reduction algorithms (e.g., ISOMAP, LLE, and LE), and LGC [47] and GFHF [49] are two classical graph-based semi-supervised learning methods. However, direct graph embedding and LGC/GFHF do not yield a method for mapping new data points that are not included in the training set. To cope with the out-of-sample problem for Direct Graph Embedding and LGC/GFHF, a linear mapping function is used in Linear Graph Embedding [37] and LapRLS/L [29], respectively. In spectral regression [8], a two-step approach is proposed to obtain the projection matrix for mapping new data points.

While the objective function of LapRLS/L (resp. Linear Graph Embedding) is not in the objective function of FME (resp. FME/U), they are still special cases of the FME framework by using different paraments μ and γ (or $\mu\gamma$). Moreover, the FME framework also reveals that the previously unrelated methods are in fact related. For example, linear graph embedding and spectral regression seem to be unrelated from their objective functions; however, they are both special cases of FME/U. Specially, FME/U reduces to linear graph embedding, when $\mu \to 0$ and $\mu\gamma \to \infty$. FME/U reduces to Spectral Regression, when $\mu \to 0$ and $\gamma = \frac{1}{\lambda}$ (i.e., $\mu\gamma \to 0$). Finally, the FME framework can be also used to develop new dimension reduction algorithms. For example, similar as in spectral regression [8], it is also possible to use the FME framework to develop a two-step approach for semi-supervised learning by setting $\mu \to 0$ and $\mu\gamma \to 0$.

5 Experimental Results

In the experiments, we use three face databases UMIST [11], CMU PIE [26] and YALE-B [10], one object database COIL-20 database [19], and one text database 20-NEWS.

Face Databases: The UMIST database [11] consists of 575 multi-view images of 20 people, covering a wide range of poses from profile to frontal views. The images are cropped and then resized to 28×23 pixels. The CMU PIE database [26] contains more than 40,000 facial images of 68 people. The images were acquired over different poses, under variable illumination conditions, and with different facial expressions. In this experiment, we choose the images from the frontal pose (C27) and each subject has around 49 images from varying illuminations and facial expressions. The images are cropped and then resized to 32×32 pixels. For the YALE-B database [10], 38 subjects are used in this work, with each person having around 64 near frontal images under different illuminations. The images are cropped and then resized to 32×32 pixels. In this work, gray-level features are used for face recognition. For each face database, ten images are shown in Fig. 3.

Object Database: The COIL-20 database [19] consists of images of 20 objects, and each object has 72 images captured from varying angles at intervals of five degrees. We resize each image to 32×32 pixels, and then extract a 1,024 dimensional gray-level feature for each image. Ten images are also shown in Fig. 3.

Text Database: The 20-NEWS database[2] is used for text categorization. The topic *rec* which contains *autos*, *motorcycles*, *baseball*, and *hockey* was chosen from the version 20-news-18828. The articles were preprocessed with the same procedure as in [17]. In total, we have 3,970 documents. We extract a 8014-dimensional tf-idf (token frequency-inverse document frequency) feature for each document.

[2] Available at http://people.csail.mit.edu/jrennie/20Newsgroups/.

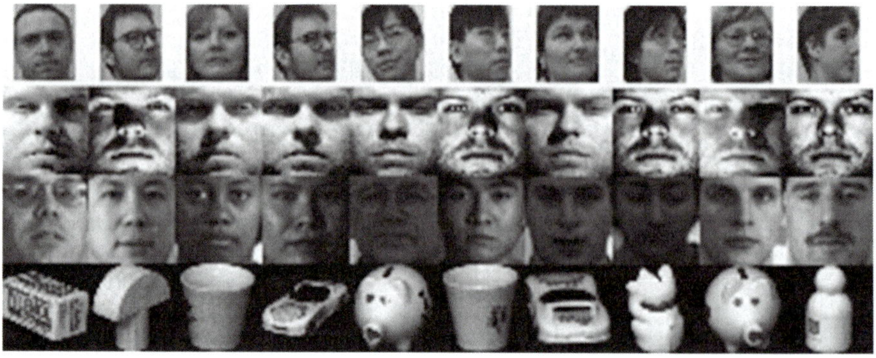

Fig. 3 Ten randomly selected image samples in each image database (From top to bottom: UMIST, YALE-B, CMU PIE and COIL-20)

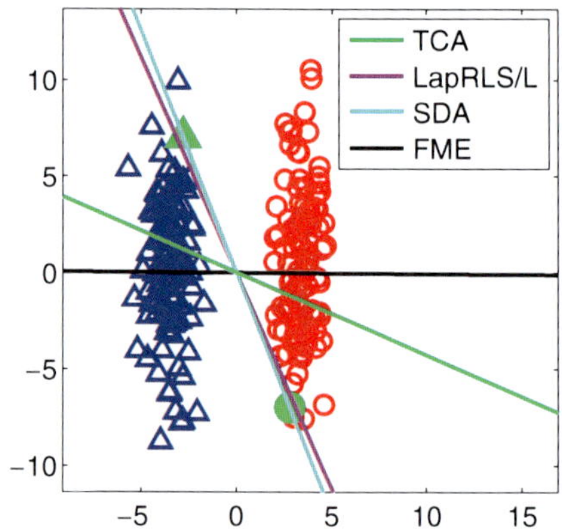

Fig. 4 The results of TCA [18], LapRLS/L [29], SDA Semi-supervised Discriminant Analysis (SDA) [7, 30] and the FME on a toy problem

5.1 Semi-supervised Learning

We firstly compare FME with other dimension reduction algorithms TCA [18], SDA Semi-supervised Discriminant Analysis (SDA) [7, 30] and LapRLS/L [29] on a toy problem. The data are sampled from two Gaussian distributions of two classes, represented by red circles and blue triangles, respectively, in Fig. 4. For each class, we label only sample (denoted by green color) and treat other samples as unlabeled data. The projection direction of TCA, SDA, LapRLS/L, and FME are shown in Fig. 4. From it, we observe that TCA, SDA, and LapRLS/L fail to find the optimal direction for all the samples. LapRLS/L does work for this toy problem, possibly because the assumption that the prediction label matrix lies in the space spanned by the training samples is not satisfied. However, FME successfully derives the discriminative direction by modeling the regression residue.

We also compare FME with LGC [47], GFHF [49], TCA [18], SDA Semi-supervised Discriminant Analysis (SDA) [7, 30], LapRLS/L [29] and MFA [37] for real recognition tasks. For dimension reduction algorithms TCA, SDA, LapRLS/L,

MFA, and FME, the nearest neighbor classifier is performed for classification after dimension reduction. For LGC and GFHF, we directly use the classification methods proposed in [47,49] for classification. For GFHF, LapRLS/L, TCA, SDA and FME, we need to determine the Laplacian matrix M (or L) beforehand. We choose the Gaussian function to calculate M or L, in which the graph similarity matrix is set as $S_{ij} = \exp(-\|x_i - x_j\|^2/t)$, if x_i (or x_j), if x_i (or x_j) is among k nearest neighbors of x_j (or x_i); $S_{ij} = 0$, otherwise. For LGC, we used the normalized graph Laplacian matrix $\tilde{L} = I - D^{-\frac{1}{2}}SD^{-\frac{1}{2}}$, as suggested in [47]. For fair comparison, we fix $k = 10$ and t is set as the method in [21]. For LGC, GFHF and LapRLS/L, the diagonal matrix U is determined according to [29,47,49], respectively. For FME, we set the the first n and the rest $m-n$ diagonal elements of the diagonal matrix U as 1 and 0, respectively, similarly as in LapRLS/L.

In all the experiments, PCA is used as a preprocessing step to preserve 95% energy of the data, similarly as in [13,37]. In order to fairly compare FME with TCA, SDA, LapRLS/L and MFA, the final dimensions after dimension reduction are fixed as c. For SDA, LapRLS/L, TCA and FME, two regularization parameters (i.e., μ and γ in FME, λ_I and λ_A in LapRLS/L, α and β in SDA and TCA) need to be set beforehand to balance different terms. For fair comparison, we set each parameter to $\{10^{-9}, 10^{-6}, 10^{-3}, 10^{0}, 10^{3}, 10^{6}, 10^{9}\}$, and then we report the top-1 recognition accuracy from the best parameter configuration. In Fig. 5, we plot the recognition accuracy variation with different parameter μ for FME, in which three labeled samples per class are used in UMIST, YALE-B, CMU PIE, and COIL-20 databases, and 30 labeled samples per class are used in 20-NEWS database. We observe that FME is relatively robust to the parameter μ when μ is small (i.e., $\mu \leq 10^{-3}$). It is still an open problem to determine the optimal parameters, which will be investigated in the future.

We randomly select 50% data as the training dataset and use the remaining 50% data as the test dataset. Among the training data, we randomly label p samples per class and treat the other training samples as unlabeled data. The above setting (referred to as semi-supervised setting) has been used in [7], and it is also a more natural setting to compare different dimension reduction algorithms. For UMIST, CMU PIE, YALE-B and COIL-20 databases, we set p as 1, 2, and 3, respectively. For the 20-NEWS text database, we set p as 10, 20, and 30, respectively, because each class has much more training samples in this database. All the training data are used to learn a subspace (i.e., a projection matrix) or a classifier, except that we only use the labeled data for subspace learning in MFA [37]. We report the mean recognition accuracy and standard deviation over 20 random splits on the unlabeled dataset and the unseen test dataset, which are referred to as *Unlabel* and *Test*, respectively, in Table 1. In table, the results shown in boldface are significantly better than the others, judged by t-test with a significance level of 0.05. We have the following observations:

1. Semi-supervised dimension reduction algorithms TCA, SDA, and LapRLS/L outperform supervised MFA in terms of mean recognition accuracy, which demonstrates that unlabeled data can be used to improve the recognition performance.

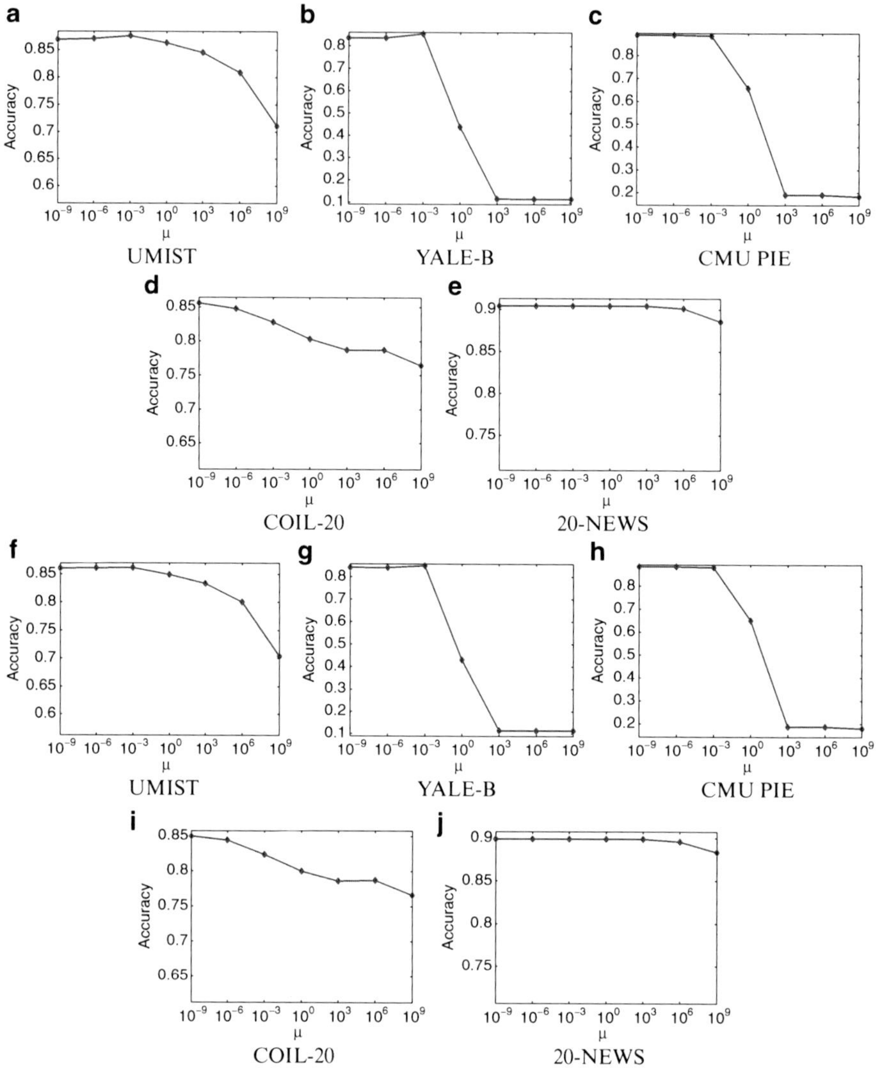

Fig. 5 Recognition accuracy variation with different parameter μ for FME. The first two rows show the results on the unlabeled dataset and the second two rows show the results on the unseen test dataset. Three labeled samples per class are used in UMIST, YALE-B, CMU PIE, and COIL-20 databases, and 30 labeled samples per class are used in 20-NEWS database

Table 1 Top-1 recognition performance (mean recognition accuracy ± standard deviation %) of MFA [37], GFHF [49], LGC [47], TCA [18], SDA semi-supervised discriminant analysis (SDA) [7, 30], LapRLS/L [29] and FME over 20 random splits on five databases

		1 labeled sample		2 labeled samples		3 labeled samples	
Dataset	Method	Unlabel (%)	Test (%)	Unlabel (%)	Test (%)	Unlabel (%)	Test (%)
UMIST	MFA	–	–	70.5 ± 4.2	70.8 ± 3.9	81.1 ± 3.9	80.6 ± 4.2
	GFHF	*63.6 ± 6.2*	–	*79.1 ± 3.9*	–	*85.8 ± 3.5*	–
	LGC	*64.5 ± 5.9*	–	*79.3 ± 3.6*	–	83.8 ± 3.9	–
	TCA	*63.2 ± 5.2*	*62.9 ± 5.8*	*78.0 ± 4.3*	*77.9 ± 4.2*	83.9 ± 4.1	83.6 ± 3.8
		$(10^3, 10^0)$	$(10^3, 10^0)$	$(10^3, 10^6)$	$(10^3, 10^9)$	$(10^3, 10^0)$	$(10^3, 10^0)$
	SDA	56.2 ± 5.4	55.6 ± 5.1	76.8 ± 4.2	76.2 ± 4.3	83.7 ± 3.8	83.2 ± 3.9
		$(10^{-9}, 10^{-6})$	$(10^{-9}, 10^{-9})$	$(10^0, 10^3)$	$(10^0, 10^3)$	$(10^{-3}, 10^3)$	$(10^0, 10^3)$
	LapRLS/L	58.1 ± 5.9	57.9 ± 6.1	74.9 ± 4.6	74.3 ± 4.5	82.1 ± 3.9	81.7 ± 3.8
		$(10^6, 10^9)$	$(10^3, 10^3)$	$(10^6, 10^9)$	$(10^3, 10^6)$	$(10^6, 10^9)$	$(10^3, 10^6)$
	FME	*63.5 ± 5.5*	*63.1 ± 5.4*	*79.7 ± 4.5*	*79.1 ± 4.2*	*86.9 ± 3.2*	*86.1 ± 3.1*
		$(10^{-9}, 10^{-6})$	$(10^{-9}, 10^{-6})$	$(10^{-9}, 10^{-5})$	$(10^{-9}, 10^{-6})$	$(10^{-3}, 10^{-6})$	$(10^{-3}, 10^{-6})$

		1 labeled sample		2 labeled samples		3 labeled samples	
Dataset	Method	Unlabel(%)	Test(%)	Unlabel(%)	Test(%)	Unlabel(%)	Test(%)
YALE-B	MFA	–	–	49.6 ± 4.6	49.1 ± 4.9	68.1 ± 2.9	68.6 ± 2.8
	GFHF	22.5 ± 2.9	–	35.9 ± 3.3	–	45.2 ± 3.9	–
	LGC	29.2 ± 3.1	–	42.1 ± 3.1	–	49.6 ± 3.5	–
	TCA	38.5 ± 3.0	39.6 ± 3.2	71.6 ± 3.3	71.4 ± 3.1	81.5 ± 2.9	81.2 ± 2.3
		$(10^0, 10^0)$	$(10^0, 10^0)$	$(10^0, 10^6)$	$(10^0, 10^6)$	$(10^3, 10^0)$	$(10^3, 10^0)$
	SDA	38.2 ± 3.0	39.0 ± 3.1	72.1 ± 3.6	71.9 ± 3.2	83.8 ± 2.1	83.0 ± 2.1
		$(10^0, 10^{-9})$	$(10^0, 10^{-9})$	$(10^0, 10^{-6})$	$(10^0, 10^{-6})$	$(10^0, 10^{-3})$	$(10^0, 10^{-3})$
	LapRLS/L	*52.9 ± 3.6*	*53.2 ± 3.1*	73.9 ± 3.2	73.6 ± 2.9	84.2 ± 2.6	83.8 ± 2.5
		$(10^{-3}, 10^{-9})$	$(10^{-3}, 10^{-9})$	$(10^0, 10^{-6})$	$(10^0, 10^{-9})$	$(10^0, 10^{-9})$	$(10^0, 10^{-9})$
	FME	*53.9 ± 3.3*	*54.2 ± 2.9*	*75.1 ± 2.6*	*75.0 ± 2.7*	*85.9 ± 2.1*	*85.6 ± 2.0*
		$(10^{-6}, 10^6)$	$(10^{-6}, 10^6)$	$(10^{-3}, 10^3)$	$(10^{-3}, 10^3)$	$(10^{-3}, 10^3)$	$(10^{-3}, 10^3)$

(continued)

Table 1 (continued)

Dataset	Method	1 labeled sample		2 labeled samples		3 labeled samples	
		Unlabel(%)	Test(%)	Unlabel(%)	Test(%)	Unlabel(%)	Test(%)
CMU PIE	MFA	–	–	72.1 ± 2.6	71.8 ± 2.3	83.0 ± 1.9	83.1 ± 1.8
	GFHF	33.9 ± 3.3	–	47.8 ± 2.6	–	55.8 ± 2.1	–
	LGC	36.2 ± 3.1	–	47.9 ± 2.3	–	55.9 ± 1.9	–
	TCA	61.3 ± 3.6	60.8 ± 3.7	78.6 ± 2.4	78.4 ± 2.3	86.9 ± 1.2	86.6 ± 1.1
		$(10^3, 10^6)$	$(10^3, 10^3)$	$(10^3, 10^0)$	$(10^3, 10^0)$	$(10^3, 10^0)$	$(10^3, 10^0)$
	SDA	59.4 ± 3.2	58.7 ± 2.8	*81.5 ± 2.1*	*81.2 ± 2.0*	*88.6 ± 1.2*	*88.8 ± 1.1*
		$(10^0, 10^{-9})$	$(10^0, 10^{-9})$	$(10^0, 10^{-3})$	$(10^0, 10^{-3})$	$(10^0, 10^{-3})$	$(10^0, 10^{-3})$
	LapRLS/L	57.9 ± 3.1	57.5 ± 2.6	79.1 ± 2.2	79.0 ± 1.8	87.8 ± 1.1	87.7 ± 1.1
		$(10^0, 10^{-9})$	$(10^0, 10^{-9})$	$(10^{-3}, 10^{-6})$	$(10^{-3}, 10^6)$	$(10^{-3}, 10^{-6})$	$(10^{-3}, 10^{-6})$
	FME	*63.2 ± 2.8*	*62.7 ± 2.6*	*81.8 ± 2.0*	*81.5 ± 1.9*	*89.1 ± 1.2*	*88.9 ± 1.0*
		$(10^{-6}, 10^3)$	$(10^{-6}, 10^3)$	$(10^{-6}, 10^3)$	$(10^{-6}, 10^3)$	$(10^{-6}, 10^3)$	$(10^{-6}, 10^3)$

Dataset	Method	1 labeled sample		2 labeled samples		3 labeled samples	
		Unlabel(%)	Test(%)	Unlabel(%)	Test(%)	Unlabel(%)	Test(%)
COIL-20	MFA	–	–	70.2 ± 2.6	70.1 ± 3.2	76.5 ± 2.5	76.2 ± 2.3
	GFHF	*78.6 ± 2.1*	–	*83.2 ± 2.2*	–	*85.6 ± 2.0*	–
	LGC	*78.5 ± 2.6*	–	*82.9 ± 2.1*	–	*85.9 ± 2.1*	–
	TCA	70.6 ± 2.9	70.5 ± 2.8	78.1 ± 2.5	77.9 ± 2.1	81.7 ± 2.2	81.5 ± 2.6
		$(10^9, 10^3)$	$(10^9, 10^3)$	$(10^9, 10^9)$	$(10^9, 10^9)$	$(10^9, 10^0)$	$(10^9, 10^0)$
	SDA	59.9 ± 2.5	59.8 ± 3.2	73.2 ± 2.7	73.3 ± 2.5	78.3 ± 2.2	78.1 ± 2.5
		$(10^{-9}, 10^0)$	$(10^{-9}, 10^0)$	$(10^3, 10^9)$	$(10^3, 10^9)$	$(10^{-9}, 10^6)$	$(10^0, 10^6)$
	LapRLS/L	60.5 ± 3.2	60.6 ± 3.5	73.5 ± 2.9	73.1 ± 2.5	78.6 ± 2.6	78.8 ± 2.5
		$(10^0, 10^6)$	$(10^0, 10^6)$	$(10^0, 10^6)$	$(10^0, 10^6)$	$(10^0, 10^6)$	$(10^0, 10^6)$
	FME	75.1 ± 3.2	*75.5 ± 3.1*	*82.2 ± 2.9*	*81.9 ± 3.1*	*86.1 ± 2.3*	*85.6 ± 2.6*
		$(10^{-9}, 10^{-6})$	$(10^{-9}, 10^{-6})$	$(10^{-9}, 10^{-6})$	$(10^{-9}, 10^{-6})$	$(10^{-9}, 10^{-6})$	$(10^{-9}, 10^{-6})$

(continued)

Table 1 (continued)

Dataset	Method	10 labeled samples Unlabel(%)	10 labeled samples Test(%)	20 labeled samples Unlabel(%)	20 labeled samples Test(%)	30 labeled samples Unlabel(%)	30 labeled samples Test(%)
20-NEWS	MFA	46.5 ± 7.2	46.2 ± 7.6	61.9 ± 5.9	61.3 ± 6.2	70.9 ± 4.5	71.5 ± 4.1
	GFHF	72.5 ± 7.5	–	83.6 ± 2.5	–	86.1 ± 1.1	–
	LGC	80.9 ± 2.3	–	83.9 ± 1.5	–	85.5 ± 1.0	–
	TCA	55.2 ± 5.3	56.6 ± 5.2	68.6 ± 3.3	67.5 ± 3.2	75.6 ± 3.2	73.9 ± 2.5
		$(10^{-3}, 10^{-3})$	$(10^{-3}, 10^{-3})$	$(10^{-3}, 10^{-3})$	$(10^{-3}, 10^{-3})$	$(10^{-9}, 10^{-9})$	$(10^{-3}, 10^{-3})$
	SDA	57.5 ± 5.9	58.1 ± 5.8	72.9 ± 3.8	73.6 ± 3.6	78.9 ± 2.7	80.5 ± 2.2
		$(10^{-9}, 10^{6})$	$(10^{-9}, 10^{6})$	$(10^{-9}\ 10^{6})$	$(10^{-3}, 10^{6})$	$(10^{-9}, 10^{6})$	$(10^{-9}, 10^{3})$
	LapRLS/L	61.9 ± 4.5	62.2 ± 4.6	75.6 ± 2.5	76.2 ± 2.6	80.9 ± 1.7	81.2 ± 1.9
		$(10^{-9}, 10^{3})$	$(10^{-9}, 10^{3})$	$(10^{-9}, 10^{3})$	$(10^{-9}, 10^{3})$	$(10^{-9}. 10^{3})$	$(10^{-9}, 10^{3})$
	FME	*83.2 ± 3.2*	*82.5 ± 3.6*	*88.2 ± 1.9*	*87.6 ± 2.0*	*90.1 ± 1.3*	*89.6 ± 1.5*
		$(10^{-9}, 10^{-6})$	$(10^{-9}, 10^{-6})$	$(10^{3}, 10^{-6})$	$(10^{-9}, 10^{-6})$	$(10^{3}, 10^{-6})$	$(10^{3}, 10^{-6})$

For each dataset, the results shown in boldface are significantly better than the others, judged by t-test (with a significance level of 0.05). The optimal parameters are also shown in parentheses (μ and γ in FME, λ_I and λ_A in LapRLS/L, α and β in SDA and TCA). Note that we do not report the results for MFA when only one sample per class is labeled because at least two samples per class are required in MFA. Considering that LGC and GFHF cannot cope with the unseen samples, the results for LGC and GFHF on the test dataset are not reported

2. When comparing TCA, SDA, and LapRLS/L, we observe that there is no consistent winner on all the databases. Among the three algorithms, TCA achieves the best results on UMIST and COIL-20 databases, LapRLS/L is the best on YALE-B and 20-NEWS databases, and SDA is generally better on CMU PIE database, in terms of mean recognition accuracy.
3. The mean recognition accuracies of LGC and GFHF are generally better than TCA, SDA, and LapRLS/L on the unlabeled dataset of UMIST, COIL-20 and 20-NEWS databases, which demonstrate the effectiveness of label propagation. But we also observe that the recognition accuracies from LGC and GFHF are much worse than TCA, SDA, and LapRLS/L on the unlabeled dataset of CMU PIE and Yale-B databases, possibly because of the strong light variations of images in the two databases. The labels may not be correctly propagated in this case, which significantly degrades the performances of LGC and GFHF.
4. The FME method outperforms MFA and semi-supervised dimension reduction methods TCA, SDA and LapRLS/L in all the cases in terms of mean recognition accuracy. Judged by t-test (with a significance level of 0.05), FME is significantly better than MFA, TCA, SDA, and LapRLS/L in 20 out of 30 cases. On unlabeled dataset, FME significantly outperforms GFHF and LGC in 9 out of 15 cases. While GFHF/LGC is significantly better than FME in one case on COIL-20 database, LGC and GFHF cannot cope with the unseen data.

5.2 *Unsupervised Learning*

We also compare FME/U with the unsupervised learning algorithms LPP [13] on three face databases UMIST, CMU PIE, and YALE-B. We also report the results from LPP-SR, in which the Spectral Regression method [8] is used to optimize the projection matrix in the objective function of LPP. The nearest neighbor classifier is used again for classification after dimension reduction. Five images per class are randomly chosen as the training dataset and remaining images are used as the test dataset. Again, PCA is used as a preprocessing step to preserve 95% energy of the data in all the experiments. The optimal parameters μ and γ in FME/U are also search from the set $\{10^{-9}, 10^{-6}, 10^{-3}, 10^{0}, 10^{3}, 10^{6}, 10^{9}\}$, and we report the best results from the optimal parameters. For LPP-SR, we use a more dense set $\{10^{-9}, 10^{-8}, \ldots, 10^{8}, 10^{9}\}$ for the parameter λ and report the best results. For PCA, LPP, LPP-SR, and FME/U, we run all the possible lower dimensions and choose the optimal one. We also report the mean recognition accuracy and standard deviation over 20 random splits in Table 2. Figure 6 plots the recognition accuracy with respect to the number of features and Fig. 7 plot the recognition accuracy variation with different parameter μ for FME/U.

We have the following observations: (1) LPP outperforms PCA on CMU PIE and YALE-B databases, which is consistent with the prior work [13]. We also observe that LPP is slightly worse than PCA on UMIST database, possibly because the limited training data cannot correctly characterize the nonlinear manifold structure

Table 2 Top-1 recognition performance (mean recognition accuracy $\pm$ standard deviation %) of PCA [32], LPP [13], LPP-SR [8] and FME/U over 20 random splits on three face databases

method	UMIST	YALE-B	CMU PIE
PCA	82.6 ± 3.2 (43)	43.1 ± 1.2 (60)	53.4 ± 1.5 (55)
LPP	80.4 ± 3.5 (31)	53.3 ± 3.1 (60)	85.0 ± 1.2 (55)
LPP-SR	81.5 ± 3.2 $(10^6, 35)$	60.5 ± 3.0 $(10^{-3}, 60)$	81.7 ± 2.1 $(10^{-5}, 55)$
FME/U	*86.3* $\pm$ *2.8* $(10^3, 10^0, 35)$	*68.2* $\pm$ *2.5* $(10^{-9}, 10^9, 60)$	*89.1* $\pm$ *1.0* $(10^{-9}, 10^9, 55)$

For each dataset, the results shown in boldface are significantly better than the others, judged by t-test (with a significance level of 0.05). Note the last numbers in parentheses are the optimal dimensions after dimension reduction. The first number in LPP-SR is the optimal λ and the first two numbers in FME/U are the optimal parameters μ and γ

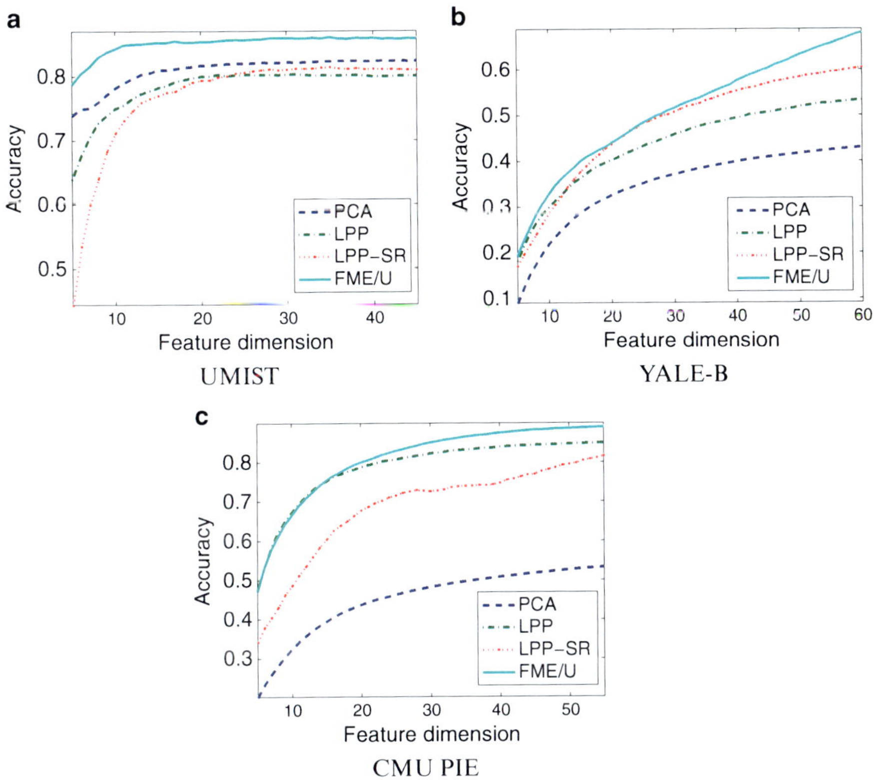

Fig. 6 Top 1 recognition rates (%) with different feature dimensions on the UMIST, YALE-B and CMU PIE databases

in this database; (2) When compared LPP and LPP-SR, there is no consistent winner on all three databases; (3) The FME/U achieves the best results in all the cases, which demonstrates that FME/U is an effective unsupervised dimension reduction method.

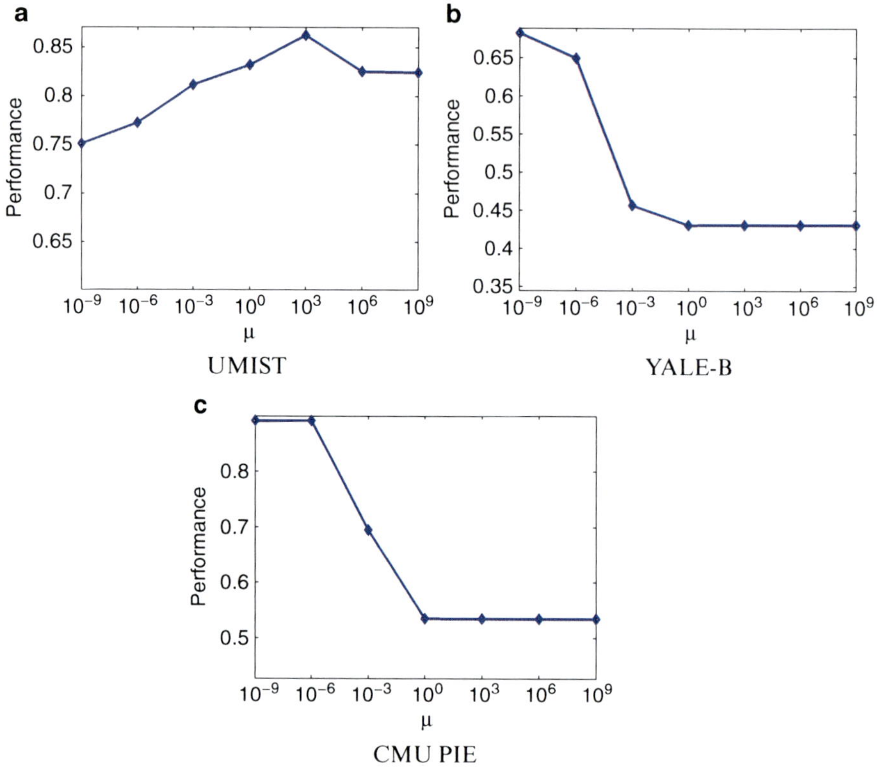

Fig. 7 Performance vs. parameter μ for FME/U

6 Summary

Linearazation technique is an important method for subspace learning and out-of-sample extension to the transductive learning method. In this chapter, we introduce three linearization methods, rigid constrained method, two-step method with regression and a flexible method. The rigid constrained method imposes $F = X^{\mathrm{T}}W + \mathbf{1}b^{\mathrm{T}}$, which is overstrict in practice. The two-step method first computes the F and then fits F with a linear model by regularized linear regression, which relies heavily on the F of the first step. The flexible linearization method overcome these drawbacks by optimizing the prediction labels F, the linear regression function $h(X)$ and the flexible regression residue F_0 simultaneously, where $h(X) = X^{\mathrm{T}}W + \mathbf{1}b^{\mathrm{T}}$ is a regression function for mapping new data points and $F_0 = F - h(X)$ is the regression residue modeling the mismatch between F and $h(X)$. We show that the first two linearization methods are the two extreme cases of the flexible linearization method.

Based on the flexible linearization method, we introduce a new framework for semi-supervised and unsupervised subspace learning, referred to as Flexible Manifold Embedding (FME), and FME/U, respectively. FME can effectively utilize label information from labeled data as well as the manifold structure from both labeled and unlabeled data, and can better deal with the data which reside on a nonlinear manifold. FME(FME/U) is a general framework, many prior works, including the general graph embedding framework are special cases of this framework.

References

1. Abernethy J, Chapelle O, Castillo C (2008) Web spam identification through content and hyperlinks. In: Proceedings of the international workshop on adversarial information retrieval on the web, pp 41–44
2. Belhumeur P, Hespanha J, Kriegman D (1997) Eigenfaces vs. fisherfaces: Recognition using class specific linear projection. IEEE Trans Pattern Anal Mach Intell 19(7):711–720
3. Belkin M, Niyogi P (2001) Laplacian eigenmaps and spectral techniques for embedding and clustering. In: Advances in neural information processing systems
4. Belkin M, Niyogi P, Sindhwani V (2006) Manifold regularization: A geometric framework for learning from labeled and unlabeled examples. J Mach Learn Res 12:2399–2434
5. Blum A, Mitchell T (1998) Combining labeled and unlabeled data with co-training. In: Proceedings of the annual conference on learning theory
6. Cai D (2009) Spectral regression: A regression framework for efficient regularized subspace learning. PhD thesis, Department of Computer Science, University of Illinois at Urbana-Champaign
7. Cai D, He X, Han J (2007a) Semi-supervised discriminant analysis. In: Proceedings of the IEEE international conference on computer vision
8. Cai D, He X, Han J (2007b) Spectral regression for efficient regularized subspace learning. In: Proceedings of the IEEE international conference on computer vision
9. Chu W, Sindhwani V, Ghahramani Z, Keerthi SS (2006) Relational learning with Gaussian processes. In: Advances in neural information processing systems
10. Georghiades A, Belhumeur P, Kriegman D (2001) From few to many: Illumination cone models for face recognition under variable lighting and pose. IEEE Trans Pattern Anal Mach Intell 23(6):643–660
11. Graham DB, Allinson NM (1998) Characterizing virtual eigensignatures for general purpose face recognition. NATO ASI Series F, pp 446–456
12. Guo Z, Zhang Z, Xing E, Faloutsos C (2008) Semi-supervised learning based on semiparametric regularization. In: Proceedings of the SIAM international conference on data mining
13. He X, Yan S, Hu Y, Niyogi P, Zhang H (2005) Face recognition using laplacianfaces. IEEE Trans Pattern Anal Mach Intell 27(3):328–340
14. Huang Y, Xu D, Nie F (2012) Semi-supervised dimension reduction using trace ratio criterion. IEEE Trans Neural Networks Learn Syst 23(3):519–526
15. Jia Y, Nie F, Zhang C (2009) Trace ratio problem revisited. IEEE Trans Neural Networks 20(4):729–735
16. Li X, Lin S, Yan S, Xu D (2008) Discriminant locally linear embedding with high order tensor data. IEEE Trans Syst Man Cybernet B 38(2):342–352
17. Liu C, Shum HY, Freeman WT (2007) Face hallucination: Theory and practice. Int J Comput Vision 75(1):115–134
18. Liu W, Tao D, Liu J (2008) Transductive Component Analysis. In: Proceedings of the IEEE international conference on data mining
19. Nene SA, Nayar SK, Murase H (1996) Columbia object image library (coil-20). Technical report, Columbia University

20. Nie F, Huang H, Cai X, Ding C (2010) Efficient and robust feature selection via joint $\ell_{2,1}$-norms minimization. In: NIPS
21. Nie FP, Xiang XM, Jia YQ, Zhang CS (2009a) Semi-supervised orthogonal discriminant analysis via label propagation. Pattern Recogn 42(11):2615–2627
22. Nie F, Xiang S, Song Y, Zhang C (2009b) Orthogonal locality minimizing globality maximizing projections for feature extraction. Opt Eng 48:017202
23. Nie F, Xu D, Wai-Hung Tsang I, Zhang C (2010) Flexible manifold embedding: A framework for semi-supervised and unsupervised dimension reduction. IEEE Trans Image Process 19(7):1921–1932
24. Nie F, Xu D, Li X, Xiang S (2011) Semisupervised dimensionality reduction and classification through virtual label regression. IEEE Trans Syst Man Cybernet B 41(3):675–685
25. Roweis S, Saul L (2000) Nonlinear dimensionality reduction by locally linear embedding. Science 290(22):2323–2326
26. Sim T, Baker S (2003) The cmu pose, illumination, and expression database. IEEE Trans Pattern Anal Mach Intell 25(12):1615–1617
27. Sindhwani V, Melville P (2008) Document-word co-regularization for semi-supervised sentiment analysis. In: Proceedings of the IEEE international conference on data mining
28. Sindhwani V, Niyogi P, Belkin M (2005a) Beyond the point cloud: from transductive to semi-supervised learning. In: Proceedings of the international conference on machine learning
29. Sindhwani V, Niyogi P, Belkin M (2005b) Linear manifold regularization for large scale semi-supervised learning. In: Workshop on learning with partially classified training data, international conference on machine learning
30. Song Y, Nie F, Zhang C, Xiang S (2008) A unified framework for semi-supervised dimensionality reduction. Pattern Recogn 41(9):2789–2799
31. Tenenbaum J, Silva V, Langford J (2000) A global geometric framework for nonlinear dimensionality reduction. Science 290(22):2319–2323
32. Turk M, Pentland A (1991) Face recognition using eigenfaces. In: Proceedings of the IEEE computer society conference on computer vision and pattern recognition
33. Vapnik V (1998) Statistical learning theory. Wiley-Interscience, New York
34. Xiang S, Nie F, Zhang C (2008) Learning a mahalanobis distance metric for data clustering and classification. Pattern Recogn 41(12):3600–3612
35. Xiang S, Nie F, Zhang C, Zhang C (2009) Nonlinear dimensionality reduction with local spline embedding. IEEE Trans Knowl Data Eng 21(9):1285–1298
36. Xiang S, Nie F, Zhang C (2010) Semi-supervised classification via local spline regression. IEEE Trans PAMI 32(11):2039–2053
37. Yan S, Xu D, Zhang B, Zhang H, Yang Q, Lin S (2007) Graph embedding and extensions: A general framework for dimensionality reduction. IEEE Trans Pattern Anal Mach Intell 29(1):40–51
38. Yang X, Fu H, Zha H, Barlow JL (2006) Semi-supervised nonlinear dimensionality reduction. In: Proceedings of the international conference on machine learning, pp 1065–1072
39. Yang Y, Nie F, Xu D, Luo J, Yunhe Pan YZ (2012) A multimedia retrieval framework based on semi-supervised ranking and relevance feedback. IEEE Trans Pattern Anal Mach Intell, pp 723–742
40. Zhang C, Nie F, Xiang S (2010) A general kernelization framework for learning algorithms based on kernel PCA. Neurocomputing 73(4–6):959–967
41. Zhang T, Popescul A, Dom B (2006) Linear prediction models with graph regularization for web-page categorization. In: Proceedings of the ACM SIGKDD international conference on knowledge discovery and data mining, pp 821–826
42. Zhang T, Tao D, Yang J (2008a) Discriminative locality alignment. In: Proceedings of the European conference on computer vision, pp 725–738
43. Zhang T, Tao D, Li X, Yang J (2008b) A unifying framework for spectral analysis based dimensionality reduction. In: IEEE international joint conference on neural networks, pp 1671–1678

44. Zhang T, Tao D, Li X, and Yang J (2009) Patch alignment for dimensionality reduction. IEEE Trans. Knowledge Data Eng, 21(9):1299–1313
45. Zhang Z, Zha H (2005) Principal manifolds and nonlinear dimension reduction via local tangent space alignment. SIAM J Sci Comput 26:313–338
46. Zhao D (2006) Formulating lle using alignment technique. Pattern Recogn 39(11):2233–2235
47. Zhou D, Bousquet O, Lal TN, Weston J, Schölkopf B (2004) Learning with local and global consistency. In: Advances in neural information processing systems
48. Zhu X (2007) Semi-supervised learning literature survey. Technical report, University of Wisconsin-Madison
49. Zhu X, Ghahramani Z, Lafferty J (2003) Semi-supervised learning using gaussian fields and harmonic functions. In: Proceedings of the international conference on machine learning

A Multi-graph Spectral Framework for Mining Multi-source Anomalies

Jing Gao, Nan Du, Wei Fan, Deepak Turaga, Srinivasan Parthasarathy, and Jiawei Han

1 Introduction

Anomaly detection refers to the task of detecting objects whose characteristics deviate significantly from the majority of the data [5]. It is widely used in a variety of domains, such as intrusion detection, fraud detection, and health monitoring. Today's information explosion generates significant challenges for anomaly detection when there exist many large, distributed data repositories consisting of a variety of data sources and formats. While traditional anomaly detection approaches focus on identifying objects that are dissimilar to most of the other objects from a single source [4, 8, 9, 15, 16, 18, 21], we aim at detecting objects that have "inconsistent behavior" among multiple information sources, which we refer to as "horizontal anomaly detection." Distinction between traditional anomaly detection and horizontal anomaly detection is shown in Fig. 1 where traditional anomaly

J. Gao (✉) • N. Du
Computer Science and Engineering Department, State University of New York at Buffalo, 338 Davis Hall, Buffalo, NY 14260-2500, USA
e-mail: jing@buffalo.edu; nandu@buffalo.edu

W. Fan
Huawei Noah Ark's Lab, 525-530, 5/F, Core Building 2, Hong Kong Science Park, Shatin, Hong Kong
e-mail: david.fanwei@huawei.com

D. Turaga • S. Parthasarathy
IBM T.J. Watson Research Center, 1101 Kitchawan Road, Route 134, 10598, Yorktown Heights, NY, USA
e-mail: turaga@us.ibm.com; spartha@us.ibm.com

J. Han
Computer Science Department, University of Illinois, Urbana-Champaign, 201 N. Goodwin Ave, 61801, Urbana, IL, USA
e-mail: hanj@illinois.edu

Y. Fu and Y. Ma (eds.), *Graph Embedding for Pattern Analysis*, DOI 10.1007/978-1-4614-4457-2_9,

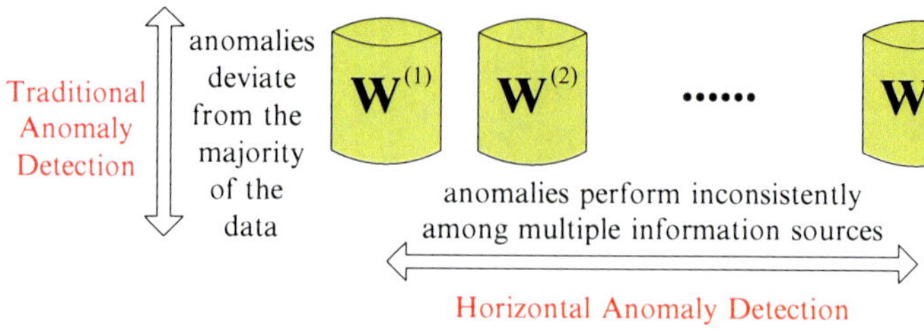

Fig. 1 Horizontal anomaly detection

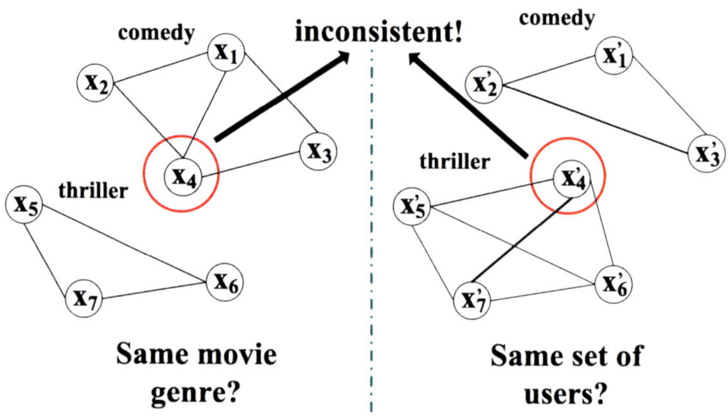

Fig. 2 A toy example illustrating multi-source inconsistency detection

detection explores the single information source vertically and horizontal anomaly detection explores horizontally the inconsistencies among multiple information sources instead. In the following discussions, we will give a few practical examples.

Nowadays, there are usually several sources of information that describe different properties or characteristics of individual objects. For example, we can learn about a movie from its basic information including genre, cast, plots, etc., or the tags users give to the movie, or the viewing histories of the users who watched the movie. On each of the information source, a relationship graph can be derived to characterize the pairwise similarities between objects where the edge weight indicates the degree of similarity. As an example, Fig. 2 shows the similarity relationships among a set of movies derived from two information sources: Movie genres and users. The genre information may indicate that two movies that are both "animations" are more similar than two other movies where one is an "animation" and the other is a "romance" movie. Similarly, movies watched by the same set of users are likely to be more similar than movies that are watched by completely different sets of users.

Clearly, objects form a variety of clusters or communities based on individual similarity relationship. For example, two clusters can be found from both of the similarity graphs in Fig. 2. One cluster represents the movies that are animations, which are loved by kids; while the other cluster represents romance movies, which are liked by grown-ups. Most of the movies belong to the same cluster even though different information sources are used. However, there are some objects that fall into different clusters with respect to different sources. In this example, the animated movie "Wall-E" by genre is expected to be liked by kids, but it is liked by grown-ups based on user viewing history. Finding such "inconsistent" movies can help film distributors better understand the expected audiences of different movies and make smarter marketing plans.

Some other example scenarios of horizontal anomaly detection are listed as follows. (1) In social networks, detecting people who fall into different social communities with respect to different online social networks would be interesting for user behavior analysis; (2) In bioinformatics, inconsistencies across different gene–gene interaction similarity graphs derived from patients with and without a certain disease represent the genes which are critical to the disease; (3) For better business marketing, one wants to find out the person who bought quite different items compared with his peers in the same social community based on the two information sources drawn from user purchase history and friendship networks; and (4) Inconsistencies across multiple module interaction graphs derived from different versions of a software project can be used to assist programmers. Besides the examples discussed above, identifying horizontal anomalies can find applications in many other fields including smarter planet, Internet of things, intelligent transportation systems, marketing, banking, etc.

In this chapter, we propose to detect objects that have "inconsistent behavior" among multiple information sources, which we refer to as *horizontal anomaly detection*. To the best of our knowledge, this is the first work on identifying anomalies by exploring the inconsistencies among multiple sources. Traditional anomaly detection [5] approaches focus on identifying objects that are dissimilar to most of the other objects from a single source [4, 8, 9, 15, 16, 18, 21]. On the other hand, most of the existing work on mining multiple information sources concerns merging and synthesizing models, rules, patterns obtained from multiple sources by reconciling their differences, such as multi-view learning [3], emerging or contrast patterns [6], multi-view clustering [2, 30], and consensus clustering [10,25]. As for multi-source anomaly detection, the studies focus on how to identify anomalies within a specific context where the pre-defined contextual attributes include spatial attributes [23], neighborhoods in graphs [26], social communities [11], and contextual attributes [24, 28]. Although these studies take two types of attributes (behavioral and contextual [5]) into consideration, they cannot be easily generalized to horizontal anomaly detection spanning multiple sources. The reason is that they simply detect anomalies from the behavioral attributes while the contextual attributes only provide the context in which the anomalies are detected. In some sense, these contextual anomalies are still extracted from one source, whereas the proposed method can identify objects with inconsistent behavior across multiple sources.

We assume that each individual information source captures some similarity relationships between objects that may be represented in the form of a similarity matrix (whose entries represent the pairwise quantitative similarity between objects). Note that although in the example shown in Fig. 2, the horizontal anomalies can be found by checking if its direct neighbors are different in the two graphs, this simple solution cannot work in real practice. The reason is that the clustering structures are much more complicated and noisy in real problems, and thus a global method that can detect both the underlying clustering structure and horizontal anomalies is needed. In this chapter, we propose a systematic approach to identify horizontal anomalies from multiple similarity matrices. A summary of this chapter is as follows:

- We combine the input matrices into one large similarity matrix that not only captures the information from each source, but also ensures that individual object relationships are preserved. We then adopt spectral techniques to identify the key eigenvectors of the graph Laplacian of the combined matrix, and identify horizontal anomalies by computing cosine distance between the components of these eigenvectors.
- We give theoretical interpretations of the proposed method from both spectral clustering and random walk perspectives. The method can be regarded as conducting spectral clustering on multiple information sources simultaneously with a joint constraint that the underlying clustering structures should be similar, and objects that are clustered differently are categorized as horizontal anomalies. The horizontal anomalies can also be regarded as those having long commute time in the random walk defined over the graph.
- We validate the proposed algorithm on both synthetic and real datasets, and the results demonstrate the advantages of the proposed approach in finding horizontal anomalies. For example, we find "One Flew Over the Cuckoo's Nest" and "Pulp Fiction" as the most anomalous, while "Star Wars" as the least anomalous among the top 20 most popular movies from the experimental results on a set of 7,595 movies.

In this chapter, we present the proposed spectral method that detects horizontal anomalies in Sect. 2, and demonstrate the effectiveness of the method in Sect. 3, and finally summarize the chapter in Sect. 4.

2 Multi-graph Spectral Framework

Suppose we have a set of N objects $X = \{x_1, x_2, \ldots, x_N\}$ and there are P information sources that describe different aspects of these objects. Let $W^{(t)}$ denote the similarity matrix derived from the t-th information source where the ij-th entry $w_{ij}^{(t)}$ represents the similarity between objects x_i and x_j $(i \neq j)$ with respect to the t-th information source. We let $W_{ii}^{(t)} = 0$ for all t and i. The objective is to assign an anomalous score s_i to each object x_i, which represents how likely the object is anomalous when its behavior differs among the P different information sources. In the simple example shown in Fig. 2, there are two matrices that describe pairwise similarities among the 7 objects, and we expect that x_4 will have the highest anomalous score.

In this section, we present a *HO*rizontal *A*nomaly *D*etection (*HOAD*) algorithm to solve the proposed problem. The basic idea is as follows: As discussed in Sect. 1, we assume that the available information sources on the same set of objects have similar clustering structures, and thus if an object is assigned to different clusters when using various information sources, it can be regarded as a horizontal anomaly. This suggests that we can first cluster the objects separately in each source and compare the clustering results. However, because clustering is unsupervised learning, we do not know the correspondence between clusters in different clustering solutions. We solve this problem by adding the constraint that the same object should be put into the same cluster by all the clustering solutions as often as possible. Another problem is that in reality, an object never belongs to just one cluster for sure, usually it can be assigned to several clusters with certain probabilities. Therefore, soft clustering is more desirable. In the proposed approach, we calculate the anomalous degree of an object based on how much its clustering solutions differ from each other. To simplify the notations, we start with the cases having two distinct information sources. We state the method in Sect. 2.1, give spectral clustering and random walk interpretations in Sect. 2.2, and explain how it is generalized to multiple information sources in Sect. 2.3.

2.1 HOAD Algorithm

Suppose we have two $N \times N$ similarity matrices on the N objects: A and W, where a_{ij} and w_{ij} define the similarity between x_i and x_j from different aspects. The algorithm consists of two major steps:

- Conduct soft clustering on A and W together with the constraint that an object should be assigned to the same cluster.
- Quantify the difference between the two clustering solutions to derive anomalous scores.

The details are as follows. We start from constructing two similarity graphs from A and W. In each of them, each node denotes an object. If the similarity between two objects x_i and x_j is greater than 0, we connect an edge between x_i and x_j and the edge weight equals to the similarity between them. We construct a combined graph by connecting the nodes which correspond to the same object in the two graphs with an edge weighted m. m, a large positive number, is a penalty parameter. An example of such a graph is illustrated in Fig. 3 for the toy example shown in Fig. 2. The set of nodes in the combined graph consists of two copies of the objects: $\{x_1,\ldots,x_N,x'_1,\ldots,x'_N\}$ ($2N$ nodes in total). Let M be an $N \times N$ diagonal matrix with m on the diagonal:

$$M = \mathrm{diag}(m, m, \ldots, m).$$

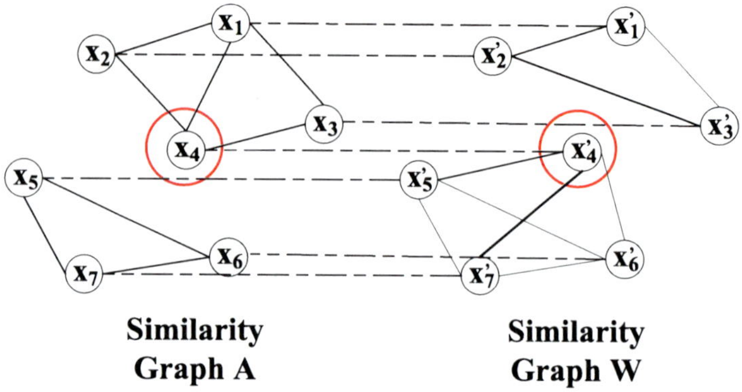

Fig. 3 Combined graph with penalty edges connecting two similarity graphs

Clearly, $M = m \cdot I$ where I is an $N \times N$ identity matrix and m represents the constraint put across the two information sources. Let Z be the adjacency matrix of the combined graph, which is a $2N \times 2N$ matrix:

$$Z = \begin{bmatrix} A & M \\ M & W \end{bmatrix}. \tag{1}$$

We cluster the nodes in the combined graph. As can be seen, there are two copies of the objects in the combined graph and with the help of the edge between the copies of the same object, we cluster the objects in the same way across different sources. In Sect. 2.2, we give a theoretical justification of this claim. First, we compute the graph Laplacian L as:

$$L = D - Z \tag{2}$$

using degree matrix D (a $2N \times 2N$ diagonal matrix):

$$D = \text{diag}\left(\left\{\sum_{j=1}^{2N} z_{ij}\right\}_{i=1}^{2N}\right). \tag{3}$$

Secondly, compute the k smallest eigenvectors of L (with smallest eigenvalues) and let $H \in \mathbb{R}^{2N \times k}$ be the matrix containing these eigenvectors as columns. We divide H into two submatrices U and V each with size $N \times k$ so that $H = [U \quad V]^{\mathrm{T}}$. Therefore, the i-th and $(i+N)$-th rows of H are represented as:

$$\mathbf{u}_i = \mathbf{h}_i, \quad \mathbf{v}_i = \mathbf{h}_{i+N}, \tag{4}$$

which correspond to two "soft clustering" representations of x_i with respect to A and W, respectively. Finally, we compute the anomalous score for object x_i using cosine distance between the two vectors:

$$s_i = 1 - \frac{\mathbf{u}_i \cdot \mathbf{v}_i}{||\mathbf{u}_i|| \cdot ||\mathbf{v}_i||}. \tag{5}$$

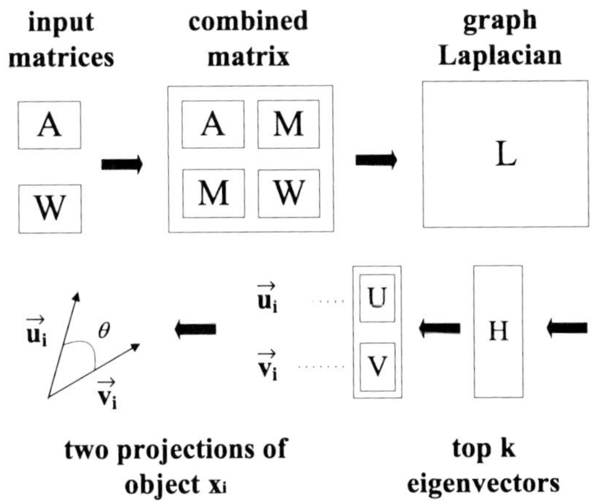

Fig. 4 Algorithm flow of HOAD algorithm

Algorithm 1: HOAD Algorithm

Input: Similarity matrices A and W, number of eigenvectors k, penalty parameter m;
Output: Anomalous score vector **s**;
1: Compute matrix Z according to (1)
2: Compute graph Laplacian L as in (2)
3: Conduct eigen-decomposition of L and Let H be the k smallest eigenvectors with smallest eigenvalues
4: Compute anomalous score of each object s_i based on (4) and (5) for $i = 1, \ldots, N$
5: **return** **s**

The algorithm flow is summarized in both Fig. 4 and Algorithm 1. We start with two $N \times N$ similarity matrices A and W and combine them together with the penalty constraint matrix M to form a combined matrix Z. After that, we compute Z's graph Laplacian L and conduct eigen decomposition on L. H contains the k smallest eigenvectors of L as column vectors, and it is divided into two $N \times k$ submatrices U and V. For each $i \in \{1, \ldots, N\}$, $\mathbf{u}_i$ and $\mathbf{v}_i$, i.e., the i-th rows of U and V can be regarded as the two clustering results of x_i. We compute the anomalous score of x_i as the cosine distance between $\mathbf{u}_i$ and $\mathbf{v}_i$, which is $1 - \cos(\theta)$ with θ representing the angle between the two vectors.

2.2 *Interpretations*

In this part, we explain the algorithm from the perspectives of spectral clustering and random walk.

Clustering on Combined Graph. As can be seen, we first perform spectral clustering on the combined graph in Fig. 3. We replace the clustering step by anomalous score computation because our goal is to detect anomalies. Now we show that the algorithm can be interpreted as conducting constrained spectral clustering on the two input similarity graphs simultaneously. The basic idea of spectral clustering is to project the objects into a low-dimensional space (defined by the k smallest eigenvectors of the graph Laplacian matrix) so that the objects in the new space can be easily separated. We call the projections as *spectral embeddings* of the objects. It has been shown that the matrix formed by the k eigenvectors (H) of L is the solution to the following optimization problem [19]:

$$\min_{H \in \mathbb{R}^{N\times k}} \quad \mathrm{Tr}(H'LH) \quad \text{s.t.} \quad H'H = I \tag{6}$$

Since we define the graph Laplacian L as $D - Z$ (2), the objective function is thus equivalent to:

$$\min_{H \in \mathbb{R}^{2N\times k}} \quad \mathrm{Tr}(H'DH) - \mathrm{Tr}(H'ZH) \quad \text{s.t.} \quad H'H = I \tag{7}$$

Let $f(H) = \mathrm{Tr}(H'ZH)$ and $g(H) = \mathrm{Tr}(H'DH)$. H is a $2N \times k$ matrix, and again we divide it into two submatrices U and V: $H = [U \quad V]^{\mathrm{T}}$. Also, from the definition of Z (1), we have:

$$\begin{aligned} f(H) &= \mathrm{Tr}\left([U' \quad V'] \begin{bmatrix} A & M \\ M & W \end{bmatrix} \begin{bmatrix} U \\ V \end{bmatrix} \right) \\ &= \mathrm{Tr}(U'AU + V'WV + V'MU + U'MV) \\ &= \mathrm{Tr}(U'AU) + \mathrm{Tr}(V'WV) + 2m \sum_{i=1}^{N} \sum_{j=1}^{k} u_{ij} v_{ij} \end{aligned} \tag{8}$$

Suppose the degree matrices for A and W are D^a and D^w, respectively:

$$D^a = \mathrm{diag}\big(\{\sum_{j=1}^{N} a_{ij}\}_{i=1}^{N}\big), \quad D^w = \mathrm{diag}\big(\{\sum_{j=1}^{N} w_{ij}\}_{i=1}^{N}\big).$$

The row sum of M is always m. Based on the definition of D (3), we have

$$\begin{aligned} g(H) &= \mathrm{Tr}\left([U' \quad V'] \left(\begin{bmatrix} D^a & 0 \\ 0 & D^w \end{bmatrix} + mI \right) \begin{bmatrix} U \\ V \end{bmatrix} \right) \\ &= \mathrm{Tr}(U'D^aU + V'D^wV + mU'U + mV'V) \end{aligned} \tag{9}$$

Also,

$$H'H = I \Leftrightarrow [U' \quad V'] \begin{bmatrix} U \\ V \end{bmatrix} = I \Leftrightarrow U'U + V'V = I$$

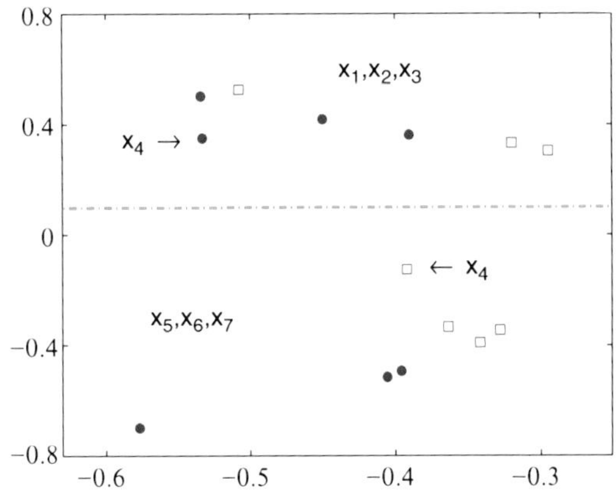

Fig. 5 The two smallest eigenvectors of the toy example in Fig. 3

By putting $f(H)$ and $g(H)$ together and ignoring the constant term $m \cdot \text{Tr}(U'U + V'V)$, we have an equivalent formulation of the problem in (6):

$$\min_{U,V \in \mathbb{R}^{N\times k}} \quad \text{Tr}(U'(D^a - A)U) + \text{Tr}(V'(D^w - W)V) - 2m \sum_{i=1}^{n} \sum_{j=1}^{k} u_{ij} v_{ij}$$

$$\text{s.t.} \quad U'U + V'V = I \tag{10}$$

Clearly, each of the first two terms in (10) corresponds to the spectral clustering problem using A or W alone. The third term acts as the constraint that the two clustering solutions should be similar (cosine similarity). Different from spectral clustering, we didn't actually perform the clustering procedure. Our method can be regarded as embedding the objects into two eigenspaces with respect to the two information sources while putting the constraint that the two projections should be similar. The parameter m controls how much we impose the constraint.

Note that in Algorithm 1, the i-th row vector in U (the first N rows of H) and V (the last N rows of H) contain the projections of object x_i. Due to the principle of spectral clustering, if the spectral embeddings $\mathbf{u}_i$ and $\mathbf{v}_i$ are close to each other, the corresponding object x_i is more likely to be assigned to the same cluster with respect to the two different sources. Therefore, the cosine similarity between the two vectors $\mathbf{u}_i$ and $\mathbf{v}_i$ quantifies how similar the clustering results of object x_i on the two sources are, and thus represents its "normal" degree. In turn, the cosine distance as defined in (5) gives the "anomalous" degree of x_i with respect to the two sources. The higher the score s_i is, the more likely x_i is a horizontal anomaly.

We show how the algorithm works using the example shown in Fig. 3. After computing the graph Laplacian L of the combined graph's adjacency matrix according to (2), we extract the two smallest eigenvectors of L and put it into H, and thus H is a 14×2 matrix. The first seven rows correspond to the spectral embeddings of the seven objects with respect to the first source whereas the remaining ones are those with respect to the second source. In Fig. 5, we plot these row vectors in a two-dimensional space where blue circles indicate the projections on the first source and red squares are results on the second source. Clearly, no matter which source we use, objects x_1, x_2, and x_3 are always projected on the top region of the space, whereas x_5, x_6 and x_7 are located at the bottom part. The biggest difference in the projections can be found in x_4, and thus it has the highest anomalous score among the seven objects. In fact, the output of the proposed algorithm is a vector: $s = (0.4626, 0.7157, 0.7736, 0.8349, 0.7013, 0.6614, 0.5587)^{\mathrm{T}}$ where each entry denotes the degree of being "horizontally" anomalous. Clearly, x_4 has the highest anomalous score.

Random Walk. In this part, we give some intuitions of the proposed method from the random walk perspective. Let z_{ij} be the edge weight between two nodes x_i and x_j in the graph, and $\mathrm{vol}(X) = \sum_{i=1}^{2N} \sum_{j=1}^{2N} z_{ij}$ be the sum of all the edge weights in the graph. Suppose we define a random walk over the combined graph, where the transition probability from node x_i to node x_j is proportional to the edge weight in the graph: $p_{ij} = z_{ij}/d_i$. If the combined graph is connected and non-bipartite, then the random walk always has a unique invariant distribution $\pi = (\pi_1, \ldots, \pi_{2N})$, where $\pi_i = d_i/\mathrm{Vol}(X)$. Now we look at the commute distance between x_i and x_i', two copies of the same object in the combined graph. Commute distance is the expected time it takes for the random walk to travel from x_i to x_i' and back. Instead of looking for one shortest path, the commute distance looks at all the possible paths. Therefore, only when there are many short paths from x_i to x_i', their commute distance is small.

It is proven that commute distance on a graph can be computed with the help of the eigenvectors of the graph Laplacian L as defined in (2). Suppose L has eigenvalues $\lambda_1, \ldots, \lambda_{2N}$, and U and V are two $N \times N$ matrices containing all the eigenvectors for the two copies of the objects respectively. Let $\mathbf{u}_i$ and $\mathbf{v}_i$ denote the i-th row of U and V. We define γ as a length-$2N$ vector with each entry γ_l equal to $(\lambda_l)^{-0.5}$ if $\lambda_l \neq 0$, and 0 otherwise. Now we divide γ into two length-N vectors: $\gamma = [\gamma_u \quad \gamma_v]$. Suppose we map x_i and x_i' into a new feature space where they are represented as $\mathbf{u}_i \cdot \gamma_u$ and $\mathbf{v}_i \cdot \gamma_v$ respectively. It can be derived that the commute distance c_i between x_i and x_i' is:

$$c_i = \mathrm{vol}(X)||\mathbf{u}_i \cdot \gamma_u - \mathbf{v}_i \cdot \gamma_v||^2, \tag{11}$$

which is the Euclidean distance between the nodes in the new feature space scaled by $\mathrm{vol}(X)$.

Recall that we compute the anomalous score of x_i as $1 - \frac{\mathbf{u}_i \cdot \mathbf{v}_i}{||\mathbf{u}_i|| \cdot ||\mathbf{v}_i||}$. We can see that both the anomalous score and the commute distance can be represented as *a distance function applied on the spectral embeddings of the two copies of the object*. The difference is: (1) The embeddings are scaled by $(\lambda_l)^{-0.5}$ in the commute distance; (2) All the eigenvectors are used in the commute distance whereas only the k smallest are used in the anomalous score computation; and (3) Euclidean distance is used instead of cosine distance in the commute distance function.

Although the connection is loose, commute distance can be a helpful intuition to understand the anomalous scores. If it takes longer time to commute between the two copies of object x_i in the graph, x_i is more likely to be a horizontal anomaly.

By the definition of commute distance, it means that it is hard to find many paths to travel between them. In fact, in the combined graph, the only way to travel from the left side to the right side is through the constraint edge with weight m. Therefore, a horizontal anomaly is the object that is categorized into different clusters with respect to different information sources, which is consistent with our definition. As an example, it is hard to travel between x_4 and x_4' in the graph shown in Fig. 3 because they link to different sets of objects in the two sources, and thus x_4 is a horizontal anomaly. On the contrary, besides the constraint edge connecting x_1 and x_1', x_1 can travel to x_1' through many other paths because its neighbors in the cluster maintain the same in the two graphs. Therefore, its commute distance is small and x_1 is a normal object.

2.3 Multiple Sources

We can adapt Algorithm 1 to handle more than two information sources as follows. Suppose we have similarity matrices $\{W^{(1)}, W^{(2)}, \ldots, W^{(P)}\}$ as the input.

Graph Construction. The combined graph is constructed in a similar fashion as discussed before: Duplicate the objects for P copies, in each copy retain the similarity information from each source, and connect each pair of the nodes corresponding to the same object with an edge weighted m. The adjacency matrix Z is thus an $NP \times NP$ matrix with $\{W^{(1)}, W^{(2)}, \ldots, W^{(P)}\}$ on the diagonal and the diagonal matrix encoding constraints $M = \text{diag}(m, m, \ldots, m)$ on all the off-diagonal blocks. We can make the framework more flexible by allowing for different m values for different pairs of information sources. m is a user-provided parameter, which characterizes the similarity between information sources in their clustering structures. Therefore, one principle to set m is to assign a larger value to it if the two information sources are more likely to share the same clustering results. In the experiments, to reduce the number of parameters, we use the simple version where we set m a uniform value. However, how to set m is still a tricky problem because m can take any value between 0 and infinity. In Sect. 3, we give some discussions on how to transform the problem of setting m to an easier task.

Eigenvectors of Graph Laplacian. After Z is obtained, we calculate its graph Laplacian and the k smallest eigenvectors following exactly the same procedure as in Algorithm 1. One concern is that, when the number of information sources increases, the size of the matrix L grows quadratically and this leads to higher computation and storage cost. However, the graph Laplacian of Z is a matrix with special structures where most entries are 0, and also, we only need the k smallest eigenvectors instead of the full eigenspace. Therefore, this problem is much easier than the general eigen-decomposition problem on any matrix. In fact, efficient packages such as ARPACK [17], have been developed to compute a few eigenvectors of large-scale sparse matrix. In Sect. 3, we show that the proposed method implemented based on ARPACK can scale well even when there are more than two information sources. Furthermore, we can use some parallel computing frameworks to process large matrices. For example, large scale top k eigensolver is available [14] using highly scalable MapReduce framework.[1] Another practical issue is how to choose the appropriate k, i.e., the number of eigenvectors we extract from the combined matrix. Choosing k is a general problem for all clustering algorithms, and a variety of methods have been developed. In particular, eigengap heuristic is proposed to choose k such that the first to the k-th eigenvalues are very small, but the $(k+1)$-th is relatively large. This heuristic works for spectral clustering methods as justified by spectral theory and perturbation theory. A brief discussion about these methods can be found in [19]. In Sect. 3, we show how the performance of the proposed method varies with respect to the value of k.

Anomalous Score Computation. The anomalous score is defined based on the distance between two vectors in (5). With P information sources ($P > 2$), we should calculate the anomalous degree of an object x_i based on the following P vectors: $\{\mathbf{h}_i, \mathbf{h}_{i+N}, \mathbf{h}_{i+2N}, \ldots, \mathbf{h}_{i+(P-1)N}\}$. There are various ways to define a distance measure among multiple vectors. In the experiment, we simply use the average pairwise distance as the measure:

$$s_i = \frac{1}{P(P-1)} \sum_{a=0}^{P-1} \sum_{b=0}^{P-1} \mathbb{1}_{a \neq b} \cdot \left[1 - \frac{\mathbf{h}_{i+aN} \cdot \mathbf{h}_{i+bN}}{||\mathbf{h}_{i+aN}|| \cdot ||\mathbf{h}_{i+bN}||} \right]$$

Similarity Computation. We need similarity matrices derived from multiple sources as the input to the algorithm. The notion of "similarity" between objects varies with the types of information sources. In real practice, the set of objects can be represented in different incompatible formats by the available information sources. For example, webpages on the Internet can be represented as bag of words feature vectors (webpage contents), or a huge graph (hyperlink relationships). We discuss how to compute the similarity matrix W for different data types as follows:

- *Graph*: $w_{ij} = 1$ if there exists an edge connecting x_i and x_j and 0 otherwise.
- *Continuous Data*: If each x_i is a feature vector of continuous values, we can use a Gaussian kernel applied on Euclidean distance: $w_{ij} = \exp(-||x_i - x_j||^2/\sigma^2)$ where σ^2 is the parameter used in the kernel.
- *Binary Data*: If each x_i is a feature vector with each entry 0 or 1, we can use Jaccard Index to compute the similarity: $w_{ij} = |x_i \cap x_j|/|x_i \cup x_j|$.
- *Documents*: Each document x_i is usually represented as a feature vector: $\{x_{i1}, x_{i2}, \ldots, x_{iT}\}$ where x_{il} denotes the number of times the l-th word is found in the document. The cosine similarity between two documents x_i and x_j is defined as: $w_{ij} = (x_i \cdot x_j)/(||x_i|| \cdot ||x_j||)$.

Note that these are simply some examples showing how the pairwise similarity can be computed. More discussions on the similarity computation can be found in [12].

[1] http://mahout.apache.org/

3 Experiments

We evaluate the HOAD algorithm on synthetic data to show its detection accuracy, as well as real datasets including DBLP, MovieLens, and NCBI to validate its ability of identifying meaningful horizontal anomalies.

3.1 Synthetic Data

The concept of "horizontal anomaly" is new, and thus there are no benchmark datasets for it. Therefore, we propose a method to convert a classification problem into a horizontal anomaly detection problem, and then apply this procedure on several UCI machine learning datasets.

Data Generation. Recall that horizontal anomalies represent the objects that have inconsistent behavior among multiple information sources. Therefore, the basic idea of the transformation is to simulate the "inconsistencies" by swapping the feature values of objects from different classes. Suppose we have a training set from a classification problem where each object consists of feature values and a class label. Suppose there are N objects in the training set: $\{x_1, \ldots, x_N\}$, and the features X can be partitioned into two views $X^{(1)}$ and $X^{(2)}$. We use $(x_i^{(1)}, x_i^{(2)})$ to denote an object x_i's feature values of the two views, and use y_i to denote its class label. We assume that each of the feature sets is correlated with the class label, and thus objects within the same class share similar feature values in each feature set. Therefore, for two objects x_i and x_j from different classes, if their feature values are swapped in one view but remain unchanged in the other, they have "inconsistent" behavior among these two views, and thus represent horizontal anomalies. In this way, we generate r pairs of horizontal anomalies, as shown in Algorithm 2. We apply the above method on four datasets obtained from UCI machine learning repository[2]: Zoo, Iris, Letter, and Waveform. On each dataset, we randomly split the feature set into two subsets with equal number of features. We repeat the transformation procedure 50 times and at each time, we generate a dataset with around 10% anomalies. We evaluate the HOAD algorithm on the 50 datasets and report the average accuracy.

Evaluation Measure and Baseline Methods. For anomaly detection problems, one of the most widely used evaluation approaches is ROC analysis, which represents the trade-off between detection rate and false alarm rate. A good algorithm would produce an ROC curve as close to the left-top corner as possible, and thus the area under ROC curve (AUC), which is in the range [0,1], is a good evaluation metric. The higher the AUC is, the better the algorithm performs. We show the performance of the proposed HOAD algorithm with various parameter settings.

[2] http://archive.ics.uci.edu/ml

Algorithm 2: Horizontal Anomalies Generation

Input: A training set from a classification problem: $\{(x_i^{(1)}, x_i^{(2)}, y_i)\}_{i=1}^N$, the number of horizontal anomalies $2r$;
Output: Two similarity matrices: A and W, and the indices of horizontal anomalies: B;

for $l = 1$ to r **do**
  Sample two objects x_i, x_j from two different classes $(y_i \neq y_j)$ without replacement;
  Randomly select a view t ($t = 1$ or 2);
  Swap the t-th view of x_i and x_j: Let $z = x_i^{(t)}$, $x_i^{(t)} = x_j^{(t)}$, and $x_j^{(t)} = z$;
  Put i and j into B.
Compute the similarity matrix A based on $X^{(1)}$ and W from $X^{(2)}$
return A, W and B

Note that the first step of the proposed algorithm is a constrained soft clustering on multiple information sources. Instead of conducting a joint clustering, the baseline method clusters multiple sources *separately* and calculates the anomalous scores based on the difference among different clustering solutions. Specifically, in two-source problems, we conduct eigen decomposition on the graph Laplacian matrices of the two similarity matrices A and W separately. Suppose the top k eigen representation of object x_i are $\mathbf{u}_i$ and $\mathbf{v}_i$, respectively, then we use (5) to compute the anomalous score of x_i for the baseline approach. Note that the major difference between the HOAD algorithm and the baseline method is on how to compute $\mathbf{u}_i$ and $\mathbf{v}_i$. Besides evaluating the proposed HOAD algorithm shown in Algorithm 1, we also evaluated an alternative version of the algorithm where the anomalous score is computed based on the commute distance (11) as discussed in Sect. 2.2. We use HOAD-c and HOAD-r to refer to the original version and the random walk version of the proposed HOAD algorithm, respectively.

Performance. The experimental results on the four datasets are shown in Fig. 6 where we vary the values of the parameters m and k. m indicates the penalty we enforce when the two clustering solutions do not agree, and k represents the number of top eigenvectors. Neither m nor k is used in the baseline and its performance remains mostly stable except slight fluctuation due to random sampling in data generation. From the experimental results, we can see that HOAD algorithm generally outperforms the baseline, especially when k is small (e.g., $k = 3$). However, when the value of m is higher, the difference in AUC between the algorithms using different k is much smaller. Therefore, we focus on how to select the appropriate m in the following discussion. It can also be observed that the two versions of the HOAD algorithm achieve nearly the same AUC values. This supports the fact that the proposed HOAD algorithm can be explained from the random walk perspective. The anomalous scores computed in Algorithm 1 simulate the commute distance between two copies of the same object in the combined graph.

On UCI datasets, it is clear that when m increases, the proposed algorithm has a higher AUC. In the simulated study, the two feature sets are two disjoint

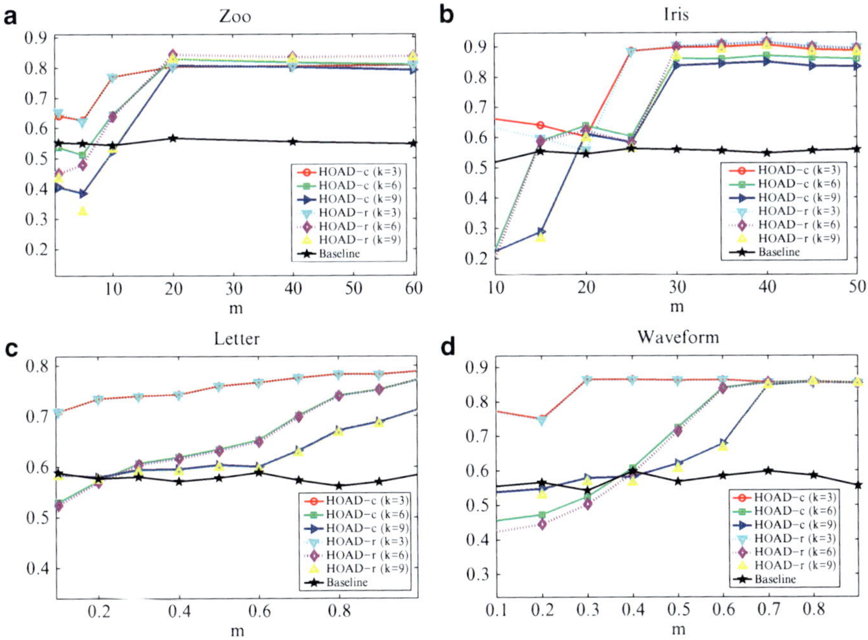

Fig. 6 Anomaly detection performance comparison on UCI data

subsets of the original features, and usually using all of the features leads to a better classification model. Hence the two views are correlated and using a large m captures this correlation well. However, this does not mean that we should assign a big number to m in all cases because this pattern may not always hold in real horizontal anomaly detection tasks. In the following experiments on DBLP datasets, we illustrate the relationship between m and the anomalous scores, and state some principles in setting m.

In Fig. 7, we show the running time of HOAD algorithm with respect to 1,000–6000 objects represented in two, three or four information sources. We conduct the experiments on synthetic datasets where we randomly generate similarity matrices for 50 trials, and report the average running time. The eigenvectors are computed using Matlab eigs function, which is based on ARPACK package [17]. As can be seen, the HOAD algorithm can scale well to large datasets when the number of objects and the number of sources both increase.

3.2 *Real World Data*

We present a set of illustrative results to highlight the concept of horizontal anomaly detection.

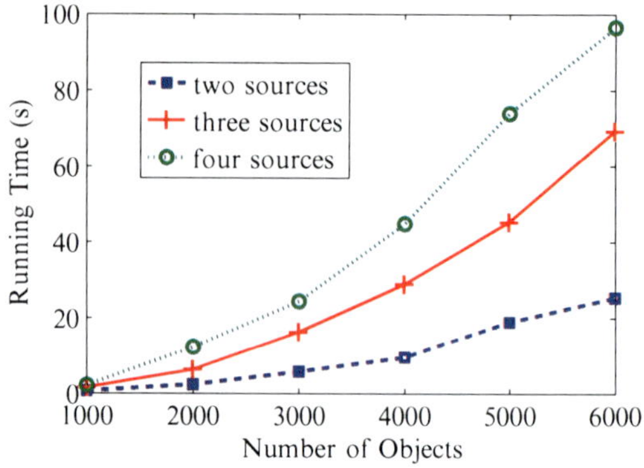

Fig. 7 Running time of HOAD algorithm on synthetic data

DBLP. DBLP[3] provides bibliographic information on major computer science journals and proceedings. We define two horizontal anomaly detection tasks based on the DBLP data where the objects are a set of conferences and authors, respectively. Let $N = 4,220$, and the set of conferences is represented as $\{x_1, x_2, \dots, x_N\}$. There are two views describing the conferences: the keywords in the conferences and the authors who published in the conferences. Specifically, each x_i has two vectors, each of which has the form $(x_{i1}, x_{i2}, \dots, x_{iT})$. In the first vector, T is the total number of words, and x_{il} is the number of times the l-th word appeared in the i-th conference profile (we concatenate the titles of papers in the conference as the conference profile). In the second vector, T is the total number of authors, and x_{il} denotes the number of times the l-th author published in the i-th conference. The pairwise similarity between two conferences x_i and x_j is defined as the cosine similarity between the corresponding vectors. Therefore, the conferences that share lots of keywords, or share lots of authors are similar.

Similarly, we select a set of 3,116 authors from data mining-related areas and extract two types of information from DBLP: the publications and the coauthorships. Each author x_i is also represented in vectors $(x_{i1}, x_{i2}, \dots, x_{iT})$ where in the first vector x_{il} denotes the occurrence of the l-th word in the authors' publications, and x_{il} corresponds to the number of times x_i and x_l collaborate in the second one. Cosine similarity is used, and similar authors will share coauthors, or keywords in their publications.

We first study the effect of m on the anomalous scores. For each m, we apply the HOAD algorithm to the datasets, and compute the mean and standard deviation of the objects' anomalous scores. The results on conferences and authors are shown in Fig. 8 where the points on the line are the average anomalous scores and the error bar denotes the standard deviation. Obviously, the average anomalous score decreases as m increases. Recall that the anomalous scores indicate the degree of differences between the spectral embeddings derived from the two similarity matrices. When

[3] http://www.informatik.uni-trier.de/~ley/db/

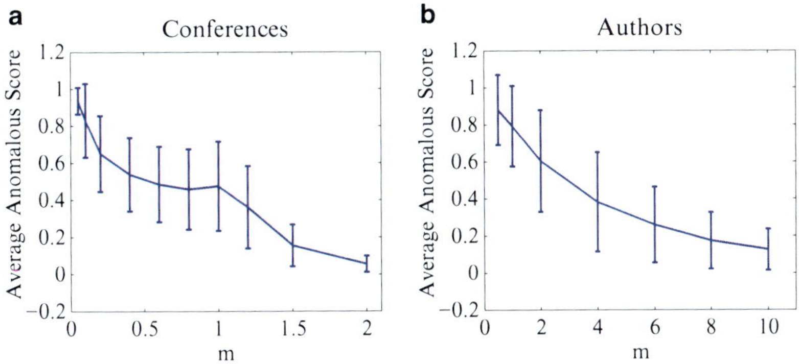

Fig. 8 Analysis of parameter m on DBLP data

Table 1 Case studies on conferences

Conf	Score	Conf	Score
IJCAI	0.3547	WWW	0.6103
AAAI	0.3748	PAKDD	0.4766
ICDE	0.7366	PODS	0.3696
VLDB	0.6299	ICDM	0.6411
SIGMOD	0.5794	ECML	0.3777
SIGIR	0.3404	PKDD	0.4205
ICML	0.5071	EDBT	0.5078
CVPR	0.1417	SDM	0.4755
CIKM	0.8211	ECIR	0.0739
KDD	0.7571	WSDM	0.0262

we give a heavy penalty on different embeddings by the two sources, we basically bias the two projections towards the ones that agree the most. Therefore, when m is larger, the spectral embeddings from the two sources are more likely to be the same, and thus the difference between them is smaller. On the contrary, when m is small, the constraint on the similarities between the two projections is often violated, so most of the objects are projected differently.

Another observation is that the variance among the anomalous scores goes up first and then goes down as m increases. When m is quite large or quite small, the two projections of all the objects would be very similar or very different, and thus the objects receive similar anomalous scores. There exists a large variability among the anomalous scores only when m is in the middle of the spectrum. Although m can be drawn from $(0,\infty)$, the average anomalous scores are within a fixed range: $[0,1]$. Therefore, we can choose m which leads to an average anomalous score around 0.5 because the variance of the anomalous scores usually reaches the highest point here and this helps us identify the horizontal anomalies.

We show the anomalous scores for selected conferences and authors output by the algorithm in Tables 1 and 2, respectively. As can be seen, the conferences that have high anomalous scores (e.g., CIKM) are those attract people from certain fields but

Table 2 Case studies on authors

Author	Score
Philip S. Yu	0.6751
Jiawei Han	0.6162
Christos Faloutsos	0.8516
Rakesh Agrawal	0.7631
H. V. Jagadish	0.8022
Divesh Srivastava	0.7808
Hans-Peter Kriegel	0.3308
Hector Garcia-Molina	0.7061
Surajit Chaudhuri	0.6905
Raghu Ramakrishnan	0.7898

the actual content can be categorized into a different area. Instead, if the authors and publications of a conference are from a somewhat pure community, its anomalous degree is low (e.g, WSDM). Similarly, the author who has different behavior in terms of the publications and the coauthorships are likely to be a horizontal anomaly.

MovieLens. We use the Movielens dataset[4] with movies as objects. There are three sources of information to capture their relationships:

- Genre: Movies are classified as being of one or more of 18 genres, such as Comedy and Thriller, which can be treated as binary vectors.
- User viewing history: Each movie has a list of users that watched the movie. This may also be represented as a vector (per movie) across all users.
- User tagging history: Movies are tagged by different users. Looking across all users, we can determine a vector per movie.

In all three cases, we compute the pairwise similarity using cosine similarity across the vectors. The dataset contains 10 million ratings and 100,000 tags for 10,681 movies by 71,567 users. We choose a set of 7,595 movies, each of which has both ratings and tags. We then have three similarity matrices, corresponding to these three different sources. To evaluate the performance of the HOAD algorithm on MovieLens dataset, we label some movies as "horizontal anomalies" based on the list of weirdest movies.[5] There are 572 movies listed as weirdest movies and among them 127 are found in the MovieLens dataset. These 127 movies are labeled as "anomalous" and the remaining 7,468 movies are "normal." Based on these labels, we calculate the area under ROC curve (AUC) of both HOAD and the baseline method based on their computed anomalies scores for the 7,595 movies. HOAD algorithm achieves a better AUC (0.4879) compared with that of the baseline method (0.4423). This demonstrates the ability of the proposed HOAD algorithm in detecting inconsistent movies across various information sources.

[4] http://www.grouplens.org/node/73

[5] http://366weirdmovies.com/the-weird-movie-list

Table 3 Anomalous scores of 20 popular movies from MovieLens

Movie	Score	Movie	Score
One Flew Over the Cuckoo's Nest	0.8079	Seven Samurai	0.6404
Pulp Fiction	0.7713	Fight Club	0.6364
Casablanca	0.7205	City of God	0.6278
The Shawshank Redemption	0.6949	The Lord of the Rings: The Return of the King	0.3512
The Godfather: Part II	0.6822	The Lord of the Rings: The Fellowship of the Ring	0.3478
The Godfather	0.6770	The Good, the Bad and the Ugly	0.3194
Goodfellas	0.6768	Raiders of the Lost Ark	0.3181
Schindler's List	0.6755	Rear Window	0.3095
12 Angry Men	0.6713	Star Wars	0.2982
The Dark Knight	0.6535	Star Wars: Episode V-The Empire Strikes Back	0.2562

Moreover, we present the anomalous scores for the 20 most popular movies[6] as shown in Table 3. We focus on these well-known movies so that our results can be easily interpreted. As may be seen, the movies "One Flew Over the Cuckoo's Nest" and "Pulp Fiction" are identified as horizontal anomalies, as they tend to show strong disagreement between the genre classification and the sets of users that watched and tagged them. Intuitively, this is expected as these two movies do not really fit into one classification or user category. Borrowing reviews from Wikipedia,[7] "Pulp Fiction" is known for its rich, eclectic dialogue, ironic mix of humor and violence, and nonlinear storyline, which make it different and anomalous among movies. For "One Flew Over the Cuckoo's Nest," the review says "it is a comedy that can't quite support its tragic conclusion." These tell us the reasons why these two movies are detected as being inconsistent. On the other hand, "Star Wars" receives the lowest anomalous score as it attracts a particular set of audiences.

NCBI. In this part, we use two NCBI databases[8] with genes as objects. Selecting informative genes that are the most predictive with respect to its corresponding class label (disease or control) from microarray experiments is one of the most important research problem in bioinformatics. One example of the benefits of such studies is that global gene expression profiling of human tumors can reveal distinct tumor subtypes not evident by traditional histopathological methods. By comparing the gene expression levels of cancerous with normal tissues, we can select those genes that might anticipate the clinical behavior of cancer. In the following experiments, we demonstrate how to detect informative genes of two kinds of cancer, breast

[6] As listed on http://www.imdb.com/chart/top on November 2010.

[7] http://en.wikipedia.org

[8] http://www.ncbi.nlm.nih.gov/

Table 4 The top 10 informative genes for breast cancer from NCBI

Gene	Description
SFN	Stratifin
FAM203A	Family with sequence similarity 203, member A
ANAPC13	Anaphase promoting complex subunit 13
RPL3	Ribosomal protein L3
RPS21	Ribosomal protein S21
HSP90AA2	Heat shock protein 90kDa alpha (cytosolic), class A member 2
E2F4	E2F transcription factor 4, p107/p130-binding
TUBBP2	Tubulin, beta pseudogene 2
SDCBP	Syndecan binding protein (syntenin)
CGGBP1	CGG triplet repeat binding protein 1

and kidney cancer, by using the proposed horizontal anomaly detection method. Specifically, the informative gene detection problem can be defined as finding the most anomalous genes from two views: the gene profiling of the cancerous and the normal samples, respectively. After two similarity matrices are constructed for genes with respect to samples with or without cancer, we detect the genes with the most obvious inconsistent behavior across the two views, and these are the informative genes which best explain the effects of a particular kind of cancer.

We use the microarray gene expression data obtained from the GEO database of NCBI [7] for both the breast and kidney cancer. The breast cancer dataset consists of 43 breast cancer and 43 normal samples. We filter out genes that are not differentially expressed and there are 4,739 genes used in our experiments. We calculate the anomalous scores for each gene based on the inconsistency degree between the cancer and the normal view. The top 10 genes with the largest anomalous scores are presented in Table 4. Among these informative genes, *SFN* has been reported to have significant association with breast cancer [22]. Levels of mRNAs coding for *HSP90* family is significantly higher in cancer tissues than in non-cancer tissues [29], suggesting that *HSP90* family is associated with the proliferation of human breast cancer. Also, *E2F4* has been reported to be related to breast cancer [20].

The statistics of the kidney cancer datasets are as follows. It consists of 69 kidney cancer and 23 normal samples, and we choose 8,710 genes in the experiment. We show the top 10 genes with the largest anomalous scores in Table 5. Among these informative genes, [1] suggested that *IGFBP3* may contribute to renal diseases via effects on podocytes and proximal tubule cells. The finding of [27] indicated that *ALDOA* is a useful biomarker for monitoring the clinical course of patients with renal cell carcinoma, which is a kidney cancer that originates in the lining of the proximal convoluted tubule. In addition, *UMOD* has been proved that it is associated with the kidney disease [13].

Note that in the presented results of these two experiments, the relationship between the other presented genes and cancer has not been revealed by existing studies yet. However, since some of the genes we identified has been verified to

Table 5 The top 10 informative genes for kidney cancer from NCBI

Gene	Description
IGFBP-3	Insulin-like growth factor binding protein 3
HUWE1	HECT, UBA and WWE domain containing 1, E3
ANXA2P1	Annexin A2 pseudogene 1
RPL41	Ribosomal protein L41
ALDOA	Aldolase A, fructose-bisphosphate
RPS10P7	Ribosomal protein S10 pseudogene 7
ACTG1	Actin, gamma 1
UMOD	Uromodulin
HLA-C	Major histocompatibility complex, class I,C
RPL8P2	Ribosomal protein L8 pseudogene 2

be informative, our study provides a promising direction to search for informative genes related to a variety of diseases.

4 Summary

In many applications, there are usually multiple information sources that describe some common set of objects. From each source, one can derive a similarity matrix describing the relationship between pairs of objects. When the relationships among multiple information sources are consistent, it is expected that similar objects form a cluster and the underlying clustering structure ought to be shared by these multiple sources. However, there exist some objects which perform inconsistently, and it is important to detect such objects for decision support in many applications. In this chapter, we introduced an effective horizontal anomaly detection approach to identify the objects that have inconsistent behavior across multiple sources. The proposed algorithm has two intrinsic steps. In the first step, we construct a combined similarity graph based on the similarity matrices and compute the k smallest eigenvectors of the graph Laplacian as spectral embeddings of the objects. After that, we calculate the anomalous score of each object as the cosine distance between different spectral embeddings. The physical meaning of the proposed algorithm is explained from both constrained spectral clustering and random walk points of view. Experimental results on several UCI as well as DBLP, MovieLens and NCBI datasets demonstrate the effectiveness of the proposed approach.

References

1. Bach L (2012) The insulin like growth factor system in kidney disease and hypertension. Curr Opin Nephrol Hypertens 21(1):86–91
2. Bickel S, Scheffer T (2004) Multi-view clustering. In: Proceedings of the IEEE international conference on data mining (ICDM'04), pp 19–26

3. Blum A, Mitchell T (1998) Combining labeled and unlabeled data with co-training. In: Proceedings of the annual conference on computational learning theory (COLT'98), pp 92–100
4. Breunig MM, Kriegel H-P, Ng RT, Sander J (2000) Lof: identifying density-based local outliers. In: Proceedings of the ACM SIGMOD international conference on management of data (SIGMOD'00), pp 93–104
5. Chandola V, Banerjee A, Kumar V (2009) Anomaly detection: A survey. ACM Comput Surv 41(3):15:1–15:58
6. Dong G, Li J (1999) Efficient mining of emerging patterns: Discovering trends and differences. In: Proceedings of the the ACM SIGKDD international conference on knowledge discovery and data mining (KDD'99), pp 43–52
7. Edgar R, Domrachev M, Lash A (2002) Gene expression omnibus: NCBI gene expression and hybridization array data repository. Nucleic Acids Res 30(1):207–210
8. Eskin E (2000) Anomaly detection over noisy data using learned probability distributions. In: Proceedings of the international conference on machine learning (ICML'00), pp 255–262
9. Fan W, Miller M, Stolfo S, Lee W, Chan P (2001) Using artificial anomalies to detect unknown and known network intrusions. In: Proceedings of the IEEE international conference on data mining (ICDM'01), pp 123–130
10. Fern XZ, Brodley CE (2004) Solving cluster ensemble problems by bipartite graph partitioning. In: Proceedings of the international conference on machine learning (ICML'04), ACM, New York, NY, pp 281–288
11. Gao J, Liang F, Fan W, Wang C, Sun Y, Han J (2010) On community outliers and their efficient detection in information networks. In: Proceedings of the the ACM SIGKDD international conference on knowledge discovery and data mining (KDD'10), pp 813–822
12. Han J, Kamber M (2006) Data mining: Concepts and techniques, 2nd edn. Morgan Kaufmann, Los Altos
13. Hart T, Gorry M, Hart P, Woodard A, Shihabi Z, Sandhu J, Shirts B, Xu L, Zhu H, Barmada M, Bleyer A (2002) Mutations of the UMOD gene are responsible for medullary cystic kidney disease 2 and familial juvenile hyperuricaemic nephropathy. J Med Genet 39(12):882–892
14. Kang U, Meeder B, Faloutsos C (2011) Spectral analysis for billion-scale graphs: Discoveries and implementation. In: Proceedings of the Pacific-Asia conference on knowledge discovery and data mining (PAKDD'11), pp 13–25
15. Khoa N, Chawla S (2010) Robust outlier detection using commute time and eigenspace embedding. In: Proceedings of the Pacific-Asia conference on knowledge discovery and data mining (PAKDD'10), pp 422–434
16. Knorr EM, Ng RT, Tucakov V (2000) Distance-based outliers: Algorithms and applications. VLDB J 8(3–4):237–253
17. Lehoucq R, Sorensen D, Yang C (1998) ARPACK users' guide: Solution of large-scale eigenvalue problems with implicitly restarted arnoldi methods. SIAM, Philadelphia, PA
18. Liu F, Ting K, Zhou Z (2008) Isolation forest. In: Proceedings of the IEEE international conference on data mining (ICDM'08), pp 413–422
19. Luxburg U (2007) A tutorial on spectral clustering. Stat Comput 17(4):395–416
20. Macaluso M, Cinti C, Russo G, Russo A, Giordano A (2003) pRb2/p130-E2F4/5-HDAC1-SUV39H1-p300 and pRb2/p130-E2F4/5-HDAC1-SUV39H1-DNMT1 multimolecular complexes mediate the transcription of estrogen receptor-alpha in breast cancer. Oncogene 22(23):3511–3517
21. Markou M, Singh S (2003) Novelty detection: A review–part 1: statistical approaches. Signal Process 83(12):2481–2497
22. Mirza S, Sharma G, Parshad R, Srivastava A, Gupta S, Ralhan R (2010) Clinical significance of Stratifin, ERalpha and PR promoter methylation in tumor and serum DNA in Indian breast cancer patients. Clin Biochem 43(4–5):380–386

23. Shekhar S, Lu C-T, Zhang P (2001) Detecting graph-based spatial outliers: Algorithms and applications (a summary of results). In: Proceedings of the the ACM SIGKDD international conference on knowledge discovery and data mining (KDD'01), pp 371–376
24. Song X, Wu M, Jermaine C, Ranka S (2007) Conditional anomaly detection. IEEE Trans Knowl Data Eng 19(5):631–645
25. Strehl A, Ghosh J (2003) Cluster ensembles–a knowledge reuse framework for combining multiple partitions. J Mach Learn Res 3:583–617
26. Sun J, Qu H, Chakrabarti D, Faloutsos C (2005) Neighborhood formation and anomaly detection in bipartite graphs. In: Proceedings of the IEEE international conference on data mining (ICDM'05), pp 418–425
27. Takashi M, Zhu Y, Nakano Y, Miyake K, Kato K (1992) Elevated levels of serum aldolase A in patients with renal cell carcinoma. Urol Res 20(4):307–311
28. Wang X, Davidson I (2009) Discovering contexts and contextual outliers using random walks in graphs. In: Proceedings of the IEEE international conference on data mining (ICDM'09), pp 1034–1039
29. Yano M, Naito Z, Yokoyama M, Shiraki Y, Ishiwata T, Inokuchi M, Asano G (1999) Expression of hsp90 and cyclin D1 in human breast cancer. Cancer Lett 137(1):45–51
30. Zhou D, Burges C (2007) Spectral clustering and transductive learning with multiple views. In: Proceedings of the international conference on machine learning (ICML'07), pp 1159–1166

Graph Embedding for Speaker Recognition

Z.N. Karam and W.M. Campbell

1 Introduction

This chapter presents applications of graph embedding to the problem of text-independent speaker recognition. Speaker recognition is a general term encompassing multiple applications. At the core is the problem of speaker comparison—given two speech recordings (utterances), produce a score which measures speaker similarity. Using speaker comparison, other applications can be implemented speaker clustering (grouping similar speakers in a corpus), speaker verification (verifying a claim of identity), speaker identification (identifying a speaker out of a list of potential candidates), and speaker retrieval (finding matches to a query set).

Text-independent speaker recognition has advanced considerably over the last ten years with dramatic performance and computational improvements. One of the first successful text-independent models was the Gaussian mixture model (GMM) universal background model (UBM) [1]. This technique used Bayesian *maximum a posteriori* (MAP) adaptation and likelihood-ratio scoring to perform speaker recognition.

A follow-on to this method was the introduction of the discriminative SVM into speaker recognition using polynomial kernels [2]. The SVM technique introduced the *vector representation* of an utterance into speaker recognition. The top-level view of this technique is to map a time-variable length utterance to a fixed dimensional vector and then perform channel compensation and classification with this representation.

Z.N. Karam (✉)
University of Michigan, MI, USA
e-mail: zahi@umich.edu

W.M. Campbell
MIT Lincoln Laboratory, Lexington, Massachusetts 02420, USA
e-mail: wcampbell@ll.mit.edu

Y. Fu and Y. Ma (eds.), *Graph Embedding for Pattern Analysis*,
DOI 10.1007/978-1-4614-4457-2_10,

The convenience and geometric intuition of vector-based techniques in speaker recognition resulted in multiple vector representations. Maximum likelihood linear regression (MLLR) adaptation parameters were used by Stolcke [3]. Also, high-level phonetic and lexical features were used in an information-retrieval style representation with SVMs for speaker recognition [4,5]. Methods for compensation for channel variation were explored in this context [6].

Vector methods were then applied to the GMM UBM model. In [7], adapted GMM model parameters were used as a vector representation with a distance motivated by the KL divergence. Alternate GMM-model distance metrics can be found in [8].

A set of methods which attempted to model the manifold structure of the vectors as a linear subspace slowly emerged. Early techniques applying PCA and KPCA to obtain subspaces to model speakers were examined in [9, 10] but did not yield performance improvements. Applying subspaces to model the local variation of a speaker due to nuisance effects was the first success of such methods with techniques such as nuisance attribute projection (NAP) [7, 11] and factor analysis [12]. The goal in this case was to remove or attenuate unwanted within-class variation. Later methods in this area are within-class covariance normalization (WCCN) [13].

Subsequent methods to model aspects of the vector representation of utterances in subspaces and the associated coordinates (factors in statistical parlance) have been quite successful. One advantage of these methods is they provide a compact low-dimensional representation. Joint-factor analysis provided the first successful modeling of both speaker and nuisance subspaces [14]. A reformulation of the joint factor analysis into a total-variability subspace resulted in the iVector method [15]. Many of these methods can be viewed in a common framework known as inner product discriminant functions (IPDF) [16].

Using graph embedding methods is a recent successful attempt to apply more advanced methods to analyze the manifold of the vector set of speaker utterances. In the paper by Karam [17], graph embedding is proposed as a method of visualizing large data sets and also as a starting point for speaker comparison (1-1 comparison). A subsequent paper [18] refines the speaker comparison method using relational learning.

This chapter examines graph embedding of speech recordings for visualization, manifold embedding, and for two speaker recognition classification problems—speaker comparison and speaker retrieval. Basic methods of vector representation of speech utterances are discussed. Graph embedding and visualization of speech corpora using these vector representations are then presented. Multiple techniques and features for speaker comparison, including geodesics, random walks, and relational learning, are examined. Finally, speaker retrieval using random walks is explored.

2 Vector Representation and Compensation of Speech Data

As discussed in the introduction, the vector representation of speech data is a critical part of the graph-embedding process. In this section, representing speech data using Gaussian mixture models (GMMs) and GMM supervectors is reviewed. Although this particular approach is used for concreteness in the graph-embedding experiments, the proposed graph-based methods are general and can be applied to any vector expansion such as iVectors [15, 19], the polynomial (GLDS) vector [9], high-level term frequency vectors [4], etc. The vector representation can further be refined to better capture speaker similarity through nuisance compensation. Though many compensation techniques exist [7, 11–13], this section will briefly present weighted nuisance attribute projection (WNAP) [20], the method used in the graph-embedding experiments.

2.1 *Vector Representation*

The basic strategy for comparing speech utterances is shown in Fig. 1. Given two speech utterances x and y to compare, the first step is to extract feature vectors, $\{\mathbf{x}_i\}$ and $\{\mathbf{y}_j\}$, respectively, representing the utterances. Feature vectors are typically cepstral coefficients with associated smoothed first- and second-order derivatives.

A GMM universal background model (UBM) is adapted to each of the sets of feature vectors separately. Intuitively, the UBM is the prior for distinct acoustic sounds across a large population. The GMM UBM is given by

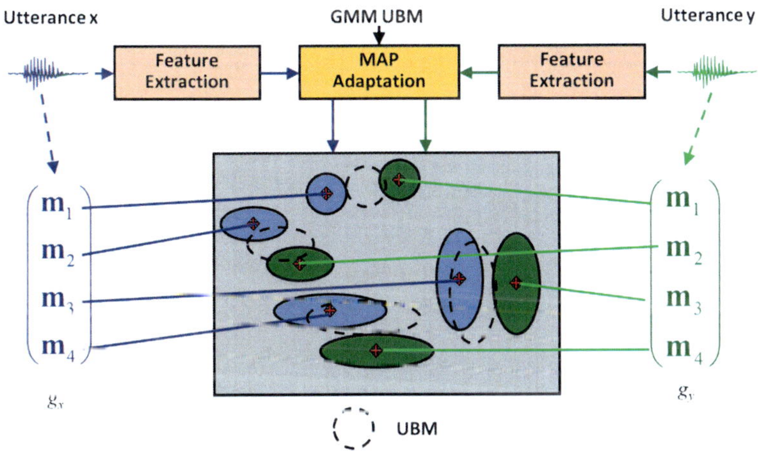

Fig. 1 Comparing speech utterances using GMM UBM MAP adaptation. The ovals in the figure indicate different mixture components and their corresponding covariance matrix. The *plus* indicates the mean of the mixture component. *Colors* indicate the results of MAP adaptation

$$g(\mathbf{x}) = \sum_{i=1}^{N_\mathrm{m}} \lambda_i \mathcal{N}(\mathbf{x}|\mathbf{m}_i, \Sigma_i). \tag{1}$$

In (1), λ_i are the mixture weights, $\mathbf{m}_i$ are the means of the individual Gaussians, Σ_i are diagonal covariances, and N_m is the number of mixture components. *Maximum a posteriori* (MAP) adaptation of the means and ML adaptation of the mixture weights using the feature vectors $\{\mathbf{x}_i\}$ and $\{\mathbf{y}_j\}$ produces two GMMs, g_x and g_y. A method of comparing the parameter vectors of stacked mixture weights and means, $\mathbf{a}_x$ and $\mathbf{a}_y$,

$$\mathbf{a}_x = \left[\lambda_x^\mathrm{T}\ \mathbf{m}_x^\mathrm{T}\right]^\mathrm{T} \tag{2}$$

$$= \left[\lambda_{x,1} \cdots \lambda_{x,N_\mathrm{m}}\ \mathbf{m}_{x,1}^\mathrm{T} \cdots \mathbf{m}_{x,N_\mathrm{m}}^\mathrm{T}\right]^\mathrm{T}. \tag{3}$$

from the two GMMs will be used for speaker comparison.

From Fig. 1, if mixture components roughly correspond to different acoustic sounds, the movement of the means away from the UBM by adaptation gives an indicator of speaker differences in acoustic production. This difference in production (e.g., from dialect and vocal tract shape) gives a method of discriminating speakers.

The two distributions g_x and g_y can be compared using KL-divergence,

$$D(g_x \| g_y) = \int_{R^n} g_x(\mathbf{x}) \log\left(\frac{g_x(\mathbf{x})}{g_y(\mathbf{x})}\right) \mathrm{d}\mathbf{x}, \tag{4}$$

but the result is not computable in closed form. An approximate symmetrized KL divergence is used as an alternative [7, 21],

$$d^2(\mathbf{a}_x, \mathbf{a}_y) = \sum_{i=1}^{N_\mathrm{m}} \left[\lambda_{x,i} \mathbf{m}_{x,i}^\mathrm{T} \Sigma_i^{-1} \mathbf{m}_{x,i} - 2\lambda_{x,i}^{\frac{1}{2}} \lambda_{y,i}^{\frac{1}{2}} \mathbf{m}_{x,i}^\mathrm{T} \Sigma_i^{-1} \mathbf{m}_{y,i} + \lambda_{y,i} \mathbf{m}_{y,i}^\mathrm{T} \Sigma_i^{-1} \mathbf{m}_{y,i}\right], \tag{5}$$

where we have used the adapted means and mixture weights. Thus, the distance (5) reflects the intuition shown in Fig. 1; i.e., the overall distance between utterances is a sum of local distances between means of the adapted GMMs appropriately weighted.

A corresponding inner product to this distance is

$$C(\mathbf{a}_x, \mathbf{a}_y) = \sum_{i=1}^{N_\mathrm{m}} \lambda_{x,i}^{\frac{1}{2}} \lambda_{y,i}^{\frac{1}{2}} \mathbf{m}_{x,i}^\mathrm{T} \Sigma_i^{-1} \mathbf{m}_{y,i}, \tag{6}$$

An alternate expression of this inner product using Kronecker notation is (see [21]),

$$C(\mathbf{a}_x, \mathbf{a}_y) = \mathbf{m}_x^\mathrm{T} (\lambda_x^{1/2} \otimes I_n) \Sigma^{-1} (\lambda_y^{1/2} \otimes I_n) \mathbf{m}_y = \mathbf{z}_x^\mathrm{T} \mathbf{z}_y \tag{7}$$

where the corresponding vector representation is

$$\mathbf{z}_i = \mathbf{b}(\mathbf{a}_i) = \left[\lambda_{i,1}^{\frac{1}{2}} \Sigma_i^{-1} \mathbf{m}_{i,1}^{\mathrm{T}} \cdots \lambda_{i,N_{\mathrm{m}}}^{\frac{1}{2}} \Sigma_i^{-1} \mathbf{m}_{i,N_{\mathrm{m}}}^{\mathrm{T}}\right]^{\mathrm{T}} \\ = (\lambda_i^{1/2} \otimes I_n)\Sigma^{-1/2}\mathbf{m}_i. \tag{8}$$

The vector representation (8) and corresponding inner product and distance will be used for constructing content graphs. Note the inner product has also been used extensively in SVM-based speaker recognition [21].

The inner product and expansion shown belongs to a more general class of comparisons called inner product discriminant functions (IPDF) [16]. In the more general framework, a comparison function of the form

$$C(\mathbf{a}_x, \mathbf{a}_y) = (L_x\mathbf{m}_x)^{\mathrm{T}} D^2 (L_y\mathbf{m}_y) \tag{9}$$

is used. In the equation, L_x, L_y are linear transforms which perform compensation and dimension reduction. The matrix D is a positive definite matrix inducing a distance metric. As shown in [16], approximate KL divergence (8) is of this form. In addition, methods such as iVector and joint factor analysis can be written in this form, since factor analysis can be written as a linear transform—see [16] for more details.

2.2 *Vector Compensation*

As mentioned, L_x and L_y can also be used for channel mismatch. Channel mismatch occurs in many forms between two recordings—different acoustic environments, different microphone types, speaker session variation, etc. A common assumption in speaker recognition is that a subspace in the vector space corresponds to nuisances to the recognition.

A framework for eliminating nuisances (nuisance attribute project, NAP) in the parameter vector based on projection was shown in [11, 20]. The basic idea is to assume that nuisances are confined to a small subspace and can be removed via an orthogonal projection, $\mathbf{m}_x \mapsto Q_{U,D}\mathbf{m}_x$. One justification for using subspaces comes from the perspective that channel classification can be performed with inner products along one-dimensional subspaces. Therefore, the projection removes channel-specific directions from the parameter space.

Alternate methods for compensation in the IPDF framework include within-class covariance normalization (WCCN), variability-compensated support vector machines (VCSVM), and Wiener filtering. WCCN [13] attenuates rather than removes nuisance directions. VCSVM [22] also attenuates nuisance directions, but does this in the context of SVM training. Finally, Wiener filtering [23] uses subspace models of speaker and channel and performs matrix-vector Wiener filtering to optimally attenuate noisy subspaces. All methods are closely related but provide different perspectives on subspace-based compensation.

3 Content Graphs

3.1 Constructing Content Graphs

Our approach to constructing graphs is based on the semi-supervised learning literature [24, 25]. A set of vectors $M = \{\mathbf{z}_i\}$, $\mathbf{z}_i = b(\mathbf{a}_i)$ obtained as a vector mapping of a corresponding set of speech signals is given, as in (8). The set of vectors occupies some manifold. A fundamental idea is to create a graph G which reflects the local connectivity and distances of points on the manifold. An approximate analogy is that the graph represents a sampling process (e.g., band-limited sampling) on the manifold. After creating this graph, fundamental operations such as finding nearest neighbors of points or computing global properties such as geodesic distances using graph methods (Dijkstra's shortest path algorithm [26]) can be performed.

The construction of a speaker content graph proceeds as follows. First, each node n in the graph corresponds to a single vector $\mathbf{z}$ from a speech signal. Edges in the graph represent speaker-similarity between a pair of recordings. Ideally if two nodes are connected, the speakers from the two nodes should be the same. When performing the graph embedding, a sparse version of the similarity matrix is first computed with the only valid entries being those corresponding to the existing edges. Note that the summarized matrix and the resultant graph are two ways to represent the same information. Figure 2 sketches out the embedding process for four utterances $\{\mathbf{R_A}, \mathbf{R_B}, \mathbf{R_C}, \mathbf{R_D}\}$.

In experiments, it has been found critical to sphere (normalize) the vector data (8) as $\mathbf{z}_x/\|\mathbf{z}_x\|_2$. Sphering the data partly corrects for variations in vector length due to different signal durations [27]. Sphering simplifies computation since the distance and inner product are related by the simple relation, $\|\mathbf{z}_x - \mathbf{z}_y\|^2 = 2 - 2\mathbf{z}_x^T\mathbf{z}_y$.

As shown in Fig. 2, in order for the graph to reflect the local neighborhood of a point $\mathbf{z}$, only a limited number of neighbors are connected. One possible method is to choose a fixed threshold ε and connect all points within an ε-ball using the distance function $d(\mathbf{z}, \mathbf{z}_i) < \varepsilon$. For our experiments, this method was found to yield graphs which were too dense and had unusual degree distributions. Alternatively,

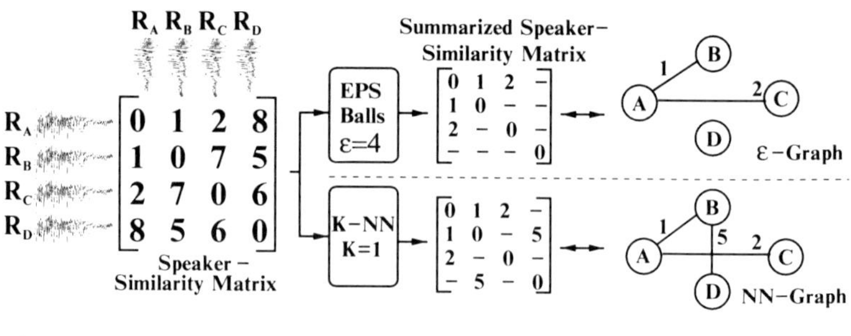

Fig. 2 Sketch of graph embedding

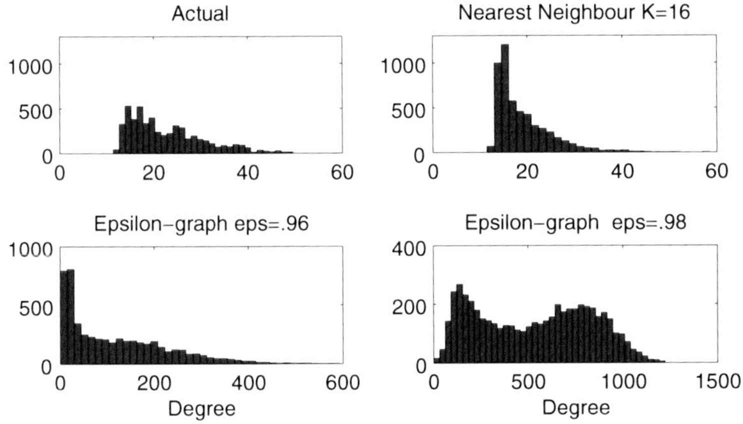

Fig. 3 Histogram of node degree for various content graph construction methods

for each point, the top-K closest neighbors can be found; i.e., for a given node n with corresponding vector $\mathbf{z}$, connect n and n_i only for the smallest K values of $||\mathbf{z}-\mathbf{z}_i||$ as $\mathbf{z}_i$ ranges over the entire vector set M. Note that this construction implies the minimum degree of each node is K, but because the edge construction is done independently for each node, the degree could be substantially larger than K. Graphs built based on an ε-ball are referred to as ε-graphs, and those based on nearest neighbors as NN-graphs.

Figure 3 shows the degree distribution for multiple types of graph construction on the NIST SRE 2004 speaker recognition corpora. The actual (true) degree distribution is derived from the graph where two nodes are connected if and only if they represent the same speaker. From the figure, it is clear to see that the ε-graph is very sensitive to the parameter setting, with a very small change in ε yielding significantly different degree distributions; furthermore, no choice of ε achieves a degree distribution similar to that of the actual graph. The nearest neighbor graph, however, is a better fit; therefore, this chapter will focus on NN-graphs.

Figure 4 shows an example of a content graph using the approximate KL distance and a top-5 nearest neighbor construction. The graph is created from four speakers and shows the corresponding clusters in the graph. The graph clearly illustrates that not all nodes within a cluster are connected (i.e., not a complete subgraph).

The graph structure has a corresponding isomorphism to a matrix representation. The weighted adjacency matrix W for the graph is defined as $W=[W_{i,j}]$ where

$$W_{i,j}=\begin{cases} f(\mathbf{z}_i,\mathbf{z}_j) & \text{if an edge exists between } i \text{ and } j \\ 0 & \text{otherwise.} \end{cases} \tag{10}$$

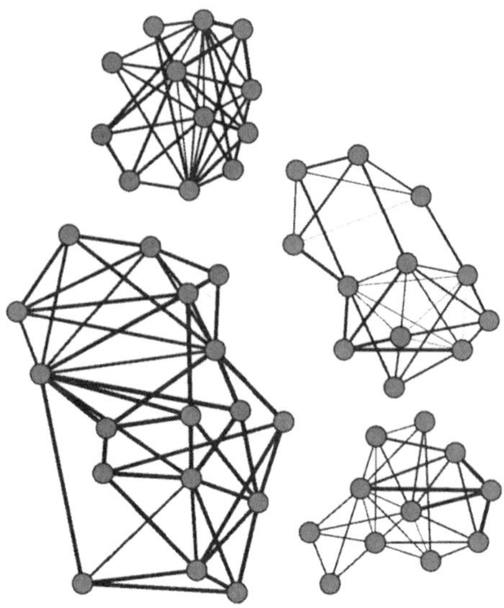

Fig. 4 Example of a content graph using four speakers from the NIST SRE 2008 male speaker set and a top-5 graph

Typical functions $f(\cdot,\cdot)$ are

$$\begin{aligned} f_0(\mathbf{z}_i,\mathbf{z}_j) &= \mathbf{z}_i^{\mathrm{T}}\mathbf{z}_j \\ f_1(\mathbf{z}_i,\mathbf{z}_j) &= \sqrt{2-2\mathbf{z}_i^{\mathrm{T}}\mathbf{z}_j} \\ f_2(\mathbf{z}_i,\mathbf{z}_j) &= \mathrm{e}^{-f_1^2(\mathbf{z}_i,\mathbf{z}_j)/\sigma^2} \end{aligned} \tag{11}$$

Note that $f_2(\cdot)$ in (11) is a standard radial basis function (also commonly used as an SVM kernel). The parameter σ controls the decay of the exponential function (i.e., the width of the basis function).

The matrix W is sparse and symmetric by construction. In fact, the process of graph construction in the matrix domain can be viewed as a sparse approximation to the dense matrix where none of the weights are zeroed out in (10).

3.2 *Incremental Construction*

Content graphs can be constructed in an incremental manner. That is, if speech data is viewed as streaming in over time, then the content graph can be updated dynamically to reflect the new data. The two steps to make this an efficient process are briefly described.

As a setup, suppose a new piece of data $\mathbf{z}$ is obtained, and an existing data set $M = \{\mathbf{z}_i\}$, content graph G, list of indices and closest K distances $\{D_i\}$, and weight matrix

W is given. Note the closest distances for vector $\mathbf{z}_i$ in increasing order are $D_{i,k}$, $k = 1, \ldots, K$. The process of adding a new piece of data to the graph is equivalent to appending a row and column to W. The steps to adding to the content graph are the following:

- Compute and store $\|\mathbf{z} - \mathbf{z}_i\|$ for $i = 1, \ldots, K$. Store the list of indices and sorted distances D with distances $D_1 \leq D_2 \leq \cdots \leq D_K$.
- For each $i > K$, compute $d(\mathbf{z}, \mathbf{z}_i)$ using an early out algorithm. First, retrieve the furthest neighbor distance $D_{i,K}$ for $\mathbf{z}_i$. Then, loop over the dimensions in the Euclidean norm in (5). The current "estimate" of the distance is monotonically increasing; if this estimate goes over D_K or $D_{i,K}$, the computation can stop and go to the next vector. Otherwise, insert the vector in the appropriate list.
- At the end, the K closest vectors to $\mathbf{z}$ in the list D are obtained. The vector $\mathbf{z}$ has been potentially inserted into the distances and lists for each of the $\mathbf{z}_i$.

Several comments on size and computation are appropriate. First, since only the top K distances and indices for each $\mathbf{z}_i$ are stored, the memory for the graph is approximately $2KN_v$ scalars where N_v is the number of vectors. Storage increases linearly with the number of vectors. Second, computation to insert an element into the graph is approximately $2cN_eN_v$ flops where N_e is the dimension of the expansion vector and c is the early-out probability (typically 0.5 in our experiments). Essentially, this computation could be viewed as an extreme form of feature reduction performed when the data is streamed into the system. Additionally, the graph construction performs most of the necessary computation up-front to avoid significant computation at query time.

4 Speaker Manifolds and Graph Visualization

This section shows how the content graphs can be used to explore the speaker manifold and for the visualization of large corpora.

4.1 Exploring the Speaker Manifold

Content graphs can be used to explore the existence of an underlying manifold upon which speech recordings lie as well as to obtain an embedding of the recordings on the manifold. To this end a popular technique, isometric feature mapping (ISOMAP) [28] can be employed. ISOMAP relies on approximate geodesic distances, distances along the manifold, computed using the content graphs. This section presents geodesic distances and their approximate computation, followed by the ISOMAP algorithm and the insights gained by applying it to speech recordings.

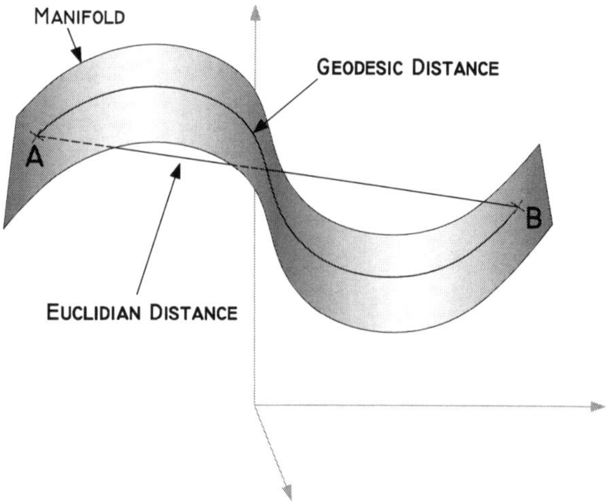

Fig. 5 Geodesic and Euclidean distances between A and B

4.1.1 Geodesic Distance

Assuming that the recordings lie on a low-dimensional manifold in the speaker-similarity space defined by the recording vector representation, then the Euclidean distance between two recordings that are far apart may not be a faithful representation of speaker similarity. A better choice may be the geodesic distance, the length of the shortest path connecting them along the manifold, between the two recordings. Figure 5 sketches the difference between the two distances for a manifold with an intrinsic dimension of two in a three-dimensional Euclidean space.

Though they differ over large distances, the Euclidean and geodesic distances are approximately equivalent for arbitrarily short distances. This equivalence can be used to approximate the geodesic distance [28] as follows. First assume that enough recordings are available such that they densely sample the manifold in the Euclidean space and create a content graph with the edge weights between two nodes being the Euclidean distance between them. The graph only connects nodes that are similar, and if the space is densely sampled one can assume the weight of the edge between two recordings is a faithful representation of their similarity. Thus, the geodesic distance between two recordings can be approximated using the graph geodesic, which is computed by summing the weights of the edges along the shortest path in the graph connecting their corresponding nodes. Figure 6 sketches this approximation for a manifold with an intrinsic dimension of two in a three-dimensional Euclidean space. The Matlab implementation [29] of the Dijkstra algorithm [26] is used to efficiently compute the shortest path between two recordings.

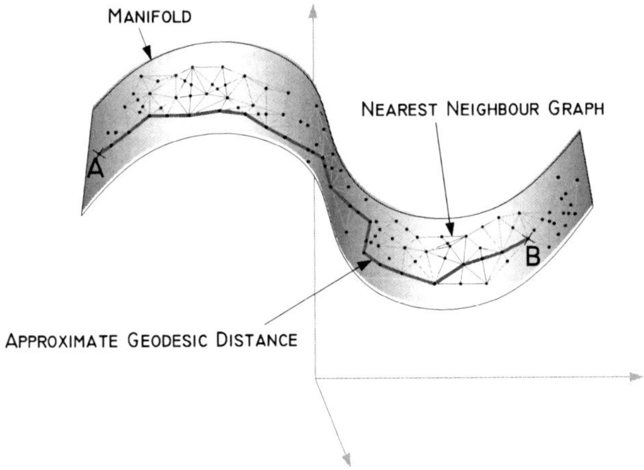

Fig. 6 Approximate geodesic distance between A and B

4.1.2 ISOMAP

ISOMAP [28] is a technique that is used to explore the existence and dimension of the manifold, as well as embed points into it. The embedding begins with approximate geodesic distances computed using the content graph and applies multidimensional scaling (MDS), a technique used for dimensionality reduction and data visualization [30], to obtain the low-dimensional embedding that best preserves these distances. The distance in the embedding space will be referred to as the ISOMAP distance. The optimal size of the lower-dimensional embedding space is, in general, not known a priori and can be estimated by examining the decay of the residual variance, the variance in the data unaccounted for by the embedding. ISOMAP embedding can be performed using the Matlab software package [29].

ISOMAP Applied to Speech Recordings: The speaker-similarity vector expansion described in Sect. 2.1 to which speech recordings are mapped has a high dimension of 19,968; however, as shown in the iVector representation of speech [15] good speaker separation can be done in a significantly smaller space of dimension 400. This smaller space is essentially the linear subspace of largest speaker variability in the original space. A question that this section attempts to answer is whether the data lies near a nonlinear manifold and if so what is its dimension. To this end, ISOMAP with an NN content graph built using $K = 6$ nearest neighbors is applied to:

- 5213 recordings of the NIST SRE 2004 data set [31], which contain 212 speakers from both genders.
- 5742 recordings, of both genders, from the 1 and 3 conversation enroll and 1 conversation test tasks of the NIST SRE 2006 [32].

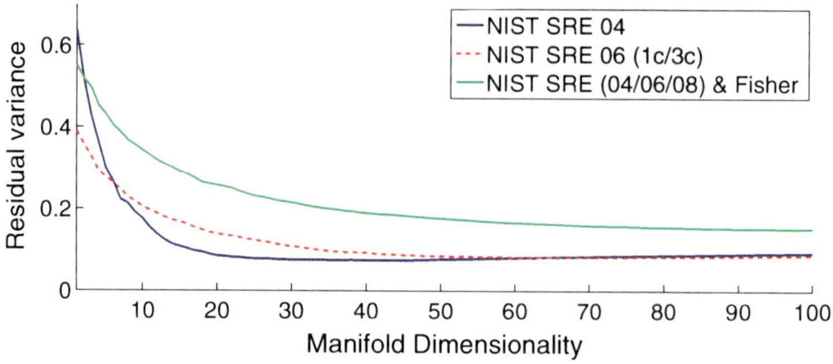

Fig. 7 Decay of residual error with increasing embedding dimension

- 23000 recordings, of both genders, sub-selected from the NIST SRE 2004/06/08 [33] evaluations as well as the Fisher corpora.

Figure 7 examines the decay of the residual error as the embedding dimension is increased. Note that most of the variability in the SRE 04 data set can be captured by a 50-dimensional manifold, and similarly for the SRE 06 data set. However, when speech recordings from multiple sources are pooled the intrinsic dimension is closer to 100 with an overall higher residual error, which seems to indicate a lack of consistency in the manifold across the data sets.

To further highlight the existence of an underlying manifold of speaker variability, Fig. 8 shows the two-dimensional embedding of 5 recordings from 10 male and 10 female speakers randomly selected from the 212 speakers from the SRE 04 data set; all recordings from the 212 speakers were used in the ISOMAP embedding. Each set of similarly colored "o"s represents recordings from a male speaker, and the set of similarly colored "x"s represents recordings from a female speaker. It is interesting to note that both speaker and gender separation can be observed in this two-dimensional embedding.

4.2 Graph Visualization of Speaker Recordings

Content graphs can also be used to visualize the efficacy of various vector representations and nuisance compensation methods, which compensate the vector representations such that they better capture speaker similarity. When visualizing content graphs, the locations of the vertices are not important; however, the existence and weights of the edges between them are. The graph can, therefore, be "laid out" (the process of choosing vertex locations) in a manner that would result in good visualization. The GUESS [34] software package can be used to perform both the visualization and the layout using the GEM algorithm [35]. An example of such a layout is presented in Fig. 9 which shows the layout of the content graph,

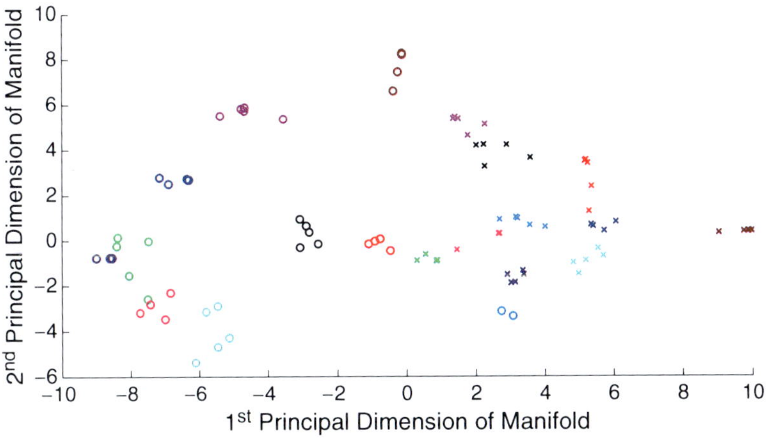

Fig. 8 Five recordings each from 20 speakers embedded on the estimated two-dimensional manifold—"o" for males and "x" for females

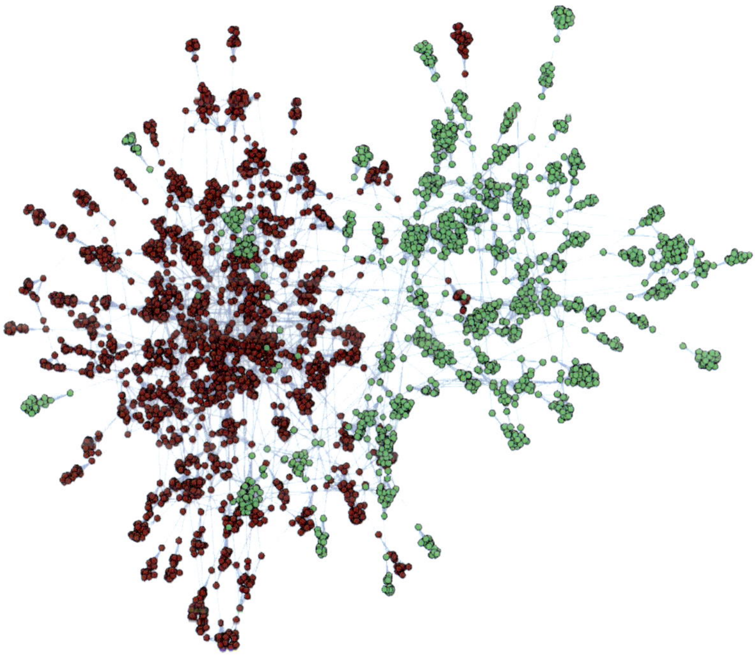

Fig. 9 NIST SRE 2004 NN-graph $K = 6$ male (*red*) and female (*green*) recordings

built using nearest-neighbors with a $K = 6$, of the SRE 04 telephony data [31]. Male and female recordings are represented by red and green nodes, respectively, and the visualization clearly shows the gender separation.

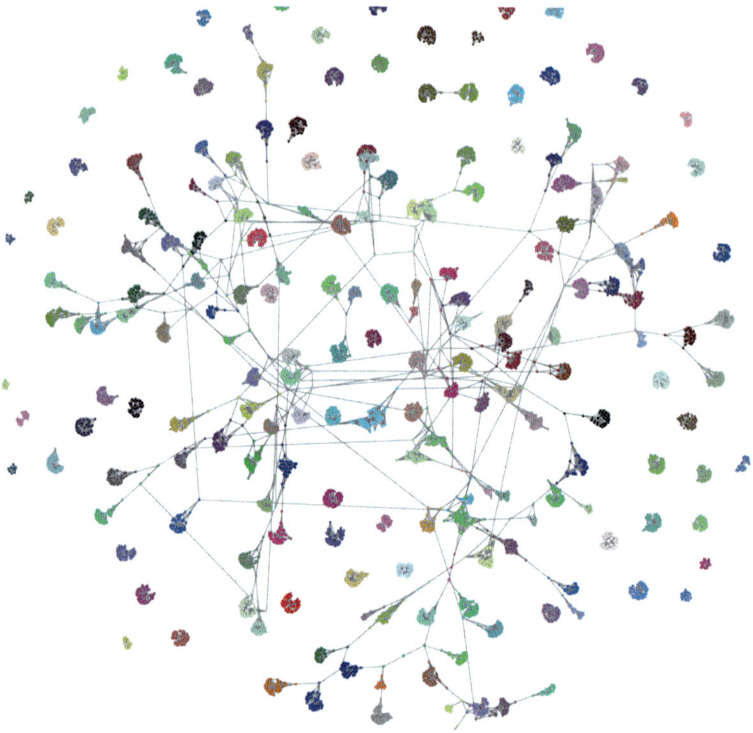

Fig. 10 Graph visualization of all NIST SRE 2010 male recordings using the full channel-blind system and with speaker metadata overlaid

Such visualizations can also be used to explore and uncover structure in a given data set. To highlight this, a case study is presented. In [19] a channel-blind system was proposed that could be used across the different tasks of the 2010 NIST speaker recognition evaluation (SRE) [36]. These tasks include recordings of telephony speech as well as various microphone recordings collected from interviews conducted in two separate rooms. This system is based on the iVector system [15], which employs with within-class covariance normalization (WCCN) [13] and linear discriminant analysis (LDA) [37] to perform the crucial role of removing the channel variability. Using the visualization of the content graphs, one can highlight the effect of the channel compensation in the system as well as explore the NIST SRE 2010 recordings. Only male recordings are presented since similar results are observed with female recordings. The graph visualizations show all male recordings in the 2010 extended NIST SRE, where the number of nearest neighbors is set to $K = 3$.

First, the efficacy of the channel-blind system is shown by building the content graph using the distance defined by the system, without using any channel or speaker information. Figure 10 shows the resultant visualization with speaker metadata

Fig. 11 Graph visualization of all NIST SRE 2010 male recordings using the channel-blind system without WCCN/LDA channel compensation and with speaker metadata overlaid

overlaid such that recordings of the same speaker are colored alike. Clusters of similar color, representing clusters of recordings of the same speaker, show that the system does indeed perform well at capturing speaker similarity.

Next, the importance of the channel compensation performed by the combination of WCCN/LDA is explored. To do this, a content graph is built using the channel-blind system without the WCCN/LDA step, without the use of any speaker or channel information. The corresponding visualization with speaker metadata overlaid is shown in Fig. 11. Notice that the speaker clustering observed with the full channel-blind system is no longer visible; however, interestingly there does seem to be some structure to the graph. Further exploration, by overlaying channel metadata, shows that the structure can be attributed to channel variability. Figure 12 shows this layout with colors representing different telephone and interview microphone channels, and the node shape representing the two different rooms the interview data was collected in. Upon careful inspection of the graph, one notices that the room accounted for more variability than the interview microphones, specifically for the far-talking microphones: MIC CH 05/07/08/12/13. Another worthwhile observation is that recordings from the two landline phones located in each of the two rooms cluster near the interview data of the corresponding room, and more specifically near the close-talking and desk microphones: MIC CH 02/04.

This ability to visualize and explore the dominant variability within a data set may prove to be a useful tool when dealing with newly collected data sets. In this particular case study, the greater effect of the room variability over that of the microphones seems to suggest that future NIST SREs should include tasks that test for robustness across varying recording rooms.

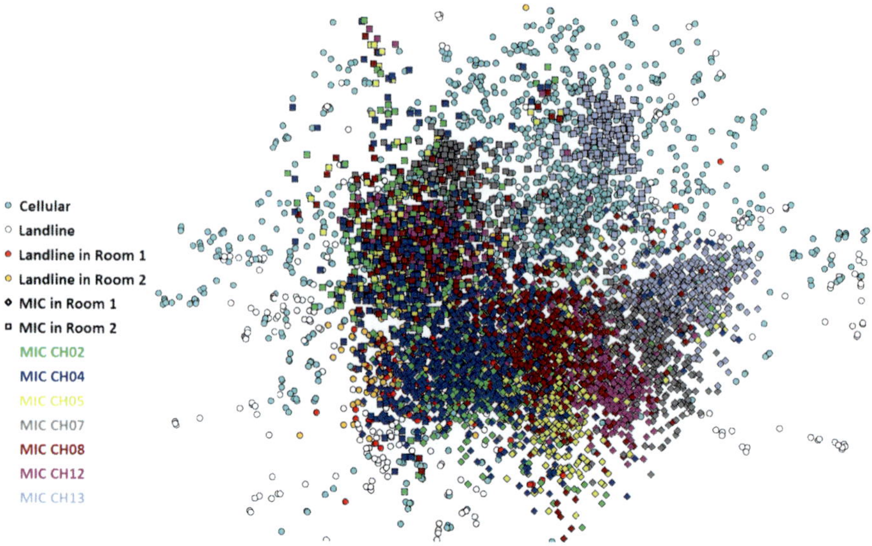

Fig. 12 Graph visualization of all NIST SRE 2010 male recordings using the channel-blind system without WCCN/LDA channel compensation and with channel metadata overlaid

Another useful aspect of visualization, which will only be mentioned here, is to help identify errors in the metadata provided with a data set. For example, an error in the speaker or gender label would manifest as a node or group of nodes not clustering with their same speaker/gender labeled counterparts.

5 Speaker Comparison

5.1 *Definition, Metrics, and Data sets*

Speaker comparison is given by the following scenario. Given two speech signals, produce a score of speaker similarity. More positive scores indicate a higher likelihood of match; more negative scores indicate a lower likelihood of match. By varying a threshold on the comparison score, different performance levels can be achieved.

Speaker comparison performance is typically measured over a large number of target and non-target trials: the former consists of two recordings from the same speaker, while the latter from two different speakers. Performance can be measured with a variety of different metrics calculated in terms of miss probability P_{m} and false alarm probability P_{fa}. Standard metrics are equal error rate (EER) which is the operating point where $P_{\mathrm{m}} = P_{\mathrm{fa}}$ and minimum decision cost function (minDCF) which is defined as the minimum of a cost function,

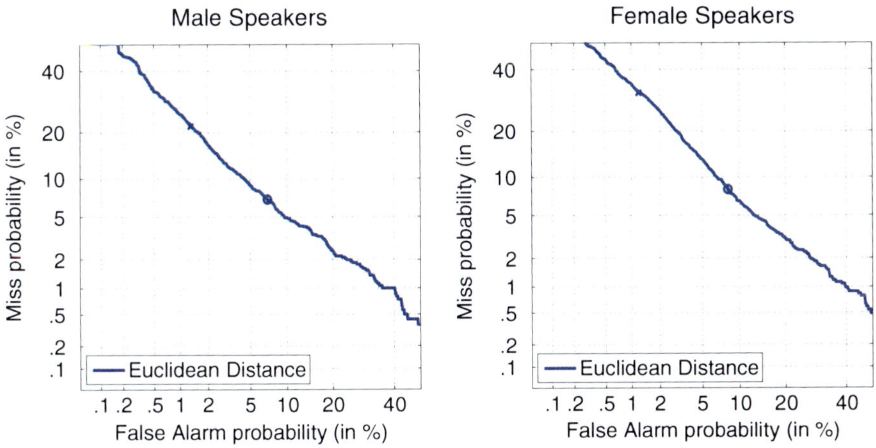

Fig. 13 DET plot showing speaker comparison performance of the baseline system on NIST SRE 2006

$$C(T) = C_m P_m(T) P_{tgt} + C_{fa} P_{fa}(T)(1 - P_{tgt}), \tag{12}$$

where T is the threshold. Typical values are $C_m = 10$, $C_{fa} = 1$, and $P_{tgt} = 0.01$. Additionally, performance can also be shown over a range of thresholds in the form of detection error trade-off (DET) plots—a type of ROC curve [38]. Figure 13 shows an example DET plot for our baseline system. To summarize overall performance, the area under the ROC (AUC) curve is used; however for most speaker comparison applications, the region above the EER operating point is of most interest.

Experiments for speaker comparison were performed on the NIST SRE 2006 [32] and NIST SRE 2008 [33] speaker recognition evaluation (SRE) data sets. All telephony data from both evaluations was used. For SRE 06, this resulted in 4,951 male utterances and 2,790 female utterances. For SRE 08, 4,951 male utterances and 6,393 female utterances were used. The SRE 06 data will be used as a training/validation set, and the SRE 08 will be reserved for testing.

5.2 *Baseline System*

The baseline system is used both for setting a performance baseline for speaker comparison and to build the content graphs used in our proposed graph-based systems.

The vector expansion used is given by (8). Standard feature extraction with 20 MFCCs (including c0) plus deltas was performed along with SAD and 0/1 feature normalization. A GMM UBM with 512 mixtures was trained using a large set of Switchboard 2 [39] and Fisher data. For MAP adaptation, a relevance factor of 0.01

was selected. To compensate the expansions, WNAP [20] was used and trained using a combined NIST SRE 2004–2006 [31, 32, 40] list with the dimension of the nuisance subspace fixed at 64. Finally the compensated expansions are sphere normalized to obtain the resultant vector representation $\mathbf{z}_x$ of recording $\mathbf{x}$. The Euclidean distance between two recordings $\mathbf{x}$ and $\mathbf{y}$ is:

$$E(\mathbf{x},\mathbf{y}) = ||\mathbf{z}_x - \mathbf{z}_y|| = \sqrt{2 - 2\mathbf{z}_x^{\mathrm{T}}\mathbf{z}_y}. \tag{13}$$

This distance is used to build the content graphs using NN construction, as described in Sect. 3.

In addition to graph construction, the negative of the Euclidean distance between two recordings will serve as the baseline system. A DET plot is used to present the performance of this system on the SRE 06 data, the 'x' and the 'o' represent the minDCF and EER operating points, respectively. Throughout this chapter, performance on male and female speakers are presented separately.

5.3 Exploiting Content Graphs for Speaker Comparison

Even though the content graph is built using the baseline Euclidean distance system, it contains a wealth of additional information which can be used to improve speaker comparison. This section derives three sets of features from the content graph, each motivated by a different application of graphs in the literature. The properties and performance of each set will be examined on the SRE 06 data set and a comparison, between them, on the test set is presented in Sect. 5.3.4.

5.3.1 Speaker Comparison Using Geodesic Distance

Section 4 showed that speech recordings do indeed lie near a manifold, therefore taking the manifold structure into consideration when computing speaker similarity should improve over the Euclidean distance baseline. To this end, the negative of the approximate geodesic distances computed via nearest neighbor content graphs can be used to compare two recordings. These are computed by applying Dijkstra's [26] shortest path algorithm to NN graphs with edge weights $f_1(\mathbf{z}_i, \mathbf{z}_j)$ (11). This geodesic measure is sensitive to the tuning parameter, the number of NNs (K); this is highlighted in Fig. 14, which shows the variability of each of the three metrics as a function of K. In the figure, the performance on the SRE 06 data is presented using $EER/2$, $1 - AUC$ and $minDCF$ to match their respective dynamic ranges. The figure shows that no one choice of K simultaneously minimizes each of minDCF/EER/(1-AUC). This phenomenon is similarly observed for the other choices of graph-based speaker comparisons presented in the following sections.

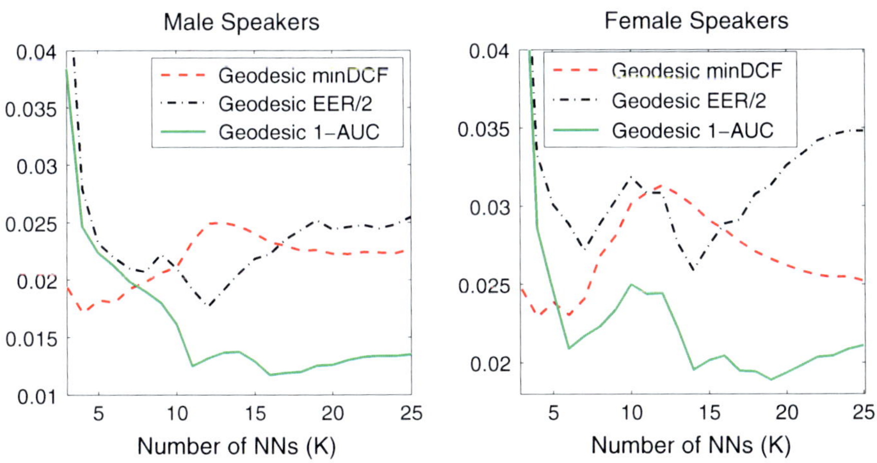

Fig. 14 Performance, as measured by minDCF/EER/AUC, on the NIST SRE 2006 data set

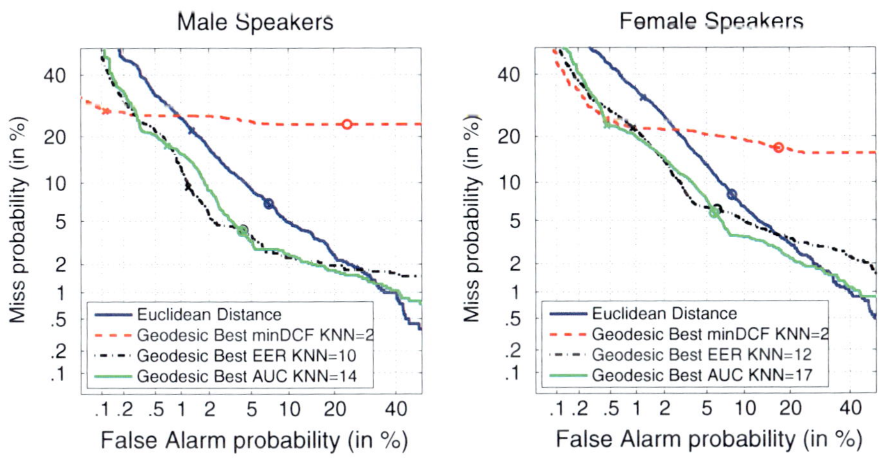

Fig. 15 DET plots using Geodesic distance for the choices of K that achieve the best minDCF, EER, and AUC on NIST SRE 2006

Figure 15 visualizes this trade-off in system performance over the full range of operating points for each optimal, as chosen by one of the three metrics, choice of K. Overall, the Geodesic distance does improve over the Euclidean baseline. Low $K = 2$ achieves a dramatic improvement, specifically for male speakers, in the low false-alarm regime of the DET curve, however at a significant reduction in performance for the remaining operating points. For larger choices of K, the improvement over the baseline is more consistent. An interesting property of DET curves using the Geodesic distance is the step-like structure, most clearly observed on the female speakers. The reason for this is that the dynamic range of scores between speakers

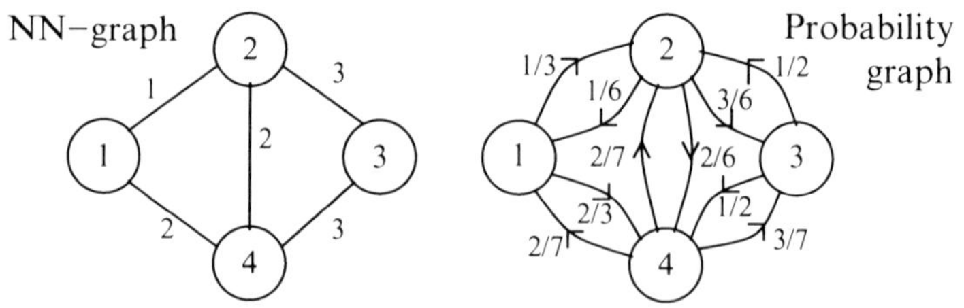

Fig. 16 NN graph and its corresponding probability graph

that are one-hop away, directly connected, and two-hops away, not connected but share a common neighbor, is large enough that the histograms of target and non-target scores of such a system are multi-modal.

5.3.2 Speaker Comparison Using K-Step Markov

The Geodesic distance only considered the shortest path connecting two nodes, however, one could consider all paths between them. K-step Markov (KSM) is an asymmetric feature that achieves that goal and is typically used to quantify the relative importance between two nodes in a graph [41]. KSM provides the aggregate probability that a random walk started at node i will visit node j after S steps are taken; the notation S was chosen so as not to be confused with K, the number of NNs in the graph.

Computing KSM requires first transforming an NN graph with edge weights given by $f_2(\mathbf{z}_i, \mathbf{z}_j)$ (11) into a probability graph with directed edges representing the probability of transitioning from a node to each of its neighbors with each step taken. This is done by dividing each outward edge from a node by the sum of all outward edges from that node. The transformation is shown in Fig. 16; note that f_2 represents similarity rather than distances; therefore, higher similarity will be transformed into a higher probability. The probability graph can be transformed into a probability-of-transition matrix $\mathbf{A}$ where the $\mathbf{A}(i,j)$ represents the probability-of-transition from node i to node j; for example in Fig. 16, $\mathbf{A}(2,4) = 2/6$ and $\mathbf{A}(4,2) = 2/7$. For the experiments in this section, the decay parameter σ in f_2 is set to 1.

Given the probability-of-transition matrix $\mathbf{A}$, the KSM score for S-steps can be computed as follows:

$$\mathrm{KSM}(i,j) = \{\mathbf{A}\mathbf{e}_i + \mathbf{A}^2\mathbf{e}_i + \mathbf{A}^3\mathbf{e}_i + \cdots + \mathbf{A}^S\mathbf{e}_i\}_j, \tag{14}$$

where $\mathbf{e}_i$ is a vector whose ith entry is 1 and all remaining entries are zero, and $\{\mathbf{v}\}_j$ represents taking the jth entry of the vector $\mathbf{v}$. For the experiments in this section a symmetric form of the KSM measure is used:

$$\mathrm{KSM}_{\mathrm{sym}}(i,j) = \frac{1}{2}\{\mathrm{KSM}(i,j) + \mathrm{KSM}(j,i)\}. \tag{15}$$

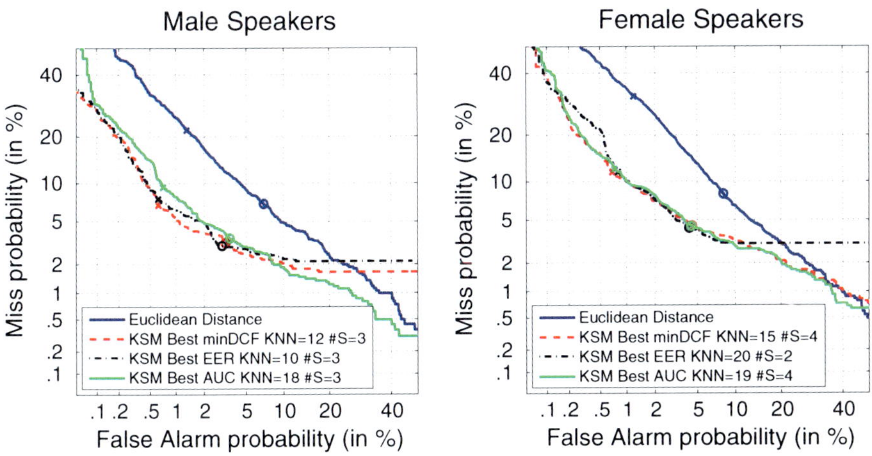

Fig. 17 DET plots using KSM for the choices of K and S that achieve the best minDCF, EER, and AUC on NIST SRE 2006

KSM is parametrized by K, the number of nearest neighbors used to build the graph, and S the number of steps used to compute the score in (14). For each of the metrics considered the best choice of $K \in \{1:25\}$ and $S \in \{1:15\}$ on SRE 06 is selected, and the corresponding DET plots shown in Fig. 17; a maximum choice of $S = 15$ is sufficient as the contributions from additional steps beyond that is negligible. The figure shows that KSM significantly improves over the baseline and is more consistent over the choice of the optimization metric than the Geodesic distance. This is due to KSM taking into consideration the multiple paths connecting two nodes rather than just the best.

5.3.3 Speaker Comparison Using Neighborhood Locality

Both the Geodesic distance and KSM were measures of flow through the graph, this section will focus on local neighborhoods. The premise being that if two recordings share a common neighborhood they are more likely to be of the same speaker. This approach was motivated by the link prediction problem [42]; however, unlike that work, a graph is not explicitly provided and must be built using the NN approach. This section shows that the link prediction measures presented in [42] can be used to quantify speaker similarity between two recordings. These measures are typically computed using binary graphs, where the edge weights are either 1 or 0; however, this section also proposes counterparts that apply to weighted graphs, whose edge weights are $f_0(\mathbf{z}_i, \mathbf{z}_j)$ (11). First, some useful notation is introduced, followed by the link prediction measures on binary graphs, and their weighted graph counterparts.

- The graph consists of T nodes (recordings).
- NN_i is the set of neighbors of node i, i.e. the nodes connected to i by an edge.

- $|X|$ is the cardinality of the set X.
- $||\mathbf{z}_i||$ is the 2-norm of the vector $\mathbf{z}_i$.
- The vectors $\mathbf{v}_i$ are, typically sparse, vectors of size $T \times 1$ that capture the interaction of i with the remaining graph nodes:
  - Zero valued entries in the vectors indicate the lack of an edge between the recording i and the nodes corresponding to the zero locations.
  - For weighted graphs, the value of the nonzero vector entries indicates the weight of the edge between i and the corresponding graph nodes.
  - For binary graphs, all nonzero entries have a value of one and indicate edges between i and the corresponding graph nodes.

Binary graphs: For binary graphs the speaker similarity metrics are based on the ones proposed in [42] for link prediction:

- *Common neighbors*$=|NN_j \cap NN_i|$ counts the number of common neighbors between j and i.
- *Jaccard's coefficient*$=\frac{|NN_j \cap NN_i|}{|NN_j \cup NN_i|}$ normalizes the common neighbor score by the total number of nodes connected to both j and i. An example scenario where the normalization would be useful is where a particular recording j shares the same number of common neighbors with two separate recordings i_1 and i_2, however $|NN_{i_2}| \gg |NN_{i_1}|$ and thus the Jaccard coefficient would penalize i_2.
- *Adamic*$=\Sigma_{z \in NN_j \cap NN_i} \frac{1}{log|NN_z|}$ a measure that combines the size of the intersection set with how highly connected the nodes in the intersection are. This could be thought of as another form of normalized common neighbors.

Weighted graphs: For weighted graphs, counterparts to the binary graph measures are used:

- *Inner product*$=v_j^{\mathrm{T}} v_i$ is based on the common neighbors measure.
- *Normalized inner product I*$=\frac{v_j^{\mathrm{T}} v_i}{||v_j||.||v_i||}$ which is inspired by Jaccard's coefficient.
- *Normalized inner product II*$=\frac{v_j^{\mathrm{T}} v_i}{||v_j||+||v_i||}$ also inspired by Jaccard's coefficient.
- *Adamic Weighted*$=\Sigma_{z \in NN_j \cap NN_i} \frac{1}{log||v_z||}$, based on the binary Adamic feature.
- *Landmark Euclidean distance*$=||v_j - v_i||$, a measure that considers the recordings in the graph as landmarks and that the vectors v_j and v_i represent the coordinates of j and i in the space defined by the landmarks.

The SRE 2006 is once again used to set the parameter $K \in \{1:25\}$, the number of nearest neighbors, for each of the different metrics. Figure 18 shows the DET plot of the best choice of K and the best performing binary-graph and weighted-graph local-neighborhood similarity measure on SRE 06. Once again, these measures outperform the baseline, with the normalized inner product I performing consistently better than the others over the different choices of optimization metric. The step-like shape of the common neighbors measure is because this measure results in integer similarity scores.

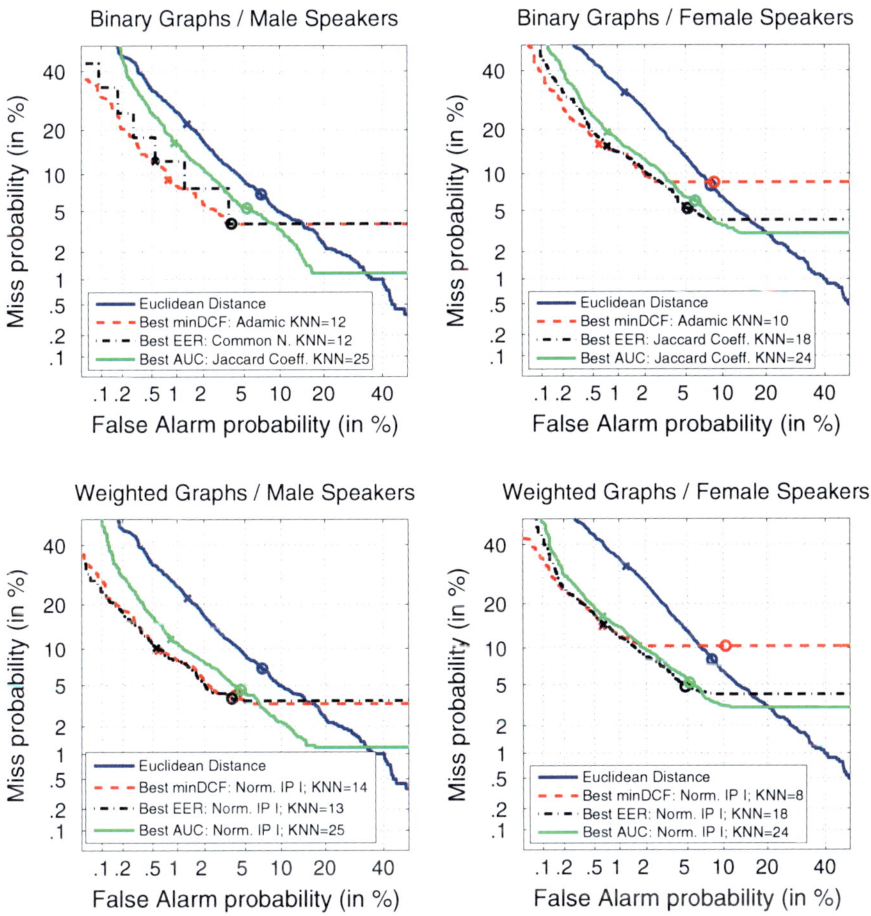

Fig. 18 DET plots of the local-neighborhood similarity measures that achieve the best minDCF, EER, and AUC on NIST SRE 2006

5.3.4 Generalization to the Test-Set

This section evaluates the best performing content-graph-based similarity measures on the NIST SRE 2008 test set. Rather than examining each of the optimization metrics (EER/minDCF/AUC) separately, this section will focus on the EER; this is because, of the three, it provided the most consistent results on the evaluation set over the region of the DET curve of most interest for speaker comparison, the region above the EER operating point.

Figure 19 presents the results on the test set for the best choice of parameters for each of the four measures: Geodesic, KSM, local-neighborhood (L-N) Binary, and L-N Weighted. The best parameters are chosen to optimize EER performance on the SRE 06 evaluation set. As seen in Fig. 19, all of the measures, except for the L-N

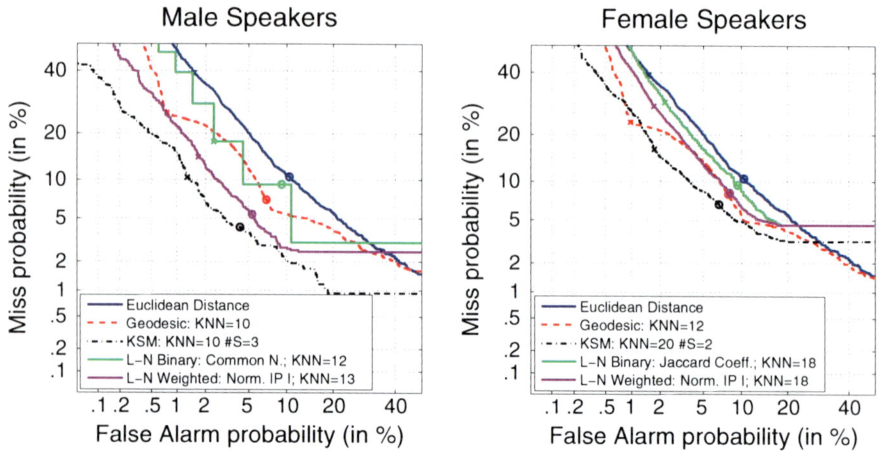

Fig. 19 DET plots showing the generalization of the different measures to the NIST SRE 2008 test set

Binary common neighbors and L-N Binary Jaccard's coefficient, generalized well to the SRE 08 test set, and significantly outperformed the baseline; with the best performing and most consistent measure being KSM.

5.4 Graph-Relational Features for SVM Speaker Comparison

Speaker comparison using SVMs has proven popular in the speaker verification scenario [2–5]. There the goal of these approaches was to build an SVM classifier, per speaker of interest, that separates the vector representation of the speaker of interest from all others. This section presents an alternate approach to SVM speaker comparison that learns a single classifier, per gender, that operates on pairs of recordings and separates those pairs of recordings of the same speaker from those of different speakers. The graph-based similarity measures presented in the previous section along with the Euclidean distance serve as graph-relational features for this SVM. An important question, however, is which subset of measures and parameter values to include in the graph-relational features. This section explores two such choices. The first, only includes the best performing subset (BEST) from each class of measures, with the selection based on minDCF, EER, and AUC performance on the validation set; these are listed in Table 1. The second uses all the features (ALL) over a wide range of parameter choices.

When the best subset along with the Euclidean distance are used, the resultant feature vectors are of length 13. Alternatively, the feature vectors for ALL are of length 300 and include the Euclidean distance and all the different measures over the range $K \in \{1:2:25\}$ for NN graph construction and $S \in \{1:15\}$ for the

Table 1 The best performing choice of parameters from each type of measure for each of the minDCF, EER, and AUC metrics

Measure	Metric	Male	Female
Geodesic	minDCF	KNN=2	KNN=2
Geodesic	EER	KNN=10	KNN=12
Geodesic	AUC	KNN=14	KNN=17
KSM	minDCF	KNN=12 #S=3	KNN=15 #S=4
KSM	EER	KNN=10 #S=3	KNN=20 #S=2
KSM	AUC	KNN=18 #S=3	KNN=19 #S=4
L-N Binary	minDCF	Adamic; KNN=12	Adamic; KNN=10
L-N Binary	EER	Common N.; KNN=12	Jaccard Coeff.; KNN=18
L-N Binary	AUC	Jaccard Coeff.; KNN=25	Jaccard Coeff.; KNN=24
L-N Weighted	minDCF	Norm. IP I; KNN=14	Norm. IP I; KNN=8
L-N Weighted	EER	Norm. IP I; KNN=13	Norm. IP I; KNN=18
L-N Weighted	AUC	Norm. IP I; KNN=25	Norm. IP I; KNN=24

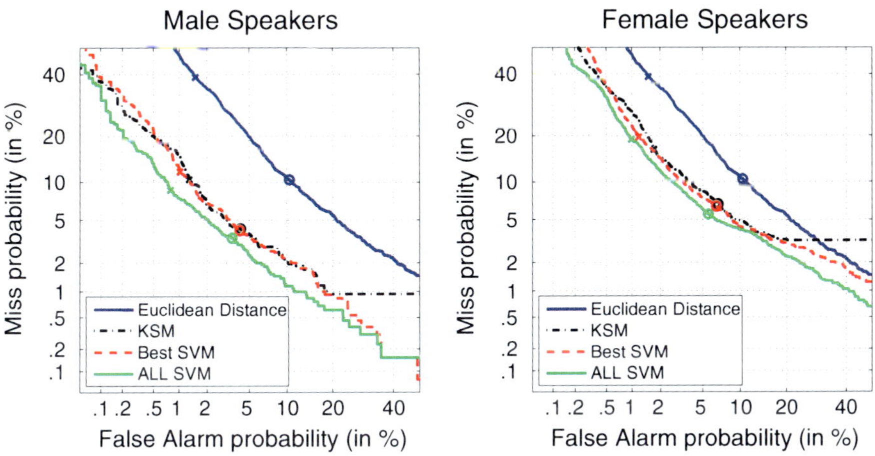

Fig. 20 DET plots comparing SVM with BEST and ALL features to KSM and the baseline on the NIST SRE 2008 test set

KSM measure. The feature vectors are mean and variance normalized based on the training set. Covariance normalization based on the training set is also explored, since the individual measures are similar across parameter choices and types; this improved performance only when the BEST choice of features were used; therefore, the results presented for the BEST features includes covariance normalization. SRE 06 is used to train the classifier, with the SVM cost parameter selected to optimize for the EER using threefold cross validation on SRE 06. Cross validation also showed that it was crucial to use an AUC-maximizing SVM [43], this work uses the SVM^{perf} Matlab implementation [43].

Figure 20 compares the SVM classifier with the two feature vector choices, BEST and ALL, to the baseline measure and KSM, the single best performing graph-based measure. The SVM with the ALL features consistently outperformed the others, while the performance of the BEST features was on par with KSM. This is an interesting observation, because even though the different graph-based measures and the different parameter choices may not perform well individually, they provide complimentary information that can be exploited by the SVM.

6 Speaker Retrieval

6.1 *Definition and Metrics*

Although no standard definition exists, we define speaker retrieval as follows. A query set of utterances is given. The goal is to find a list of utterances in a larger corpus that match the speakers in the query set. This problem is similar to the standard multi-utterance enrollment multi-speaker detection task. Two differences are that speaker retrieval does not require speaker labeling of the query set. Also, since a result list is returned (e.g., top 10 matches), speaker retrieval does not require scoring and ranking all utterances in the larger corpus. Typical applications for speaker retrieval are recommender-systems based on a user's browsing history or query-by-example.

For speaker retrieval, a typical metric is average precision at a given result list of length N. For a given query, find the top N scoring nodes and then find the percentage of relevant speakers (true trials) on the list—the precision. The mean of the precision across multiple queries is the average precision. For this work, the assumption is made that recall will not be an issue; i.e., there are enough relevant speakers to fill the results list.

6.2 *Approaches to Speaker Retrieval*

Suppose a set of query speech signals represented as vectors Q contained in a larger set of vectors M with top-K NN-graph G and corresponding weighted adjacency matrix W is given. The speaker retrieval problem is to find a list of N candidate matches from $M \setminus Q$ (set difference) to the query set Q.

The baseline approach to speaker retrieval using standard speaker recognition techniques is a stack detector [44]. First, assume that the query set has S labeled speakers. Then, for each of the speakers construct an SVM model, $\mathbf{w}_j$. Next, apply the models $\{\mathbf{w}_j\}$ as a stack detector to the vectors $M \setminus Q$. The stack detector score for vector $\mathbf{m}_i$ is $s_i = \max_j \mathbf{w}_j^t \mathbf{m}_i$. Typically, the stack detector score would be compared to a threshold for verification. In speaker retrieval, the scores $\{s_i\}$ are sorted, and the largest N values and corresponding indices are found returned as the result of the query.

6.2.1 Speaker Retrieval with Random Walks

Speaker retrieval can also be performed with a random walk (Markov process) on the content graph constructed in Sect. 3. Suppose again a set of query vectors, Q is given. Our goal is to set up a random process on the content graph with walks from the query nodes to good candidates with high probability.

The random walk method presented is inspired by approaches typically used for text processing. Random walks are commonly used in semi-supervised learning for inference, see [45,46]. Additionally, random walk methods can be viewed as related to using homophily, relational autocorrelation, or label propagation techniques in relational learning [47]. In both cases, the goal is to exploit the fact that nodes "close" to the query nodes will have similar labels.

Speaker retrieval is accomplished by computing the probability of arriving at the jth node after L steps from the query nodes. This calculation can be computed iteratively using a matrix multiply. If the vector P_t is defined with entries $P_{t,i} = p(v_t = i)$, then by the Markov property $P_{t+1} = TP_t$. Thus, $P_L = T^L P_0$ where P_0 is the starting (prior) distribution. For the query process, set $P_{0,i} = 1/|Q|$ if q_i is in the query set and zero otherwise ($|Q|$ is the number of elements in Q).

In addition to the random walk on the graph, a restart is included that returns to the query nodes. This is equivalent to adding edges from every node back to the query node set. With this modification, our update equation becomes

$$P_{t+1} = \alpha P_0 + (1-\alpha)TP_t, \tag{16}$$

where $0 \leq \alpha \leq 1$. For α greater than 0, the restart term reinforces the query.

Note that since the random walk process is only performed for a limited number of steps, it may be the case that not all of the nodes in the graph can be reached (depending on the diameter of the graph). Thus, random walks essentially classify unreachable nodes as "not similar" if used in speaker comparison mode. Another way of viewing this is that a random walk is a speaker comparison detector operating in the low probability of false alarm region.

6.3 *Experiments*

6.3.1 Baseline System

Experiments were performed on the NIST 2006 and NIST 2008 speaker recognition evaluation (SRE) data sets as described in Sect. 5.1. SRE 06 was used as a parameter tuning set, and SRE 08 was used as a final evaluation.

For both data sets, queries were taken from enrollment lists for speaker models taken from the eight conversation training scenario from the SRE. That is, a typical query would be formed by taking the utterances from S models, where S is the stack size.

The vector expansion used is given by (8). Standard feature extraction with 20 MFCCs (including $c0$) plus deltas was performed along with SAD and 0/1 feature normalization. A GMM UBM with 512 mixtures was trained using a large set of Switchboard 2 and Fisher data. For MAP adaptation, a relevance factor of 0.01 was selected. WNAP [21] was trained using a combined NIST SRE 2004–2006 list with the dimension of the nuisance subspace fixed at 64.

SVM speaker models were trained using the kernel in (6) and the methods in [7]. A subset of the Fisher corpus (approximately 4,000 utterances) was used as an SVM background. The SVM models were applied using a stack detector with stack sizes S of 10, 50, and 100. For each stack size, 500 combinations of models were selected at random and scored against all of the NIST SRE test data.

6.3.2 Speaker Retrieval System Tuning on NIST SRE 2006

A random walk system was implemented using the methods described in Sect. 6.2.1 and applied it to the NIST data set using the same enroll/test protocol as the stack models for the baseline system. Content graphs per gender were constructed with all of the NIST data. Then, the probabilities of random walks from the utterances of a query set to other nodes on the graph were computed with (16).

To facilitate understanding hyperparameters, the values of $\sigma = 0.1, 1, 10$ in (10), $\alpha = 0, 0.1, 0.25, 0.5, 0.75$ in (16), $K = 2, 5, 10, 25, 50, 100$ for the top-K distances described in Sect. 3, and $L = 1, \ldots, 5$ as the number of steps taken in the random walk in (16) are swept. The best average precision is tuned on the SRE 06 data set and applied to the SRE 08 data set in the next section.

Initially the number of steps L in the random walk was considered. Table 2 shows the optimal number of steps to achieve the best average precision broken out by gender. The trend seen is that for a longer query result list length (N), more steps are needed to obtain the best precision. In addition, if the stack size is large with respect to the query result list length, then less steps are needed; i.e., we can choose a few high scoring nearest neighbors.

Next, the top-K value for NN content graph construction was considered. For the different experiments (18 total) in Table 2, the optimal value was $K = 10$ for 16 out of the 18 cases. A possible reason for the system preferring a smaller K is that it reduces the number of false alarm connections. The random walk process can recover from this by using more than one step to find a good candidate.

The values of α and σ were also tested for male speakers. For σ, the parameter achieving the best average precision for multiple S and K was $\sigma = 1$; this value was optimal for 17 out of the 18 possible cases. For alpha, values of 0 (5 times), 0.1 (11 times), 0.5 (1 time), 0.75 (1 time) were found to produce optimal precision. This indicates that very little restarting is needed in the process.

Table 2 Number of steps L with best average precision at N with the random walk method on the NIST SRE 2006 telephone data

Stack size S	Query result length N	Male best L	Male AvgP (%)	Female best L	Female AvgP (%)
10	5	1	99.08	1	98.68
50	5	1	100	1	100
100	5	1	100	1	100
10	10	4	98.4	3	97.94
50	10	5	99.48	2	99.96
100	10	1	100	1	100
10	20	4	97.09	5	96.08
50	20	5	99.16	1	99.22
100	20	1	99.72	1	99.89

Table 3 Results on the NIST SRE 2008 data set; comparison with optimal L and average precision in %

Stack Size S	Query result length N	Best L	Tune AvgP (%)	Best AvgP (%)	Stack AvgP (%)
10	5	1	98.4	99.24	96.46
50	5	2	98.4	100	95.5
100	5	1	92.92	100	95.72
10	10	3	93.72	95.34	92.82
50	10	5	98.98	100	95.71
100	10	2	94.54	100	95.21
10	20	2	72.63	72.77	72.66
50	20	5	98.5	99.91	96.66
100	20	3	95.32	100	95.53

6.3.3 Application to the NIST SRE 2008 Data Set

From the previous analysis, two possibilities were explored, a fixed parameter setting and the optimal setting. For the fixed parameter setting, $\alpha = 0.1$, $\sigma = 1$, $K = 5$ were used based on consideration of the optimal choices for the NIST SRE 2006 data set; for L, a compromise value of 3 was chosen based on Table 2. Results for the SRE 06 tuned system applied to the SRE 08 data, the optimal system (best mean average precision across all hyperparameters), and the stack system with SVM models are shown in Table 3.

Table 3 shows that the random walk is relatively robust to parameter settings. The tuning parameters were applied in the algorithm to an unseen data set. Also, the tuned and optimal average precision are close in many cases. Additional work could be done on using alternate walk retrieval strategies—variable hyperparameters based on query length or stack size, or alternate walk scores. Finally, the random walk is performing well in comparison with a standard stack approach. This fact is significant since the random walk approach uses much less computation at query time.

7 Summary

This chapter has explored the interaction between graph embedding and speaker classification for multiple applications. Methods of vector-embedding of audio based upon parameters of GMM statistical models were reviewed. Then, representing data sets of audio using graphs derived from the vector embedding was presented in detail. The resulting graph structures visually showed significant clustering of speakers. Applications of both speaker comparison (1-1 classification) and speaker retrieval (finding speakers in a large data set) were presented.

For speaker comparison, individual methods such as random walks, geodesics, neighborhood features, and landmark features were explored. Ultimately, this exploration led to the combination of features with relational learning methods via an SVM. Performance for speaker comparison was shown to considerably outperform the baseline and generalized well. Our exploration of techniques provides guidance for future applications.

For speaker retrieval, random walks were explored as a viable low-complexity method. For different query set sizes, the precision of graph-based methods was shown to be similar to SVM vector-based retrieval methods. Also, the graph structure was shown to provide a natural structure for encoding information for arbitrary data queries.

For future research, graph embedding of audio data has many potential areas of application and exploration because of its natural representation of large data sets. Possible areas of research are—optimal graph construction, clustering, advanced classification methods on graphs, compensation, data set summarization, and visualization. Overall, the interaction of graph embedding and speaker recognition is a rich source of future research and applications.

Acknowledgements This work was sponsored by the Department of Defense under Air Force Contract FA8721-05-C-0002. Opinions, interpretations, conclusions, and recommendations are those of the authors and are not necessarily endorsed by the United States Government.

References

1. Reynolds DA, Quatieri TF, Dunn R (2000) Speaker verification using adapted Gaussian mixture models. Digital Signal Process 10(1–3):19–41
2. Campbell WM (2002) Generalized linear discriminant sequence kernels for speaker recognition. (IEEE) In: Proceedings of ICASSP, pp 161–164
3. Stolcke A, Ferrer L, Kajarekar S, Shriberg E, Venkataraman A (2005) MLLR transforms as features in speaker recognition. In: Proceedings of Interspeech, pp 2425–2428
4. Campbell WM, Campbell JP, Reynolds DA, Jones DA, Leek TR (2004) High-level speaker verification with support vector machines. Curran Associates, Inc. (IEEE) In: Proceedings of ICASSP, pp I–73–76
5. Shriberg E, Ferrer L, Venkataraman A, Kajarekar S (2004) SVM modeling of SNERF-grams for speaker recognition. Curran Associates, Inc. In: Proceedings of interspeech, pp 1409–1412

6. Solomonoff A, Quillen C, Campbell WM (2004) Channel compensation for SVM speaker recognition. (IEEE) In: Proceedings of Odyssey-04, the speaker and language recognition workshop, pp 57–62
7. Campbell WM, Sturim DE, Reynolds DA, Solomonoff A (2006) SVM based speaker verification using a GMM supervector kernel and NAP variability compensation. (IEEE) In: Proceedings of ICASSP, pp I–97–I–100
8. You CH, Lee KA, Li H (2009) An SVM kernel with GMM-supervector based on the bhattacharyya distance for speaker recognition. IEEE Sign Process Lett 16(1):49–52
9. Campbell WM, Liu H (2001) Using feature transformation and selection with polynomial networks. In: Applications and science of computational intelligence IV, SPIE Aerosense
10. Kajarekar SS (2005) Four weightings and a fusion: A cepstral-SVM system for speaker recognition. (IEEE) In: Proceedings of ASRU
11. Solomonoff A, Campbell WM, Boardman I (2005) Advances in channel compensation for SVM speaker recognition. (IEEE) In: Proceedings of ICASSP
12. Kenny P, Dumouchel P (2004) Experiments in speaker verification using factor analysis likelihood ratios. (IEEE) In: Proceedings of Odyssey04, pp 219–226
13. Hatch AO, Kajarekar S, Stolcke A (2006) Within-class covariance normalization for SVM-based speaker recognition. (IEEE) In: Proceedings of the international conference on spoken-language processing, pp 1471–1474
14. Kenny P, Ouellet P, Dehak N, Gupta V, Dumouchel P (2008) A study of inter-speaker variability in speaker verification. In: Transactions on Audio, Speech and Language Processing
15. Dehak N, Dehak R, Kenny P, Brummer N, Ouellet P, Dumouchel P (2009) Support vector machines versus fast scoring in the low-dimensional total variability space for speaker verification. Curran Associates, Inc. In: Proceedings of interspeech
16. Campbell WM, Karam ZN, Sturim DE (2009) Inner product discriminant functions. In: Advances in neural information processing systems 22. MIT, Cambridge, MA
17. Karam Z, Campbell WM (2010) Graph embedding for speaker recognition. Curran Associates, Inc. In: Proceedings of interspeech, pp 2742–2745
18. Karam Z, Campbell WM, Dehak N (2011) Graph relational features for speaker recognition and mining. In: IEEE statistical signal processing workshop, pp 525–528
19. Dehak N, Karam ZN, Reynolds DA, Dehak R, Campbell WM, Glass JR (2011) A channel-blind system for speaker verification. (IEEE) In: Proceedings of ICASSP
20. Campbell WM (2010) Weighted nuisance attribute projection. In: Proceedings of IEEE Odyssey
21. Campbell W, Karam Z (2010) Simple and efficient speaker comparison using approximate KL divergence. Curran Associates, Inc. In: Proceedings of interspeech
22. Campbell W, Karam Z (2009) Variability compensated support vector machines applied to speaker verification. Curran Associates, Inc. In: Proceedings of interspeech
23. McCree A, Sturim D, Reynolds D (2011) A new perspective on GMM subspace compensation based on PPCA and wiener filtering. In: Proceedings of interspeech, Florence, Italy, 2011
24. Belkin M, Niyogi P (2003) Using manifold stucture for partially labeled classification. (IEEE) In: Thrun S, Becker S, Obermayer K (eds) Advances in neural information processing systems 15. MIT, Cambridge, MA, pp 929–936
25. Zhu X (2007) Semi-supervised learning literature survey. Tech Rep., University of Wisconsin, Madison
26. Dijkstra EW (1959) A note on two problems in connexion with graphs. Numerische Math Springer Berlin / Heidelberg Signal, 1:269–271
27. Wan V, Campbell WM (2000) Support vector machines for verification and identification. In: Neural networks for signal processing X, proceedings of the 2000 IEEE signal processing workshop, pp 775–784
28. Tenenbaum JB, Silva V, Langford JC (2000) A global geometric framework for nonlinear dimensionality reduction. (AAAS), Science 290
29. Tenenbaum J (xxxx) Matlab package for a global geometric framework for nonlinear dimensionality reduction. http://isomap.stanford.edu/

30. Cox TF, Cox MAA (2000) Multidimensional scaling, 2nd edn. Chapman and Hall, London
31. The NIST year 2004 speaker recognition evaluation plan (2004) Accessed date on September 30, 2012. http://www.itl.nist.gov/iad/mig/tests/sre/2004/index.html
32. The NIST year 2006 speaker recognition evaluation plan (2005) Accessed date on September 30, 2012. http://www.itl.nist.gov/iad/mig/tests/sre/2006/index.html
33. The NIST year 2008 speaker recognition evaluation plan (2008) Accessed date on September 30, 2012. http://www.itl.nist.gov/iad/mig/te
34. Adar E (2006) Guess: A language and interface for graph exploration. (ACM) In: CHI
35. Battista D, Eades P, Tamassia R, Tollis IG (2002) Graph drawing: Algorithms for visualization of graphs. Prentice Hall, Englewood Cliffs
36. The NIST year 2010 speaker recognition evaluation plan (2010) Accessed date on September 30, 2012. http://www.itl.nist.gov/iad/mig/te
37. Bishop CM (2009) Pattern recognition and machine learning. Springer, Berlin
38. Martin A, Doddington G, Kamm T, Ordowski M, Przybocki M (1997) The DET curve in assessment of detection task performance. In: Proceedings of eurospeech, pp 1895–1898
39. Linguistic Data Consortium (xxxx) Switchboard-2 corpora. http://www.ldc.upenn.edu
40. The NIST year 2005 speaker recognition evaluation plan (2005) Accessed date on September 30, 2012. http://www.itl.nist.gov/iad/mig/tests/sre/2005/index.html
41. White S, Smyth P (2003) Algorithms for estimating relative importance in networks. In: Proceedings of the ninth ACM SIGKDD international conference on Knowledge discovery and data mining
42. Liben-Nowell D, Kleinberg J (2003) The link prediction problem for social networks. (ACM) In: Proceedings of the 12th international conference on information and knowledge management
43. Joachims T (2005) A support vector method for multivariate performance measures. ACM Press. Proceedings of the 22nd International Conference on Machine Learning 377–384
44. Singer E, Reynolds DA (2004) Analysis of multitarget detection for speaker and language recognition. In: Proceedings of Odyssey, pp 301–308
45. Szummer M, Jaakkola T (2001) Partially labeled classification with random walks. In: Dietterich TG, Becker S, Ghahramani Z (eds) Advances in neural information processing systems 14. MIT, Cambridge, MA
46. Lin F, Cohen WW, (2010) Semi-Supervised Classification of Network Data Using Very Few Labels, asonam, pp. 192–199, 2010 International Conference on Advances in Social Networks Analysis and Mining, ASONAM Publisher is Institute of Electrical and Electronics Engineers (IEEE)
47. Macskassy SA, Provost F (2007) Classification in networked data: A toolkit and a univariate case study. J Mach Learn Res 8:935–983

Printed by Publishers' Graphics LLC
SO20121120.19.22.37